Tamás Matolcsi

Common Sense Relativity and Spacetime

Also of Interest

Radical Relativity
An Uncommon Way to Spacetime
Heinz Blatter, Thomas Greber, 2025
ISBN 978-3-11-150309-7, e-ISBN (PDF) 978-3-11-150359-2

Quantum Mechanics
Guo-Ping Zhang, Mingsu Si, Thomas F. George, 2024
ISBN 978-3-11-067212-1, e-ISBN (PDF) 978-3-11-067215-2

Logic and Deduction
Applications to Theoretical Physics, and Number Theory
Bertrand Wong, 2025
ISBN 978-3-11-914531-2, e-ISBN (PDF) 978-3-11-221436-7

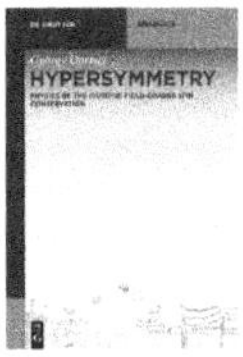

Hypersymmetry
Physics of the Isotopic Field-Charge Spin Conservation
György Darvas, 2021
ISBN 978-3-11-071317-6, e-ISBN (PDF) 978-3-11-071318-3

Data Science
Time Complexity, Inferential Uncertainty, and Spacetime Analytics
Ivo D. Dinov, Milen Velchev Velev, 2021
ISBN 978-3-11-069780-3, e-ISBN (PDF) 978-3-11-069782-7

Tamás Matolcsi

Common Sense Relativity and Spacetime

Looking Physics Right in the Eye

DE GRUYTER

Author
Tamás Matolcsi
Department of Applied Analysis and Computational
Mathematics
Eötvös Loránd University (ELTE)
Pázmány Péter sétány 1/A
1117 Budapest
Hungary
matolcsi.tamas@gmail.com

ISBN 978-3-11-914606-7
e-ISBN (PDF) 978-3-11-221955-3
e-ISBN (EPUB) 978-3-11-221963-8

Library of Congress Control Number: 2026930397

Bibliographic information published by the Deutsche Nationalbibliothek
The Deutsche Nationalbibliothek lists this publication in the Deutsche Nationalbibliografie; detailed bibliographic data are available on the Internet at http://dnb.dnb.de.

Cover image: agsandrew / iStock / Getty Images Plus
Typesetting: VTeX UAB, Lithuania

www.degruyterbrill.com
Questions about General Product Safety Regulation:
productsafety@degruyterbrill.com

I am grateful to my former students, my friends, Patroklos Benatos, Áron Szabó and Péter Ván who made several suggestions to improve the text and helped me to publish this book, to another my former student Tibor Gruber who revised the manuscript

How can you look physics right in the eye?

A special Foreword by three former students of the author

> Proceed, proceed: we will begin these rites,
> As we do trust they'll end, in true delights.
>
> Shakespeare, *As You Like It*, 5.4. 207–208

Why this foreword? While it is a true delight to look physics right in the eye, certain rites have to be performed before you will be able to do so. As former students of the author of this book, Tamás Matolcsi, we have performed these "rites" by learning a style and an approach from him that are refreshingly clear and helpful for thinking coherently and consistently about theoretical physics. The rites begin with the stage where physics "happens": *spacetime.*

Some time ago, one of us came across a 2500-year-old parable that he thought was an excellent metaphor for explaining what lies ahead.

The tale of six blind men and the elephant

Patroklos Benatos

Figure 1: Six blind men touching different parts of an elephant. Each of them comes to different conclusions based on which body part they happen to touch.

Once upon a time in a village—so the story goes in an old Indian folklore tale—six blind men learned that a strange creature called an elephant had arrived. Driven by curiosity, they set out to encounter it using one of the senses available to them—touch (see Figure 1).

Each man felt a different part of the animal: one pressed his hand against the broad side of the elephant and concluded, "An elephant is like a wall." Another grasped the pointed tusk and declared, "No, it's like a spear—sharp and smooth." The third man explored the trunk and laughed, "You're both wrong! An elephant is clearly like a thick snake." The fourth wrapped his arms around a leg and insisted, "It's like a mighty tree." The fifth fluttered his fingers across the ear and reasoned, "You're mistaken—it's like a large fan." The sixth seized the tail and argued, "No, no, it's just like a rope!"

Overhearing the debate, a sighted villager intervened. "You're all partly right," he said. "But you've each only touched a fragment of the whole. Only by combining your perspectives can you truly understand what an elephant is."

https://doi.org/10.1515/9783112219553-203

Interestingly, this is exactly how the modern science of physics started some four hundred years ago. To describe physical phenomena, the first and quite natural notion introduced was that of a *frame of reference* (with a corresponding *coordinate system*). Reference frames enable the quantitative description of physical phenomena and then the drawing of conclusions about how these interrelate with each other, in other words, the discovery of the perceived "laws of physics".

Sounds good—but remember the blind men drawing different conclusions depending on their own perspectives? Indeed: how will the conclusions, drawn in a particular frame of reference, express something *not dependent* on the particular frame of reference?

First: is there a "best" frame of reference from which you can conclude the "true" laws of Nature, while from others there is no chance to discover these laws? Current theories in physics operate under the assumption that you would suspect from the parable: there is *no* such "best" frame of reference, *all* frames of reference are on equal footing in this regard.[1] The quantitative descriptions of physical phenomena can, of course, be (very) different in different reference frames, but the eventual conclusion has to be the same: an elephant.

Great—now how should you make this principle operational, i. e. check that some conclusion is indeed independent of the particular frame of reference it was drawn in? The way to go about this is again no different from what the sighted man in the parable said: "only by combining different perspectives can you truly understand what an elephant is". In physics this translates to transforming the quantitative description of a presumed law seen in one frame of reference into another frame (in fact, all possible other frames) and, if the description changes in a certain, prescribed way—like the view of the elephant changes as you move around it—the presumed law passes the independence check and can be declared a candidate for experimental verification. This checking procedure is usually referred to as "*transformation rule*" and the prescribed way of changing as "*covariance*". Figure 2 contains a diagram I came up with during my thesis work to illustrate this.

The "equal footing" principle[2] (and the above way of filtering out subjective elements originating from particular points of view) is quite natural and, in retrospect, *trivial*: why should the laws of Nature depend on a frame of reference (a human concept) you pick to describe them in?

Moving on, notice how cumbersome the checking procedure is: each blind man has to talk to every other blind man and then not only compile what he learned into a consistent picture, but also confirm that his picture matches everybody else's. Let's put a spin on the parable: annoyed by the inefficiency of this procedure and inspired by the

1 There are frames of reference that show Nature with minimal, but still essential distortion; these are called *inertial* frames of reference.

2 The official name for this principle is "*general covariance*".

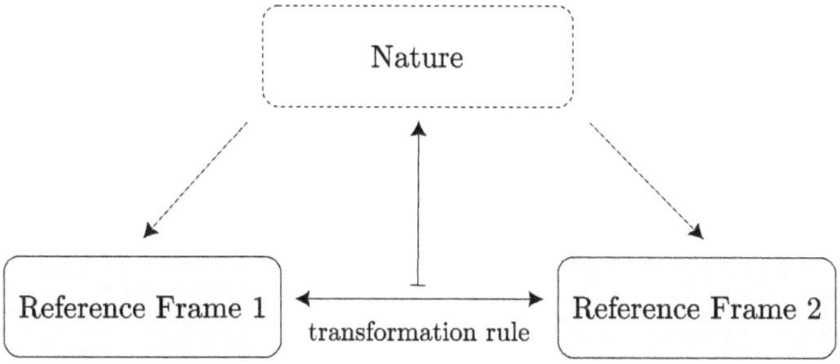

Figure 2: The transformation rule approach to covariance. Note that reference frames are primary, Nature is secondary and it gets represented only indirectly via the transformation rule.

sighted villager's remark, one of the blind men has an epiphany and exclaims: "Folks, now that we have completed this arduous process of talking-and-comparing, why don't we *save this work* for others? Let's make our combined individual perceptions tangible by sculpting a model elephant statuette! This way, our fellow blind villagers will be able to immediately explore the entire model by touch, gaining not only a full mental picture of the elephant but also the ability to reproduce each partial perspective, if necessary" (see Figure 3).

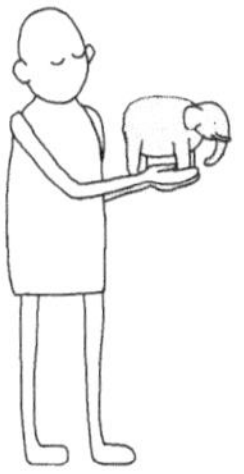

Figure 3: A blind man holding a statuette of an elephant. Such a model is useful for creating a mental model of the entirety of the animal through touch.

Analogously, since the workings of Nature are independent of any particular frame of reference they are observed in, there should be a way to formulate them *without using reference frames*; this will make them automatically "covariant". As opposed to Figure 2, I can illustrate this with the diagram in Figure 4.

As you might suspect, this fundamental change in approach has a domino effect on the way the laws of physics can be expressed: it requires going through the rather painful, but rewarding procedure of reformulating the descriptions, traditionally given in reference frames, from scratch. Note, however, that while reference frames are not needed for building theories, they play a *central role* in translating them into an *experimentally verifiable form*, as experiments (and the construction of any kind of equipment by engineers) are always carried out in some frame of reference.

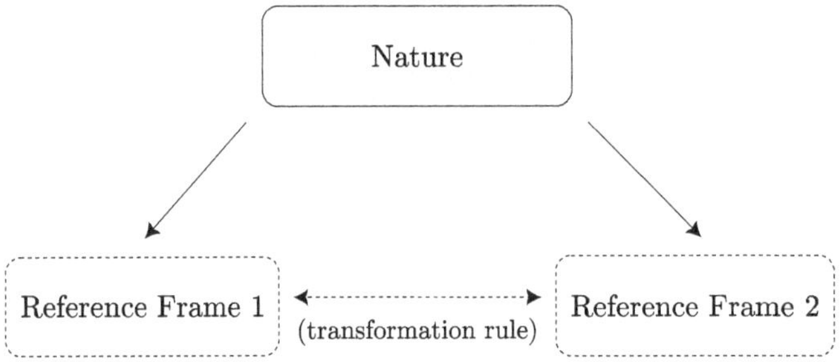

Figure 4: The spacetime approach to covariance. Note that Nature takes the primary role and reference frames are relegated to a useful secondary role. In contrast to Figure 2, Nature gets represented directly, and the transformation rule is emergent.

How to make a statuette?

Just as the blind men need some well-designed tools to build the statuette as a model of the elephant, so do theoretical physicists need some well-designed tools to build theories—except their set of tools is called *mathematics*. The book offers such a set of tools.

I distinctly remember my transformation from an interested blind man into a professional sculptor as I progressed in my studies under Tamás's supervision. It required me to do some serious soul-searching in realizing how much of the seemingly objective reasoning in the fundamental areas of physics contains tacit human assumptions that are instinctively "projected" onto Nature, even though Nature has nothing to do with them (a nice concrete example is given in the next section about "time"). These human assumptions, intermingled with the objective ones, are then wrapped in semi-consistently used mathematical tools. No surprise, this "package" eventually leads to hard-to-understand and circumspect explanations as well as contradictions big and small, especially in the modern chapters of physics, like relativity.

How can such a state of affairs be avoided? This is where wielding a well-designed toolset comes to play, i. e. using proper mathematics to build a coherent *mathematical model* for the physical phenomenon (in this case, spacetime). Additionally, you need to pair the toolset with a certain approach: after reasoning "in English" for what the important core building blocks of the phenomenon should be, you really have to force yourself to put all your assumptions into *mathematical form*. This produces remarkable clarity in the assumptions being made. I have experienced time and again that once I managed to formulate something in proper mathematical style, I felt I really knew what I was talking about (the best source of confidence!). Luckily, the mathematical machinery *preserves this mental effort* and saves you from yourself by preventing adding or changing elements haphazardly or messing something up in the "internals" of the model. Finally, by cranking the model, the rigorous rules of mathematics guide you as per the implications of your assumptions (provided, of course, that you have formulated all of them), just like the statuette guides the hands of a blind man telling him the shape of the real elephant.

I can honestly say that learning theoretical physics this way is a joy—no wonder one of the titles originally considered for this book was, actually, "The Joy of Relativity".

The mathematical modelling approach extends well beyond physics. During my doctoral program in the United States, I changed my area of study to game theory, which is (mostly) applicable to the *social* sciences. Lo and behold, I encountered the same lack of proper mathematical modeling. Not surprisingly, the "making a statuette" principles work the same way in game theory as they do in physics, and, of course, the joy remains!

Summarizing

Just as the blind men ended up transcending their particular perspectives and combined trunks, tails, tusks, legs, ears and broad sides into a statuette of the elephant, so will the book avoid using notions that are particular to a frame of reference and combine human-scale experiences right from the start into a mathematical model for spacetime.

> This is how you can look physics right in the eye:
> by eliminating the inevitable distortion in its formulation introduced by a frame of reference.

I still remember when the simplest example for "looking physics right in the eye" hit me: *motion*. I had previously read all sorts of philosophical views about how motion is supposed to be understood, starting from the ancient Greeks to the first pages of standard physics books that quantitatively describe motion with coordinate systems only to get into explanations much later that motion is, actually, not "really real". In the spacetime approach, the motion of a point-mass "disappears" and *existence* (in a technical sense) *in spacetime* takes its place. When looked at through a frame of reference, the "projection" of this existence onto the frame of reference is what becomes the perceived motion, an *entangled mix* of how the point-mass *and* the frame of reference exist in spacetime. This change of perspective felt like putting a clear-cut full stop to all the navel-gazing about motion, and I could not stop grinning for days.

Space? Time? Spacetime!

Áron Szabó

Space?

The usual intuitive assumption about space is that it is a universal background within which everything exists. This originates from planet Earth giving all of humankind a default reference. However, when the space defined by the planet is not practical to use, people naturally switch to another reference body. For example, on a long overnight flight with window shades closed, the space you find yourself in becomes defined by the plane, and it feels distinctly separate from the usual space given by Earth.

This intuition proves to be spot on from a theoretical point of view as well: the role of space is to make sense of whereabouts, and whereabouts make sense only with respect to a reference body. Correspondingly, reference bodies carry their own individual spaces. In particular, the spaces created by Earth and the airplane are *different* spaces that a priori have nothing to do with each other. However, since all spaces ultimately coexist in the same universe (spacetime), a comparison should be possible in some way. But how? See Subsections 4.5 and 9.6 in the book.

In any case, this means that space is incidental and not at all as fundamental as it used to appear to me.

Time?

The common instinctive perception of time is that it is "ticking away" in the background, outside of everything, registering *when* things happen in the universe. This perception comes from living in a finely organized society on this planet, where timetables and tax return deadlines dictate the rhythm of life. However, upon further reflection, this concept of time proves to be untenable.

What remains tenable is each physical entity carries its *own individual* "time", meaning that its own events can be ordered by "before" and "after". Such biographical ordering proves is usually called the *proper time* of the entity in question.

However, everything else is invented by people: there is no "proper time of the universe" that would give a universal drumbeat as the rhythm of existence for everything. If you want to keep track of when various events occur, you need to take drumsticks in your hands and create an administrative *system time* via "centrally issuing an official time stamp" for events (see Figure 5). Such a procedure is *synchronization* (see Subsubsection 3.5.1).

Figure 5: There is no universal drumbeat of the universe. If you want to use system time, you have to actively define it. Usually, this means making an arbitrary choice.

Since system time is a purely administrative measure for human bookkeeping of events, Nature offers no guidance on how it should be carried out. Worse, two reasonable ways of establishing system time may disagree to the extreme: even if the difference of system time between two events is finite in one particular synchronization, it can diverge to *infinity* in another.

This means that system time is highly subjective. It was eye-opening for me that Einstein's standard way of establishing system time, usually accepted without giving it any further thought, is but one choice among many (albeit a reasonable one). It is not a preoccupation with theory to insist that system time is a matter of choice. While Einstein's standard procedure works well for *inertial* reference objects, it breaks down for certain non-inertial reference objects. For these, all synchronization procedures are nonstandard, and you are *forced* to choose, on a whim, if you insist on having system time.

You could say that a system time tells a lot about those who establish it but almost nothing about Nature. Nature, of course, cares as little about the convention on system time as an elephant cares about the annual tax return deadline. If you try to understand the properties of the elephant from tax rules, you risk ending up with something beside the point—analogously, if you think in terms of system time, you might end up mistaking synchronization artifacts for genuine physical phenomena. For example, *the famous time dilation and length contraction prove to be artifacts* of the particular synchronization method used to establish system time: with a different synchronization, you can obtain time *contraction* and length *dilation*. How can this be? See concrete examples in Chapter 13.

Spacetime!

So it turns out that neither space nor system time are adequate fundamental tools for the description of Nature. I still remember how disorienting it felt to be "betrayed" by space and system time, two notions that I had considered to be indispensable parts of any framework for describing Nature. There seemed to be no way out of this as I was and would forever be bound to perceive everything in terms of space and time, and they offer only a partial and subjective perspective on physical phenomena. Like the blind men in the parable, I can experience only a part of the whole.

Luckily, there is a way out: spacetime. However, spacetime is not immediately available to our senses. Still, just as the elephant statuette proved to be useful for the blind men to make the abstract whole of the elephant comprehensible, a mathematical model can serve as a useful tool to make the abstract whole of spacetime manageable to everybody bound to space and system time.

Tamás takes this even further by establishing a general framework for flat spacetime models, based on general human-scale experiences (see Figure 6). In turn, this framework can be specialized by additional assumptions like absolute time, leading to nonrelativistic spacetime (yes, spacetime exists nonrelativistically, too!) or absolute light propagation, leading to special relativistic spacetime. In my opinion, this framework is the most important contribution of this book compared to his earlier works on spacetime.

In this way, you can capture the essence of spacetime without reference frames or coordinate systems getting in the way. Moreover, since both nonrelativistic and special relativistic spacetime share a common framework, their easy comparison helps you come to terms with the "weirdness" of special relativity. Finally, flat spacetime models offer a good conceptual preparation for general, *curved* spacetimes without getting bogged down by the heavy mathematical machinery necessary for the latter.

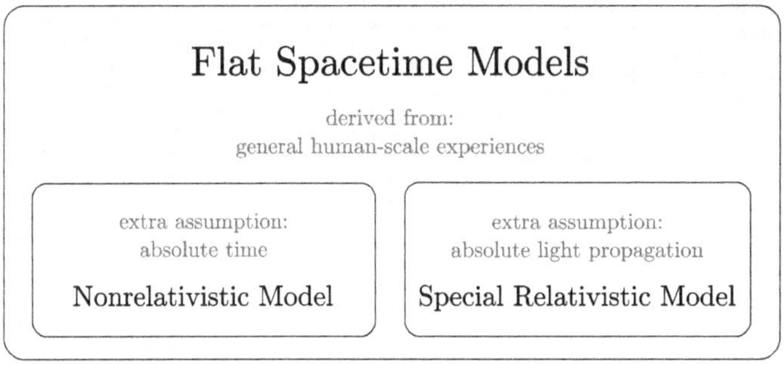

Figure 6: The general framework of flat spacetime models with two important special cases.

As nice as spacetime models might be for theory, you still live and operate within a frame of reference, inevitably perceiving spacetime as being decomposed into space and system time. Physical quantities decompose accordingly into spatial and temporal components. This is how, for example, the existence of a point-mass in spacetime becomes motion relative to a frame of reference. If you start with the decomposed quantities, glue them together, and decompose the result with respect to another reference frame, you obtain a transformation rule (see Patroklos' second diagram). I still remember the pleasant surprise when I first saw a transformation rule emerge as a consequence and not as a postulate after carrying this out (cf. Subsubsections 5.6.3 and 10.7.3).

Electromagnetism

Finally, let me share a result of spacetime models that had a profound impact on me. As you might recall, classical electromagnetism is governed by Maxwell's field equations in terms of two fields: the electromagnetic field strength and the electromagnetic displacement. The spacetime approach shows that these equations can be formulated without any reference frames, on spacetime itself. What is more, surprisingly, the *same equations* are valid both relativistically and nonrelativistically, implying that Maxwell's equations are *both* Galilean covariant and Lorentz covariant (see Subsections 7.3.2 and 12.3). This seems paradoxical since Einstein famously invented the theory of relativity to solve the problem that Maxwell's equations supposedly do not transform well under Galilean transformations.

The spacetime approach makes it easy to pinpoint the exact source of this apparent paradox: it is the link between the two electromagnetic fields. Nonrelativistically, this link is usually expressed in terms of permeability and permittivity, but a simple algebraic fact (see Subsection 7.4) shows that these quantities cannot be used in vacuum—that is, unless *a material* (with the traditional name Ether) *is postulated.*

Not so in special relativity, where the geometry of spacetime *offers a simple and natural link* between the two electromagnetic fields in vacuum (see Subsection 12.4). This means that there is no need for Ether in relativity, but on the other hand, this natural link—implicitly present in Einstein's works—inevitably introduces something manifestly relativistic, and thereby puts an end to Galilean covariance. This resolves the apparent paradox: Einstein worked with a manifestly relativistic version of Maxwell's equations, but Maxwell's equations are, in their essence, compatible with both the relativistic and the nonrelativistic frameworks.

Elephantology in thermodynamics

Péter Ván

Shortly after I started my university studies in physics, I became very interested in and eventually decided to focus on the understanding of the second law of thermodynamics. Tamás supervised my thesis research on the stability structure of ordinary thermodynamics and, in parallel, also taught me about spacetime models. As I was becoming more and more familiar with his spacetime approach, I experienced something obvious but important: that with clear concepts and notions, I can understand more about the world. As a professor today, this is what I strive to instill in my students, and this is the message of the regular summer schools I have been organizing since 2015.

Recall the method of "building a statuette" from earlier: you have to filter out the human elements from the model that have nothing to do with reality, and the best tool for this is forcing yourself to express your assumptions in mathematical form. Let me add something to this: just like a handle makes it easier to transport the elephant statuette, so can some subjective elements, even if in mathematical form, be useful for practical purposes. The best-known practical human elements are reference frames with a system time, units of measurement and all these together in a coordinate system.

Let me now outline a long-standing, open theoretical challenge where the spacetime approach can be immensely useful. To this end, something important about covariance needs to be discussed first.

General covariance—without gravity

Recall from earlier in Patroklos's part the "equal footing principle" regarding reference frames and its official name, *general covariance*. Although Einstein originally proposed this principle within his general theory of relativity, its validity extends beyond curved spacetimes and the theory of gravitation. Indeed, nonrelativistic theories should be generally Galilean covariant and special relativistic theories should generally be Lorentz covariant, that is, they should be able to be formulated on spacetime itself without reference frames. The book demonstrates this in Newtonian mechanics and classical electrodynamics.

Dissipation—objective and covariant?

My research area is continuum physics and non-equilibrium thermodynamics. Here, the apparent reference frame *dependence* of dissipation (for example energy loss due to friction, an objective physical phenomenon) is one of the most important and long-standing theoretical challenges.

Interestingly, it was not realized until the 1960s that this frame dependence is a problem in theory, and that is should be remedied. The requirement of general Galilean covariance was phrased as “the principle of material frame indifference”; however, the mathematical implementation of this principle was based only on rigid rotating frames of reference rather than on general ones. Although this first formulation is still useful for engineers today, it is unsatisfactory from both a general theoretical and a practical point of view. The goal is, of course, to obtain a generally covariant, spacetime-based formulation.

Why is this a difficult problem? In continuum theory, there is a distinguished reference body: the continuum itself. In this context, a spacetime-based formulation is far from evident, since the properties of the matter must be formulated in a way that takes into account the existence of the matter itself. This is particularly important in dissipative processes, since energy dissipation is objective, measurable, technologically significant and costly. It really isn’t the way to go to build separate theories for when a heat engine spins and then again for when your reference frame revolves around it; in any frame of reference the properties of dissipation should be derived from a single, spacetime-based law. There is a similar problem in chemical technology: in the current form of classical irreversible thermodynamics, the mechanical part of entropy production is reference frame-dependent.

A particle-based, statistical approach does not help much in any of these issues either. In the relativistic case, the very definition of a “fluid” is problematic, let alone what it means for it to *flow*. And what about a quark–gluon plasma, where fields and particles are not necessarily separable?

Furthermore, mathematical and physical problems are intertwined: the initial value problems of extended equations of heat conduction have no unique solution unless they are formulated using an appropriate “spacetime-based” time differentiation (usually called “objective time derivative”).

All these problems, and many more—as everything in physics—are connected with spacetime. I am convinced that the absolute, and therefore generally covariant, formulation of flat spacetime models presented in this book provides a concept together with a powerful set of tools for gaining insight into these problems and also offers the best line of attack. Tamás’s spacetime-based approach has been, and will be, a great source of inspiration in my research!

⋆

So, our dear reader, the book awaits you. We wish you just as much thinking, soul-searching, learning and, eventually, joy as we had while we were once traveling on this path. And don’t get scared when you ultimately do look physics right in the eye!

Thanks, Tamás, it was a privilege!

Patroklos Benatos first connected with Tamás during his undergraduate studies while searching for a more coherent framework in physics; he later completed his Master's thesis on spacetime under Tamás's supervision. During his doctoral program in the USA, he transitioned to the field of game theory. He later served as the director of a distance education program tailored for enlisted service members of the US Military. Currently, he is spearheading his cutting-edge initiative to turn academic game theory into actionable insights for business executives and policymakers and teaches game theory with applications at the Eötvös Loránd University in Budapest, Hungary. For more details, visit his LinkedIn profile.

Áron Szabó met Tamás during his graduate physics studies when, inspired by one of Tamás' courses, he got interested in mathematical physics. Áron typeset the lecture notes for a general relativity class for Tamás and this inspired him to pursue studies in mathematical physics and later a PhD in mathematics. He is a regular lecturer at the spacetime summer schools organized by Péter.

Péter Ván completed his Master's thesis under the supervision of Tamás Matolcsi and has been working in non-equilibrium thermodynamics ever since. He is interested in every aspect of thermodynamics starting from non-Fourier heat conduction through rock rheology all the way to relativistic dissipative fluids, thermodynamic gravity and quantum physics. He is the director of the Institute of Nuclear and Particle Physics of HUN-REN Wigner Research Centre for Physics, Budapest, Hungary and also works at the Department of Energy Engineering of Budapest University of Technology and Economics.

Preface

1. Special relativity arose because the naive concepts of space and time failed regarding light phenomena. Though special relativity became a well-working tool in modern physics, the usual fundamental concept of homogeneous and isotropic light propagation in every inertial frame and some conclusions contradict "common sense," which resulted and still results in several paradoxes. Since space and time play fundamental roles in every branch of physics, anything that is not completely clear concerning spacetime may yield serious errors in a theory.
   The difficulty for students is that in the theory of relativity they suddenly find themselves confronted with previously familiar concepts. Spacetime is usually introduced—if it is introduced at all—after a number of "strange" formulas. The novel approach of the present book is that employing a pedagogical strategy, it starts with spacetime explained by common sense experiences and assumptions.

   The idea of spacetime is very simple, it can be understood even in high school.

   Think of a meteor in the void; it meets a dust particle, absorbs it, then it pushes away another tiny object, and so on, and then it collides with another meteor and they blow up. If all that occurred near Earth, then we would tell that the meteor met the dust particles "here and there" at "this and that time" and the meteors blew up at "that moment" in "that place"; a space point and a time point, however, make no sense in the void.
   There are a lot of different objects in the universe. Their meetings, collisions, and other events are real physical facts, which do not determine "where" and "when" they occur. Considering such "tiny" objects pointlike in the huge universe, we can conceive that their events take place at something that unifies "where" and "when" as a whole: at a **spacetime point**. Note that the Earth, a house, a piece of stone, a soccer ball, etc. and even we are objects in the universe. A meteor hits the Earth, a ball goes into the net, I enter a house—all these are events occurring at some spacetime points.
   Spacetime itself is the background of such possible events. Since our space is three-dimensional and our time is one-dimensional, it seems evident that spacetime is a four-dimensional something.
   All these are simply understandable; further progress, however, requires some mathematical knowledge.
   The consistent mathematical formalism of the present book is as simple as possible. The affine structure of spacetime is easy and is close to everyday practice of physicists, and works even in a nonrelativistic context; moreover, it removes the gap between the usual formulations of special relativity (based on coordinate systems and transformation rules) and general relativity (based on spacetime itself, a differential manifold, and coordinates play only a practical rule).

https://doi.org/10.1515/9783112219553-204

2. Information about the world can be obtained by experimental devices. A collection of such devices is a **reference frame** with a built-in **coordinate system**, which represents results of experiments by numbers. Those results depend on the coordinate system in question. It was a natural way to eliminate this dependence by comparing results obtained in diverse reference frames by the aid of so-called **transformation rules**. According to that conventional method, a relation described in coordinates expresses a physical law only if it is "covariantly" transformed. This "coordinate-free description with coordinates," however, leads to misunderstandings.
   By now, quite a lot of knowledge has been gathered, so it is possible to apply a different method by reversing the way: starting with **absolute**, i. e., coordinate-free notions, the corresponding **relative** notions "are automatically transformed in a covariant way."
   This novel approach, achieved by constructing exact mathematical models for spacetime, has two important advantages:
   (i) Students realize that precise mathematical models are essential in understanding and eliminate errors.
   (ii) The perspective of spacetime broadens significantly the thinking of nature, which can be fruitful in effective researches in various branches of physics.
3. Usual considerations of space and time are based on
   - **Intuitive notions**, which are "well known, evident" for everyone, so there is no need to say more about them.
   - **Tacit assumptions**, which are some unspoken, "natural properties" of the intuitive notions.

   These, however, often lead to contradictions.
   This book is based on the following conceptions:
   (i) Everything must be explained in accordance with "common sense," which is extremely important from a didactic point of view.
   (ii) Relaying on the explanations, **exact mathematical models** must be constructed, i. e., all the intuitive notions must be mathematically formalized, which allows us to verify or reject tacit assumptions.
   (iii) **Only physical facts** must be taken into account in the construction of a model; coordinates are artificial objects, not physical facts.

   A spacetime model is a mathematical structure, a human picture of reality, and in fact only some aspects of reality. A model must contain everything we want to be reflected of reality and must be as simple as possible. There may be objects in a model, which do not correspond to some objects of reality.
   It took me a while to find some adequate mathematical definitions because it occurs that an intuitive notion has different names in different books or the same name in a book means different intuitive notions.
   My ideas on spacetime models has been formed in the course of university lectures. Students' comments helped me very much. For instance, at the beginning, long time age, a student told me that he found disturbing in relativity theory that "time passes

from bottom to top." I realized that, indeed, in everyday practice, in technical applications, in the elementary school, in both mathematical and physical studies at the university, even in Feynman's graphs "time passes from left to right," except special relativity. Since then, I apply "left to right" in the figures. The unfortunate "bottom to up" originated in the early formulation in which time coordinate was taken to be imaginary: *ict*.

The treatment of this book:

(i) Is arranged in such a way that the reader realizes that spacetime is a simple natural notion, which is not a special feature of relativity theory.

(ii) Explains all the principles in a way, which is palpable and natural for "common sense."

(iii) Makes clear the intuitive notions and tacit assumptions by giving them mathematical definitions.

(iv) Calls attention to the cardinal points of possible errors.

(v) Applies simple mathematical tools, which differ from the usual ones in the way of thinking; besides some elementary knowledge of set theory, continuity, differential calculus, and integral calculus only a deep knowledge of affine spaces, vector spaces, linear algebra, and a proficiency in their applications are required.

(vi) Gives mathematical definitions and statements in a colloquial manner, contrary to a strict succession of declared definitions and proved theorems what is customary in mathematics; proofs that can be skipped without any harm in understanding are typed in small size.

Part I presents all the notions—spacetime itself, world lines, future-like cones, observers, synchronizations, etc.—in a general setting of flat spacetimes with simple, natural, and clear meaning,

Part II constructs the nonrelativistic spacetime model based on absolute time progress, which corresponds to everyday thinking, and then examines in details its properties.

Part III constructs the special relativistic spacetime model based on absolute light propagation, in particular, derives the Lorentz form, and then examines in details its properties.

4. The nonrelativistic spacetime model and the special relativistic spacetime model with a number of special topics are published in the book,

   T. Matolcsi: *Spacetime without Reference Frames*
   Minkowski Institute Press, Quebec, 2020

   That book, however, does not treat flat spacetime models, does not construct the models from common sense experiences, and is somewhat hard from a didactic point of view.

Introduction

In the following, we examine the intuitive notions and tacit assumptions concerning space and time.

1. A reference frame in the literature is mostly considered to be a spatial coordinate system determined by three orthogonal axes realized, e. g., by the edges of a room. Intuitive notions ("which need not be discussed because everyone knows what they are") are

- The **right angle**.
- The **straight line**.
- The **space** in which the lines are fixed.

Tacit assumptions are that:

(i) There are straight lines in the space (how would such axes be realized in a room with rubber walls shaken by an earthquake?).

(ii) The straight lines (if they exist at all) are infinite (the edges of our room can be continued without limit in space).

(iii) Two nonparallel straight lines can have only one point in common, etc.

2. Distinguished importance is attached to inertial reference frames defined by "... the law of inertia, a material particle when left to itself will continue to move in a straight line with constant velocity" [12], "An inertial reference frame is a reference frame, with respect to which Newton's first law holds." [2] "... an inertial frame is one in which a free body ... is not accelerating." [1]

It is important to note that in the quoted definitions:

(1) A reference frame concerns only space ("straight lines" and "angles").

(2) An inertial frame implies time, too ("constant velocity" in Newton's law).

That is why it is *misleading* to say that an inertial frame is a special type of reference frames. Moreover, the definition of an inertial frame does not refer to any coordinate system, it implies only that there are straight lines in space.

Later, in Section 3.5 we shall define precisely the notion of a reference frame; for a clear distinction, in the following we refer to the intuitive notion in **1.** as a **spatial coordinate system**.

3. Constant velocity in Newton's law is based on

- The intuitive notion of **time**.
- The tacit assumption that *time passes uniformly*.

In general, velocity is based the intuitive notion of **the time during which a distance is covered**, which is highly problematic. To see the problem, let us consider an everyday situation, a foot race: how to measure the time elapsed between a racer's leaving the start and arriving at the finish line?

We mention two simple methods.

https://doi.org/10.1515/9783112219553-205

The first one requires two stopwatches and two umpires. The stopwatches are placed next to each other and started simultaneously. Then one of them is transported to the start, the other one is transported to the finish line. The umpire at the start stops the watch when the racer leaves and the umpire stops the watch at the finish line when the racer arrives. The desired duration is obtained as the difference of the values measured by the two watches.

The second method requires one stopwatch and one umpire who is equidistant from the start and from the finish line. The umpire starts the stopwatch when he/she sees the racer leave the start line and stops it when he/she sees the racer arrive at the finish line. This method is applied nowadays in a refined form: a computer starts and stops the watch by receiving ("seeing") radio signals from the start and the finish.

According to the superficial conviction, the two methods—and every other one—give the same result. Is it true or not?

The problem regarding the first method is that during the transport the watches are accelerated, shaken, and decelerated, which may influence their time keeping, even if we consider idealized devices. In everyday situations, we do not experience such an influence but it is not excluded that in extreme situations—e. g., in case of intense shaking—the influence could be significant.

The second method seems better because "it is commonly known that radio signals (electromagnetic wave, light) propagate everywhere in every direction at the same speed." But how do we know that light speed has this property? To know it, we should measure just the time during which a light signal covers a given distance. The astonished reader is asked to keep this vicious circle in mind; the exact meaning of homogeneous and isotropic light propagation will be explained in Subsection 9.1.3.

Recapitulating:

The time during which a distance is covered is a rather questionable intuitive notion.

For better understanding, we draw attention to that "the time during which a distance is covered" concerns two events occurring in two distinct space points: one of the events is the racer's starting, the other event is the racer's arrival. It is not necessary, however, for the two events to be connected to the same object; the problem is similar, e. g., if we are interested in the time elapsed in a football match between the first goals of the two teams. Summing up:

The time elapsed between two events occurring in two distinct space points is a rather questionable intuitive notion.

Note that this intuitive notion implies (by zero elapsed time) the notion of **simultaneity** of two events occurring in two different space points.

4. A usual fundamental principle is that "... all systems of inertia ... are completely equivalent in respect of our description of nature" [12]. "... all inertial frames are physically equivalent" [1]. "All inertial reference frames are equivalent for the performance of all physical experiments" [2].

It is not clear what the description of nature is and how the equivalence makes sense.

The equivalence is usually attempted to be explained by the relationship between spatial coordinate systems moving with respect to each other, as follows.

Let us consider a spatial coordinate system (X', Y', Z'), which moves at velocity $\boldsymbol{v}$ with respect to the spatial coordinate system (X, Y, Z) in such a way that the corresponding axes are parallel to each other and the axes X' and X coincide (see Figure 1).

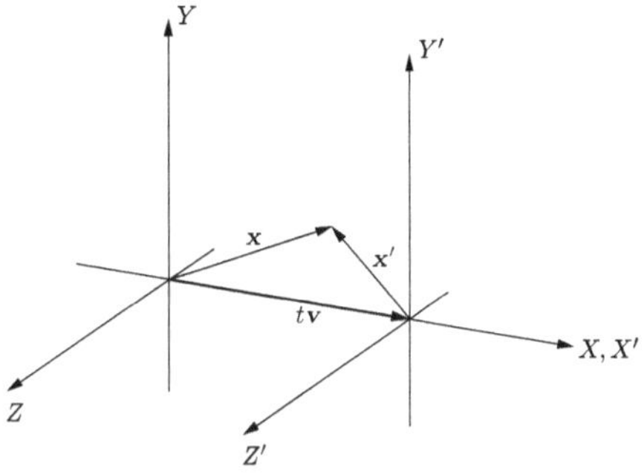

Figure 1: Galilei transformation rule.

Such a parallelism is usually taken "so evident that there is no need to speak about": we "see" in the figure and we "imagine" that a little earlier all the corresponding axes coincided.

In the "everyday" situation, one says that, according to the figure, t time after the meeting of the origins, the vector $\boldsymbol{x}$ of the "unprimed" system is observed by the "primed" system as the vector $\boldsymbol{x}' = \boldsymbol{x} - t\boldsymbol{v}$; in coordinates

$$x' = x - tv, \quad y' = y, \quad z' = z,$$

which is the well-known Galilei transformation rule.

Moreover, the figure suggests that the "unprimed" system moves with respect to the "primed" one at the velocity $-\boldsymbol{v}$.

In usual treatments of relativity theory, a similar figure (see Figure 2) shows two spatial coordinate systems having parallel axes and moving with respect to each other, but the vectors are not drawn at all, because they do not explain the Lorentz transformation rule:

$$x' = \frac{1}{\sqrt{1 - (v/c)^2}}(x - tv), \quad y' = y, \quad z' = z. \tag{1}$$

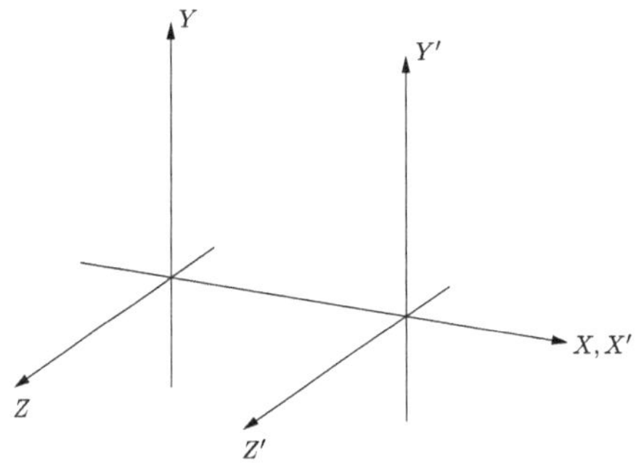

Figure 2: Parallel axes moving with respect to each other.

So, what the arrows of Figure 1 show "evidently" is not true in the relativistic case. Can we still attribute an "evident" meaning to the notion of parallelism of the corresponding axes of two spatial coordinate systems moving relative to each other? Is this notion so trivial that one does not need to clarify it?

5. Let us return to Figure 2, where we "see that earlier all the corresponding axes coincided."

This coincidence can be described as follows. Let a camera be on the axis Z of our (the "unprimed") spatial coordinate system at distance d from the origin 0. The distance of the camera from a point y of the vertical axis is $\sqrt{y^2 + d^2}$.

Let some curve C (this would correspond to the axis Y' of the moving spatial coordinate system) move in the X–Y plane at velocity v.

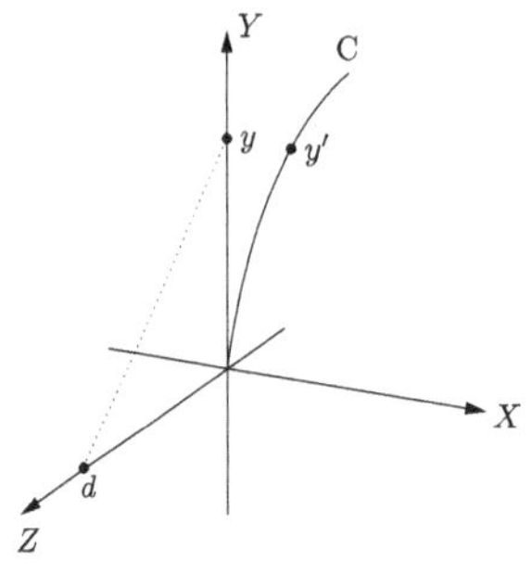

Figure 3: The curve C moving in the X–Y plane seems in a snapshot a straight line parallel to axis Y.

The camera makes a snapshot in which the images of our axis Y and C coincide. Snapshot means that the camera catches light rays from different objects at the same moment.

Let $0'$ and y' be the points of the curve C meeting the zero 0 and the point y of the axis Y. Light rays from these points arrive to the camera at the same moment.

The travel time of light from the points y and y' to the camera is $\sqrt{y^2 + d^2}/c$. As a consequence, the light from y and y' starts $\frac{1}{c}(\sqrt{y^2 + d^2} - d)$ earlier than the light from 0 and $0'$.

Therefore, when $0'$ meets 0 then y' has already left y; their distance is just $\frac{v}{c}(\sqrt{y^2+d^2}-d)$.

Consequently, at the moment of the snapshot the moving curve has coordinates

$$\left(\frac{v}{c}\left(\sqrt{y^2+d^2}-d\right), y\right)$$

in the X–Y plane: C *is not a straight line* (see Figure 3).

Our eyes (cameras) have misled us.

The attentive reader can observe that the previous reasoning applies the problematic notion of the time period between two events in two space points ("light starts earlier from y than from the origin"), and speeds v and c, which are problematic as well. This, however, does not query the validity of the reasoning: it is not evident at all what the parallelism of straight lines moving with respect to each other means.

Parallelism of straight lines moving with respect to each other is a rather questionable intuitive notion.

Remark. According to Figures 1 and 2, the "unprimed" system is usually called "rest frame" and the "primed" system is called the "moving frame." This distinction is unfortunate and misleading because neither of the spatial coordinate systems is at rest or moving in itself; *they are moving with respect to each other.*

Contents

Part III: **Absolute light propagation**

1 Heuristics of spacetime

In the following sections, we collect human-scale experiences about space, time, and motion, which serve as the basis for constructing spacetime models. It is emphasized that

a model is a mathematical structure, a human picture of reality, in fact, only of some aspects of reality.

For constructing a model, only physical facts must be taken into account, artificial object such as coordinates are avoided.

1.1 Space

Even though space is one of our simplest and most fundamental notions, it is worth taking a closer look at it when we want to build a mathematical model. Intuitively, space serves as a background with respect to which it is possible to tell *where* something happens. The following considerations help us to see the basic properties of spaces.

I am sitting in a room: a corner of the room, a spot on the carpet are points of my space, the table is a part of my space. More widely, the roads, the houses near my house constitute a space, which will be called room-space.

A car travelling along the road is a phenomenon, not a part of the room-space. On the other hand, the seats, the dashboard, etc. constitute the space for someone sitting in the car. The room-space is not a part of the car-space.

It is obvious then that the room-space and the car-space differ. We can assert:

Space is constituted by some material objects, there are different spaces; space is relative, i. e., there is no absolute space.

Different spaces can have very different properties: e. g., the space of our room differs from the space of a room with rubber walls shaken by an earthquake and differs from the space of a river (water) for a fish.

Luckily, it turns out that for constructing (a certain type of) spacetime models, we only need "well-behaving" ideal spaces whose points do not twist and whirl around:

Inertial spaces are determined by the physical fact that their space points are material points free of any action and not moving with respect to each other.

We have the following simple experiences.

($\mathcal{S}$1). *In an inertial space,* **there are straight lines**. *Between any ordered pair of space points, there is a* **vector**, *i. e., an oriented straight line segment. The set of such vectors obeys the well-known rules of addition and multiplication by real numbers.*

https://doi.org/10.1515/9783112219553-001

($\mathcal{S}$2). *An inertial space* **is three-dimensional**, *i. e., there are three essentially different directions—"right-left, ahead-back, up-down"—and these can be "combined" to obtain any other direction.*

It is a less simple experience but experiments verify (by the asymmetry of the decay rate of an elementary particle called kaon).

($\mathcal{S}$3). *An inertial space* **is oriented**, *i. e., "right-handed" order (right-ahead-up) and the "left-handed" order (left-back-down) are distinguished by nature.*

Distances and angles can be measured in space; a measuring rod for distance is physically defined, e. g., as a chain of a given number of quartz molecules free of any action. Distances and angles satisfy the well-known relations, e. g., the shortest path between two space points is a straight line segment, or the sum of the angles of a triangle is 180°, etc.; the collection of these relations is called a Euclidean structure. Then we state the following.

($\mathcal{S}$4). *An inertial space* ***has a Euclidean structure.***

We remind that the listed properties concern "human-size" experience. The question arises whether they are valid beyond human size, for instance, whether the vector between two stars of distant galaxies or the vector between two atoms of a crystal are meaningful. We do not answer such questions as we construct models extrapolating our human-size experience: a model is not the reality itself, it is a human depiction of reality.

1.2 Time

1.2.1 Time period, time point

Time, too, is one of our fundamental notions

Processes indicate that time passes: we are breathing, someone speaks, a clock ticks, the Sun moves across the sky.

In everyday speech as well as in physics, the word "time" can mean

- A time period (interval): e. g., "long time ago," "a lot of time passed."
- A time point (instant): e. g., "what is the time," "at the same time."

Moreover, the same device (clock) *measures time periods* and *indicates time points*, which helps to confuse the two distinct notions, time period, and time point, though it is extremely important to distinguish between them.

One could think that the time period between two time points is similar to the vector between two space points. This is not the case, however. Namely, both space points and the vector between two space points are "tangible" but this does not hold for time points and the time period between two time points, as we shall show below.

1.2.2 Time periods—proper times

Time passes: human beings, animals, plants, and even our tools, furniture, the rocks in the mountain—which are not living organisms—grow older.

We experience the flight of time through time periods. I perceive the time period between my breakfast and my dinner (by hunger). Similarly, you perceive the time period between your breakfast and your dinner. At risk of stating the obvious, I cannot perceive either the time period between your breakfast and your dinner or between my breakfast and your dinner.

In the same way, an elementary particle "feels" the time period between its events, e. g., between two collisions, in particular, between its creation and decay.

We can accept the following basic fact:

For every material point time progresses individually, i. e., every material point has its own time, called **proper time**.

Visualizing this fact, we imagine that a minute quartz crystal is tied to every material point and the oscillations (ticks) of that crystal measure the time passed for the material point. Such a minute device measuring the progress of time will be called a **chronometer**. We intentionally do not say "watch" or "clock" in order to avoid the possibility that the everyday properties of our usual device mislead our thinking; a chronometer has no dial, it does not show what the time is ("it is two o'clock" makes no sense in this respect), it measures only the time passed—the number of ticks—between two arbitrary events of the given material point.

We have the following fundamental experiences:

($\mathcal{T}$1). *The proper time of any material point is* **one-dimensional** *because only forward (future) and backwards (past) exist.*

($\mathcal{T}$2). *The proper time of any material point is* **oriented** *because future and past cannot be interchanged.*

It is emphasized that all these are valid for individual proper times. We have not stated anything about the relation of individual time progresses.

It may happen—why not?—that different amounts of proper time pass for two material points between their two meetings. For instance, let us consider me and my home as two material points. I leave my home, I fly to another country for holiday, then I return; it is not evident that between my departure and arrival my chronometer has ticked just as many times as the chronometer left at home. According to our everyday experience, the two proper times that elapsed between two meetings of material points are equal, or better to say, their difference does not exceed the practical error of measurements. But this experience refers to "mild" circumstances. Let us consider two chronometers resting together and then one of them being left there and the other one being revolved again and again very fast in a centrifugal machine (without causing harm to it); is it certain that the two chronometers have measured the same number of ticks between their meetings?

1.2.3 Time points—simultaneity, synchronizations

It is worth repeating:

A proper time period is a physical reality.

Such a proper time period—the number of ticks of a chronometer—is meaningful only between two events of the same material point. The time period between "my breakfast" and "your dinner" cannot be measured by a chronometer (unless we are together).

What is a time point? What is 12 o'clock? What does it mean that John in London and Pierre in Paris has dinner at 12 o'clock?

What does "at the same moment" in different space points mean? More widely: what does "a time period between two events" in different space points mean? Chronometers cannot answer these questions.

This problem has been raised in the Introduction considering a foot race where the stopwatches are chronometers according to our definition.

Now it will be examined how it can be defined, how it can be stated that two events far from each other occur at the same time point (at the same moment), i. e., simultaneously.

Our everyday experience regarding human-size phenomena is based on seeing. What is happening **now** in the street? Looking out the window, we can see it. We can have doubts, however, since what we see does not happen "now" but happened a little "earlier" because light "requires time"—even if a very little— to arrive at us from the street. What is happening **now** in a town 1200 kilometers far from us? The answer is not so immediate as previously. Two fireworks explode, one in London, the other in Paris: how can we determine whether the explosions are simultaneous or not? Saying that both occurred at 12 o'clock puts off the answer because the question remains: how is 12 o'clock established in London and in Paris?

Proper time progresses (a chronometer ticks) in each space point but these proper times are not related to each other. A quartz crystal oscillates in London, another one oscillates in Paris, measuring how time progresses; they do not give information about whether it makes sense at all when a tick in London is simultaneous to a tick in Paris. In particular, they do not tell, e. g., when 12 o'clock in the two towns is.

Simultaneity can be constructed by some procedures. A **synchronization** means establishing simultaneity in a continuous manner, creating **time points**. We give an overview of how the synchronization was developed throughout history.

The ancient method was based on the position of the Sun in the sky. In the Middle Ages, every colony established its own simultaneity. The church was the synchronization center. The bell rang at church midday, i. e., when the Sun reached its highest position in the sky, and midday at any point of the colony— the instant simultaneous with church midday—was the instant when the ringing arrived there. I will tell a story now. A man had bought a watch and set it to 12 o'clock when hearing the ringing in his house next to the church. The next day he went to work on his field 5 miles away. Hearing the bell ringing, there he saw that his watch showed more than 12 o'clock, so then he reset the clock. The next day he was again at home and saw that the watch showed less than 12 o'clock. He was angry that the merchant had sold him a faulty watch. The watch, however, was working fine. (The reader is kindly asked to put aside that this story could be true only with high precision of hearing, seeing, and setting.)

In the 19th century, Greenwich became the synchronization center in Europe. In a given day, midday was defined in Greenwich by the highest position of the Sun in the sky. Then midday in the towns of Europe—the instant simultaneous to Greenwich midday—was defined by a position of the Sun before or after its summit by a given angle corresponding to the degrees of longitude of the towns. Of course, to determine simultaneity in this way requires highly precise measurement of directions and angles. The inaccuracy of measurement results only in a practically negligible error.

In the 20th century, up to 1970, Greenwich remained the synchronization center; a radio signal (a pip) started at Greenwich midday towards the towns of the world; knowing the distance of the town from Greenwich and taking the speed of light to be 299792458 m/s, the travel time of the signal is calculated and the midday in the towns was set correspondingly.

Nowadays synchronization is established on Earth according to the position of stars, by a complicated method using hundreds of cesium chronometers on Earth and on satellites, which communicate with radio signals.

The presented development history exhibits well that simultaneity in itself does not exist, the above mentioned methods **define** simultaneity, perhaps different ones. Indeed, the simultaneity determined by the position of the Sun differs from that obtained by radio signals as it will be shown (see Subsection 10.6.1); the difference, though being practically negligible, is theoretically important:

> Synchronization is a human convention, not a physical reality. According to our basic principle, synchronizations—consequently, time points—cannot play a role in constructing a spacetime model.

Time points of a synchronization have nothing to do with the proper time periods of a chronometer, which is well shown by the story of the church bell.

We can realize in principle the time of a synchronization in such a way that a device, called **synchroindicator** is put in each space point showing continuously the instants of the synchronization.

Up to now, we have illustrated simultaneity concerning space points far from each other on Earth but the same will hold for places near each other, e. g., those in a building, too. For instance, the ringing of class bells give in every classroom the same instant of the end of a lesson.

This way we have two different devices, the chronometer **measuring** only proper time periods of material points, and the synchroindicator **showing** only time points of a synchronization. It is important that a synchronization time point cannot be measured. To emphasize this, we have spoken about instants (midday, end of lessons), which can be described without numbers.

It cannot be excluded that different proper times pass between two synchronization instants in two different space points. Between yesterday midday and today midday, the chronometer in Paris might not tick the same number of times as the chronometer in London. If this is not experienced within a practically negligible error, then a chronometer and a synchroindicator can be united into a single device, a clock which represents synchronization time points by the corresponding proper time periods passed from a reference instant. Clocks, however, must be reset occasionally.

The previous considerations show that the synchronization on Earth is a delicate matter. And we face a new problem concerning synchronization in cosmic region. Our system of synchronization establishes time points in the space points of Mars, too. Let us suppose that intelligent creatures live on Mars who establish time points with their synchronization system similar to ours. Do the two systems define the same synchronization time points on Mars or not?

The movement of the Sun in the sky—i. e., the rotation of Earth—made it evident how to determine simultaneity on Earth; this notion of simultaneity, together with that based on seeing, was developed and became a part of our everyday life long ago and simultaneity seemed a law of nature even in physics.

The fiction of absolute simultaneity was ruled out by the theory of relativity but, unfortunately, this is not evident in the formulation with coordinates and is rarely emphasized. All the paradoxes I know, formulated against the theory of relativity, involve implicitly absolute simultaneity or some of its consequences.

1.3 Movements

1.3.1 Paths of movements

We experience that a material object can move relative to us, i. e., it changes its position in our space. The same material object moves differently with respect to different spaces. A bird flies, it moves differently with respect to the road and with respect to the car traveling on the road; the bird does not move in itself: it exists. In general, we have the following.

A material body exists and its existence (history) is perceived in different spaces as different movements: motion is a relative notion.

Properties of movements give us important information about space and time.

The **path** of a movement of a material point **in a space** is the collection of the space points met by the material point.

Our first simple experience is the following.

($\mathcal{M}$1). Every straight line *in an inertial space can be the* **path of a movement**.

1.3.2 Uniform motion on a straight line

One of the most important laws of physics is Newton's law of inertia, "a material particle when left to itself will continue to move in a straight line with constant velocity." We have seen that this formulation requires a synchronization, and even a special one. Namely, it may happen that in the same movement "equal distances are covered over equal times" with a synchronization, but "different distances are covered over equal times" with another synchronization. According to our principle, synchronization cannot be used in the construction of a spacetime model, so we look for another formulation. First, we can say the following.

($\mathcal{M}$2). The path of an inertial material point *(free of forces) in an inertial space is a single point or a* **straight line**.

The uniformity of a motion can be defined without synchronization by a little but far-reaching tricky cooperation of the body and the space in which it moves. The body marks the space points that it meets one after another at the ends of a prescribed, given proper time interval and then the distance is measured between these marked space points.[1] Then we can state:

An inertial material point in an inertial space is at rest or moves along a straight line in such a way that it covers equal distances during its equal proper time periods.

which will be referred to as the following.

($\mathcal{U}$). *If an inertial material point moves in an inertial space, then there is a* **uniform relation** *between inertial proper time periods of the material point and the corresponding space distances traveled.*

1 This is the method by which a car determines its own speed: it measures how many times its wheels turn in one second of its proper time and uses the circumference of the wheels to indicate the distance traveled.

1.3.3 Faster-slower

Let us return to the foot race. Two racers leave the start line together (their leaving is the same occurrence of the start line) and run along the same path (in fact, their paths are beside each other but the racers are considered pointlike in such illustrating examples, so the paths can be considered the same) but they arrive at the finish line separately (their arrivals are different occurrences of the finish line). By proper time passing, the notion of earlier-later is meaningful at every space point, in particular at the finish line, too. The racer who arrives earlier is faster, the one who arrives later is slower. So, we know which of the racers is faster without knowing what their time results are (regardless of how it is measured). In other words, we know which of them is faster without knowing how fast they are (what their speeds are).

Then we can declare the following important statement:

For two movements along the same path in the same direction, it makes sense without synchronization which of them is faster (slower).

As regards the comparison of movements in an inertial space, we experience the following.

($\mathcal{M}$3). *For every inertial movement* **along a straight line**, *there is another,* **faster** *inertial movement* **along the same line in the same direction**.

($\mathcal{M}$4). *For every inertial movement* ***along a straight line***, *all the* **slower** *inertial movements are possible* ***along the same line in the same direction***.

The content of ($\mathcal{M}$3) is evident. ($\mathcal{M}$4) is explained as follows. Let us consider the movement of a racer1 along a path between a start and a finish. For an arbitrary (long) proper time interval $\boldsymbol{t}$, there is another racer2 who starts together with racer1, moves along the same path and arrives at the finish $\boldsymbol{t}$ time after racer1.

Lastly, we emphasize that faster-slower along the same path is independent of synchronization but can depend on spaces. Let a swallow and a sparrow fly in the same direction above a road; with respect to the road, the swallow is faster than the sparrow. Let a car travel on the road in the same direction; then the birds fly backwards with respect to the car, the sparrow faster than the swallow.

1.3.4 Round-way speed

There is a further, extremely important notion in connection with movements.

First of all, in the sequel we make a clear distinction between velocity and speed. **Velocity** will mean a vector, **speed** will mean a scalar (which can be the magnitude of a velocity but not necessarily).

Let us consider a race where the start and the finish coincide. Then the time period between the starting and the arrival of a racer is measured by the chronometer of the start-finish, and knowing the length of the path run through, we can say without any synchronization how fast the racer is. In general:

The **round-way speed** for movements on a closed path, in particular, the **two-way speed** for movements on a straight path to and from are meaningful without synchronization.

1.4 Measure lines

Treating time and space, we meet the problem of measuring time periods and distances. In usual treatments (using coordinates), time periods as well as distances are given by real numbers but 3 is neither a time period nor a distance; 3 seconds, 3 hours, and 3 years are time periods, 3 meters, 3 yards, 3 miles are distances where the units of measurement, seconds, meters, etc. are intuitive notions. Moreover, second, meter, etc. are human conventions and the use of the corresponding real numbers can result in serious

(i) Technical problems; e. g., a space rocket destroyed because a real number a meant a yards and a meters in a cooperation between English engineers and American engineers.
(ii) Theoretical problems; e. g., years of research were in vain because the basic mathematical object of quantum field theory, the C^*-algebra of physical quantities is a nonsense: different quantities such as position (meter) and momentum (kilogram × meter/second) cannot be added.

If we want to avoid such problems—and we want—then we have to go beyond the real numbers and to construct a precise mathematical model of physical dimensions (units of measurements). Of course, such a model must be based on the common properties that we have got accustomed to. Namely,

- The double, the half, etc. in general, the nonnegative multiples of a time period (distance, etc.) make sense.
- Two time periods (two distances, etc.) can be added.
- A time period and a distance cannot be added.

The way of construction can be find in Subsection 18.2. The result is the following.

($\mathcal{D}$). Units of measurement *of a physical quantity are modeled by positive elements of an oriented one-dimensional real vector space, called a* **measure line**.

Every usual rule of multiplication and division of measure line elements are well-defined in this context, e. g., m^3, $\frac{\mathrm{m}}{\mathrm{s}}$, etc. have a precise meaning, so the reader can handle measure lines without knowing their exact construction.

Part I: **Flat spacetime models**

2 Construction of flat spacetime models

In the following sections, we construct a special type of spacetime models, which we call the **flat spacetime model**, by giving exact mathematical formulas for the heuristic assertions ($\mathcal{S}$**1**)–($\mathcal{S}$**4**), ($\mathcal{T}$1)–($\mathcal{T}$2), ($\mathcal{M}$1)–($\mathcal{M}$4), ($\mathcal{U}$), using ($\mathcal{D}$).

It is emphasized again that all the assertions above are extensions of human-size experiences to arbitrarily "large" and "small" size.

Our basic requirement is that all functions involved in the construction be continuous (in a convenient sense).

2.1 Spacetime

The mathematical notions included in this subsection are treated in Section 18.5 of the mathematical supplement.

In everyday practice, we characterize events by where and when they occurred. "Where" refers to space points, "when" refers to time points. We have seen in preceding subsections that various space points and various synchronization time points can be associated to the same event. Events take place at something that unifies "where" and "when" as a whole:

> We can conceive a **spacetime point** as "here and now" or "there and then" keeping in mind that "here and now" is an **absolute entity** not determining "here" and "now."

Spacetime points are usually called "events," which, unfortunately, has led to some misunderstandings. Namely, the word "event" has a definite meaning in probability theory applied in several areas of physics. A collision is an event of two balls, the explosion is an event of a firework: all these are not events of spacetime. They illustrate a spacetime point. The same spacetime point can be illustrated by different events: the collision of two balls as well as the explosion of a firework may occur "here and now."

That is why we avoid the terminology "event" and we say spacetime point or **world point** and a physical phenomenon illustrating a world point will be called an **occurrence**.

Properties ($\mathcal{S}$1)–($\mathcal{S}$3) in Section 1.1 suggest that an inertial space is a three-dimensional oriented affine space. ($\mathcal{T}$1) and ($\mathcal{T}$2) in Subsection 1.2.3 suggest that the proper time of an inertial material point is an oriented one-dimensional affine space. Finally, we can state according to properties ($\mathcal{M}$2) and ($\mathcal{U}$) in Subsection 1.3.2 that these affine structures are not independent.

That is why we accept:

The basic notion of a flat spacetime model is **spacetime, a four-dimensional oriented affine space**.

---

https://doi.org/10.1515/9783112219553-003

The affine structure allows us to formulate our experience that there is no distinguished part of spacetime: what may occur "here and now," may occur "there and later" (spacetime is "homogeneous").

The affine spacetime and the underlying vector space will be denoted by M and **M**, respectively.

It is important that the **world points** (elements of M) are different from the **world vectors** (elements of **M**). This difference gets lost in treatments based on coordinates where both world points and world vectors are represented by quartets of numbers.

The affine spacetime M will be illustrated by the plane of the page (see Figure 2.1). Though such an illustration is useful, we have to be cautious; we must not attribute the usual properties of the plane to spacetime. For instance, two points in the plane have a distance between them but two world points do not.

Figure 2.1: World points.

The vector between two world points is illustrated by an arrow between the points (see Figure 2.2).

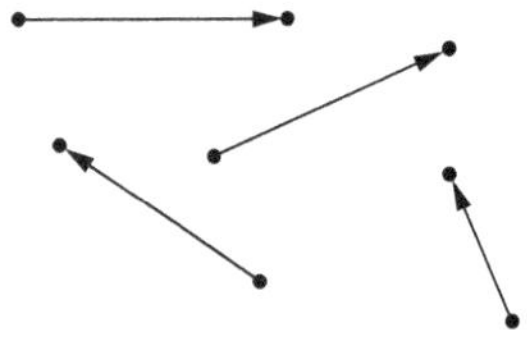

Figure 2.2: Affine structure of spacetime: There is a vector between each pair of world points.

Since we have no other choice, the vector space **M**, too, will be illustrated in the plane of the page. Then vectors are shown by arrows starting from zero (see Figure 2.3). Though such an illustration of vectors is useful, we must be careful not to be deceived; in general, the length of a world vector and the angle between two vectors are not meaningful.

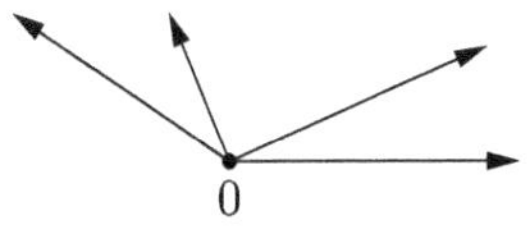

Figure 2.3: World vectors.

We indicate, if necessary, whether M or **M** is illustrated by the plane of the page.

2.2 Future-like vectors

2.2.1 World lines

The mathematical notions included in this subsection are treated in Subsection 18.7.1 of the mathematical supplement.

The life (history) of a tiny material object, a point-like body, is a continuous sequence of its occurrences, which have a definite order according to how its proper time passes: it is meaningful, which of two occurrences of the same material point is earlier or later than the other (e. g., if I am considered such a body, then "my having breakfast" is earlier than "my having dinner"). This earlier-later relation in the life (history) of a material point means an **orientation**. The proper time of the material point is one-dimensional, the occurrences of the material point are illustrated by spacetime points, so we accept:

The life (history) of a material point in spacetime is a **connected, oriented curve**, called a **world line.**

Figure 2.4 illustrates a world line, the arrow shows its orientation, i. e., the earlier-later relation is expressed by "from left to right," which corresponds to the everyday practice (beginning from the elementary school to an arbitrary technical application) that if a process is visualized then "time passes from left to right."

Figure 2.4: A world line.

Some oriented curves in spacetime are world lines, some others are not. If a curve with a given orientation is a world line, then with the opposite orientation it is not. Moreover, there are curves that cannot be world lines with any orientation. This can be made palpable as follows. Let us consider a string of lamps for decoration in an otherwise dark room. The lamps can flash "one after the other" or "at the same time." In the first case, the sequence of flashes corresponds to a world line because they can be obtained by a single lamp moving along the string and blinking. In the second case, a single lamp cannot produce the flashes.

In the following, we characterize what kind of curves can be world lines in spacetime.

First of all, according to the affine structure, there is no distinguished part of spacetime, "what may happen here and now, may happen there and then," therefore, every world line translated by an arbitrary world vector is a world line, too, as shown by Figure 2.5 and Figure 2.6.

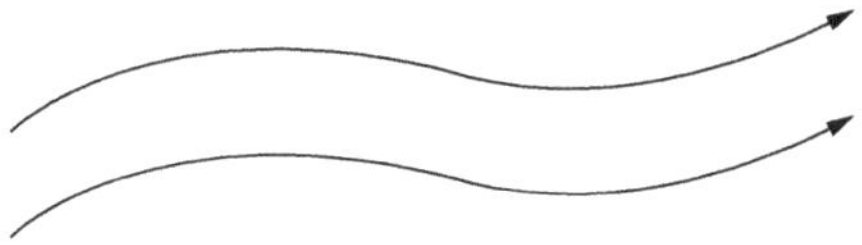

Figure 2.5: Translated world lines.

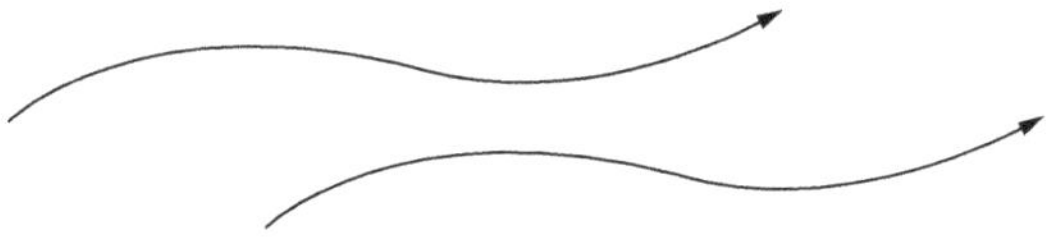

Figure 2.6: Translated world lines.

2.2.2 Inertial world lines

Properties ($\mathcal{M}$**1**) and ($\mathcal{M}$**2**), together with the affine structure of spacetime, suggest us to accept:

The world line of an inertial material point is an oriented straight line.

An oriented straight line is determined uniquely by an arbitrary point of its and a direction vector. Since the orientation of a world line expresses earlier-later, a direction vector "goes from an earlier point to a later one":

The direction vectors of inertial world lines are called **future-like**.

The translations of an inertial world line are inertial world lines, having the same direction vectors, Accordingly, in order to determine the set of possible inertial world lines, we need to specify the possible direction vectors.

We know that the orientation of a straight line means a "half line" of its direction vectors; therefore, the positive multiples of future-like vectors must be future-like (in other words, the set of future-like vectors is a cone whose apex is the zero vector) and the negative multiples of future-like vectors must not be future-like.

Furthermore, properties ($\mathcal{M}$**1**), ($\mathcal{M}$**3**), and ($\mathcal{M}$**4**) in Section 1.3 imply that the set of future-like vectors is convex and open (which will be proved in Section 2.6 after having acquired more knowledge). Therefore,

---

The structure of a flat spacetime model must contain the set of **future-like vectors**, a proper subset $T^{\rightarrow}$ of $\mathbf{M} \setminus \{\mathbf{0}\}$, which is an open and convex cone whose apex is at zero.

It is worth detailing: $T^{\rightarrow}$ is open and:

(1) If $\boldsymbol{x}$ is in $T^{\rightarrow}$ and $\alpha > 0$ is a real number, then $\alpha\boldsymbol{x}$ is in $T^{\rightarrow}$ (cone)

(2) If x, y are in $T^{\rightarrow}$ and $\alpha, \beta \geq 0$ are real numbers such that $\alpha + \beta = 1$, then $\alpha x + \beta y$ is in $T^{\rightarrow}$ (convex)

which are equivalent to

⋆ If $x, y \in T^{\rightarrow}$ and $\alpha, \beta \geq 0$ are real numbers, $\alpha + \beta > 0$, then $\alpha x + \beta y \in T^{\rightarrow}$.

The zero vector (the apex of $T^{\rightarrow}$) does not belong to $T^{\rightarrow}$.

$T^{\rightarrow}$, being convex, is a connected set.

Figure 2.7 shows three essentially different forms in three dimensions. The first one is the half-space on the right side of a plane, the second one is the part between two half planes, the third one is just a customary cone inside.

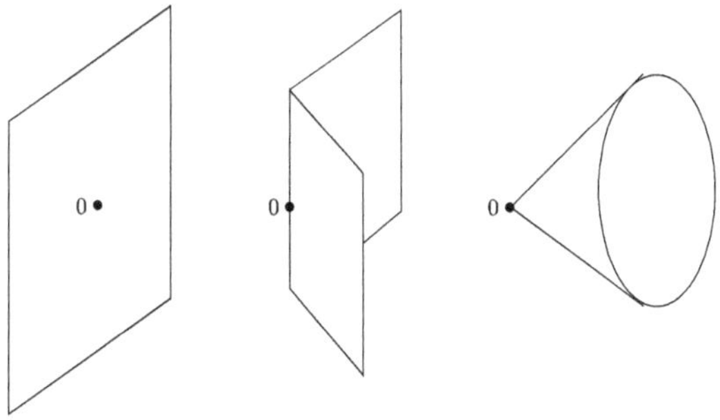

Figure 2.7: Future-like cones.

In our illustration in the plane of the page, the first and the third possibilities take the form shown in Figure 2.8. As for the second possibility depicted in Figure 2.7, different "natural" two-dimensional illustrations can be drawn, according to the point of view.

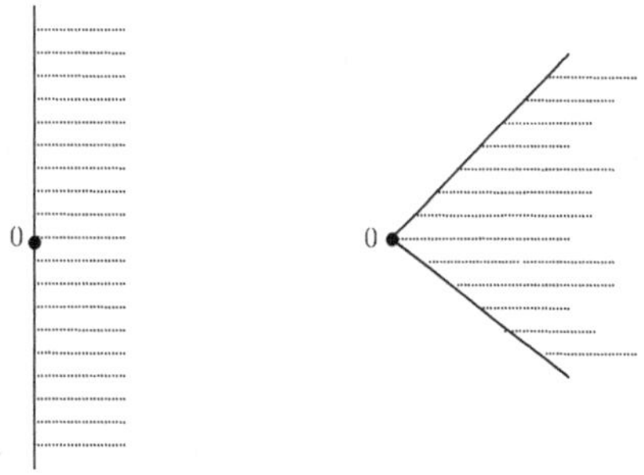

Figure 2.8: Future-like cones.

Once we have a cone of future-like vectors, we can define a few other notions. A vector opposite to a future-like vector is called **past-like**, their set, $T^{\leftarrow} := -T^{\rightarrow}$ is disjoint from $T^{\rightarrow}$. A **time-like** vector is either future-like or past-like; their set is $T^{\leftarrow} \cup T^{\rightarrow}$.

If x and y are world points and $y - x \in T^{\rightarrow}$, then y is called **future-like with respect to** x. Accordingly, $x + T^{\rightarrow} = \{x + \mathbf{x} \mid \mathbf{x} \in T^{\rightarrow}\}$ is the set of world points future-like with respect to the world point x.

A similar definition, according to the sense, holds with $T^{\leftarrow}$ and "past-like" instead of $T^{\rightarrow}$ and "future-like."

The world points x and y are called **time-like separated** if $y - x$ is a time-like vector.

2.2.3 Properties of world lines

Since "on a small scale" every sufficiently well-behaved curve is approximately straight, it is reasonable to accept the definition:

A **world line** is an oriented curve whose tangent vectors are time-like.

This means that if p is a parametrization of a world line, then $\dot{p}(a)$ is time-like (so it is not zero) for every parameter value a. The set of time-like vectors is the disjoint union of the set of future-like vectors and the set of past-like vectors; the derivative of a parametrization is continuous; thus, every $\dot{p}(a)$ is future-like or every $\dot{p}(a)$ is past-like. The parametrization is called **progressive** if $\dot{p}(a)$ is future-like.

World lines have the following fundamental property:

The vector between arbitrary two different occurrences of a world line is time-like.

In formula: if x and y are different points of a world line C, then $y - x \in T^{\leftarrow} \cup T^{\rightarrow}$. Equivalently: if x is an arbitrary point of a world line C, then any other point of C is future-like or past-like with respect to x:

$$C \setminus \{x\} \subset x + (T^{\leftarrow} \cup T^{\rightarrow}) \quad (x \in C).$$

Let p be a progressive parametrization of C and let a be an arbitrary parameter value. Then $p(b) - p(a) = \dot{p}(a)(b - a) + \mathrm{ordo}(b - a)$; $\dot{p}(a)$ is in $T^{\rightarrow}$, which is an open set, therefore, $p(b) - p(a)$ is future-like for $b > a$ in a neighborhood of a. Let us suppose that there is an s in the domain of p such that $p(b) - p(a) \in T^{\rightarrow}$ for all $a < b < s$ and $p(s) - p(a) \notin T^{\rightarrow}$. Then $p(s) - p(b) = \dot{p}(b)(s - b) + \mathrm{ordo}(s - b)$; thus, $p(s) - p(b)$ is future-like for b sufficiently near s and $b < s$. $T^{\rightarrow}$ is a convex cone, so $(p(s) - p(b)) + (p(b) - p(a)) = p(s) - p(a) \in T^{\rightarrow}$; this contradiction shows that $p(b) - p(a)$ is future-like for all $b > a$ in the domain of p.

We can argue similarly for the past-like part of C.

This assertion implies that if x and y are world points and:
(i) $y - x$ is time-like, then there is world line—e. g., a straight line (an inertial world line)—passing through them.
(ii) $y - x$ is not time-like, then there is no world line passing through them.

2.3 Progress of time

2.3.1 Inertial proper times

First of all, we accept:

The structure of a flat spacetime model must contain the **measure line of time periods**, denoted by $\mathbb{T}$.

According to ($\mathcal{U}$) in Subsection 1.3.2, we accept:

Proper time passes "uniformly" on inertial world lines.

This is formulated in the model in such a way that the proper time period between two occurrences of an inertial world line depends only on the future-like vector between them and, moreover, the proper time period corresponding to the double, triple, etc. of a future-like vector $\boldsymbol{x}$ is the double, triple, etc. of the proper time period corresponding to $\boldsymbol{x}$.

So, we have to assign a positive element $\boldsymbol{P}(\boldsymbol{x})$ of $\mathbb{T}$ to every future-like vector $\boldsymbol{x}$ in such a way that

$$\boldsymbol{P}(\alpha\boldsymbol{x}) = \alpha\boldsymbol{P}(\boldsymbol{x}) \tag{2.1}$$

for all real numbers $\alpha > 0$; $\boldsymbol{P}$ has the physical meaning that if x and y are world points, and y is future-like with respect to x, then $\boldsymbol{P}(y - x)$ is the proper time passed on the straight world line from x to y (inertial proper time elapsed from x to y).

Formula (2.1) is referred to as $\boldsymbol{P}$ is **positive homogeneous.**

Moreover, as an elementary requirement, $\boldsymbol{P}$ must be continuous. Summing up:

The structure of a flat spacetime model must contain the **inertial time progress**, a positive homogeneous and continuous function $\boldsymbol{P} : \mathrm{T}^{\rightarrow} \to \mathbb{T}^{+}$.

2.3.2 Absolute velocities

Future-like vectors are direction vectors of inertial world lines, positive multiples of such a direction vector are direction vectors, too. Time progress makes it possible to introduce some "unit" direction vectors, by dividing future-like vectors by the corresponding time period, which results in the following:

An **absolute velocity** is a "straight spacetime path covered over an inertial proper time period."

The future-like vectors are elements of $\mathbf{M}$, the time periods are elements of $\mathbb{T}$; thus, the quotient will be in the tensorial quotient $\frac{\mathbf{M}}{\mathbb{T}}$. The precise meaning of such a quotient can be found in the mathematical supplement.

Then the set of absolute velocities is

$$V(1) := \left\{ \frac{\boldsymbol{x}}{\boldsymbol{P}(\boldsymbol{x})} \;\middle|\; \boldsymbol{x} \in T^{\rightarrow} \right\} \subset \frac{\mathbf{M}}{\mathbb{T}}. \tag{2.2}$$

An element of $\frac{\mathbf{M}}{\mathbb{T}}$ is called future-like if it can be written as the quotient of a future-like vector by a positive time period. With this, the definition above can be made more palpable, extending the definition of $\boldsymbol{P}$ to the future-like elements of $\frac{\mathbf{M}}{\mathbb{T}}$ by the property of positive homogeneity (as for linear maps): if $\boldsymbol{x} \in T^{\rightarrow}$ and $\boldsymbol{t} \in \mathbb{T}^{+}$, then $\boldsymbol{P}(\frac{\boldsymbol{x}}{\boldsymbol{t}}) := \frac{\boldsymbol{P}(\boldsymbol{x})}{\boldsymbol{t}}$. Accordingly,

$$V(1) = \left\{ \boldsymbol{u} \in \frac{\mathbf{M}}{\mathbb{T}} \;\middle|\; \boldsymbol{u} \text{ is future-like, } \boldsymbol{P}(\boldsymbol{u}) = 1 \right\}. \tag{2.3}$$

This explains the number 1 in the notation of V(1).

Note that $\boldsymbol{P}(\boldsymbol{t}\boldsymbol{u}) = \boldsymbol{t}$ for $0 < \boldsymbol{t} \in \mathbb{T}$ and $\boldsymbol{u} \in V(1)$. Because of the positive homogeneity of $\boldsymbol{P}$, if the absolute velocities $\boldsymbol{u}$ and $\boldsymbol{u}'$ are parallel, then $\boldsymbol{u} = \boldsymbol{u}'$.

The word “velocity” inevitably reminds us of our everyday notion, therefore, it is worth keeping in mind that

- *There is no zero absolute velocity.*
- *An absolute velocity has no magnitude; in particular, it makes no sense for an absolute velocity to be higher or lower than another one.*
- *The angle between two absolute velocities is not meaningful.*

Remark. $T^{\rightarrow}$ is four-dimensional; if $\boldsymbol{P}$ is differentiable, then the condition $\boldsymbol{P}(\boldsymbol{u}) = 1$ reduces the dimension by one, so V(1) is a three-dimensional submanifold.

2.3.3 Proper time progress on a world line

We conceive that the proper time passed on a small piece of a world line equals approximately the inertial proper time corresponding to the straight line segment approximating the small piece in question. Approximating the world line by a broken line consisting of small straight line segments, we expect to get an approximation for the proper time progress on the world line.

In formula: if p is a progressive parametrization of the world line C and $a_1 < a_2 < \cdots < a_{n+1}$ are parameter values, then $p(a_{k+1}) - p(a_k)$ is future-like for all k, according to our previous result in Subsection 2.2.3. So, $\boldsymbol{P}(p(a_{k+1}) - p(a_k))$ makes sense, and the proper time passed on the world line between the occurrences $x := p(a_1)$ and $y := p(a_{n+1})$ is approximated by

$$\sum_{k=1}^{n} \boldsymbol{P}(p(a_{k+1}) - p(a_k)) \approx \sum_{k=1}^{n} \boldsymbol{P}(\dot{p}(a_k))(a_{k+1} - a_k)$$

where we used the fundamental equality of differential calculus, $p(a_{k+1}) - p(a_k) = \dot{p}(a_k)(a_{k+1} - a_k) + \mathrm{ordo}(a_{k+1} - a_k)$ and the continuity of $\boldsymbol{P}$.

The right side of the above expression looks like an approximation of an integral, therefore, we accept:

The proper time passed on the world line C between its occurrences x and y is

$$\boldsymbol{t}_{\mathrm{C}}(x,y) := \int_{p^{-1}(x)}^{p^{-1}(y)} \boldsymbol{P}(\dot{p}(a))da \tag{2.4}$$

where p is an arbitrary progressive parametrization of the world line.

The integral is the same for all progressive parametrizations. Let p and q be such parametrizations. We know that $S := p^{-1} \circ q : \mathbb{R} \to \mathbb{R}$ is continuously differentiable and $S' > 0$, therefore, $q(b) = p(S(b))$ and $\dot{q}(b) = \dot{p}(S(b))S'(b)$, so

$$\int_{q^{-1}(y)}^{q^{-1}(x)} \boldsymbol{P}(\dot{q}(b))db = \int_{q^{-1}(y)}^{q^{-1}(x)} \boldsymbol{P}(\dot{p}(S(b)))S'(b)db = \int_{p^{-1}(y)}^{p^{-1}(x)} \boldsymbol{P}(\dot{p}(a))da;$$

the first equality is based on the positive homogeneity of $\boldsymbol{P}$, while the second is the consequence of the well-known formula of integration by substitution.

Formula (2.4) is also accepted when y is past-like with respect to x i. e. $\boldsymbol{t}_{\mathrm{C}}(x,y)$ is defined for arbitrary two elements x and y of the world line. $\boldsymbol{t}_{\mathrm{C}}(x,y)$ is positive or negative depending on whether y is future-like or past-like with respect to x and $\boldsymbol{t}_{\mathrm{C}}(x,x) := 0$.

It is worth mentioning that our definition allows the possibility that different proper times pass on different world lines between two world points which is illustrated by tick lines in Figure 2.9.

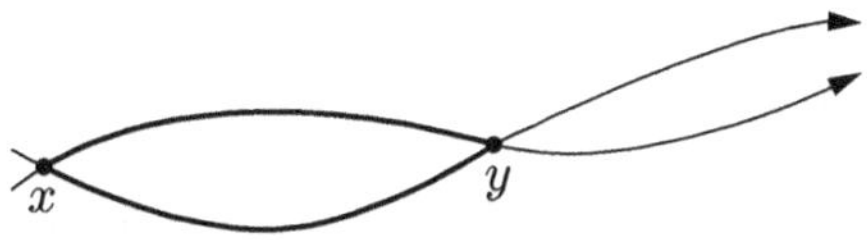

Figure 2.9: Different proper times may pass between x and y on two different world lines.

2.3.4 World line functions

The notion of proper time along a world line allows us to parametrize world lines in a "canonical" way. Indeed, choosing an arbitrary point x_0 in the world line C, the proper time as a function of parameter values is

$$b \mapsto \boldsymbol{t}_C(x_0, p(b)) = \int_{p^{-1}(x_0)}^{b} \boldsymbol{P}(\dot{p}(a))\, da.$$

It is well known from the theory of integration and differentiation that such a function is continuously differentiable and $\frac{d}{db}\boldsymbol{t}_C(x_0, p(b)) = \boldsymbol{P}(\dot{p}(b))$, which is positive since p is a progressive parametrization. The derivative being positive, the function above can be inverted in principle: the parameter can be given as a function of proper time.

This means that the world line can be parameterized by its proper time, which passes from an arbitrarily chosen point of its; this way we obtain a function $r : \mathbb{T} \to \mathrm{M}$ for which $p(b) = r(\boldsymbol{t}_C(x_0, p(b)))$ holds. Differentiation yields

$$\begin{aligned}\dot{p}(b) &= \dot{r}(\boldsymbol{t}_C(x_0, p(b)))\frac{d}{db}\boldsymbol{t}_C(x_0, p(b)) \\ &= \dot{r}(\boldsymbol{t}_C(x_0, p(b)))\boldsymbol{P}(\dot{p}(b));\end{aligned}$$

dividing by $\boldsymbol{P}(\dot{p}(b))$, we get that the values of $\dot{r}$ are absolute velocities. Based on this result, we accept the following definition:

A **world line function** is a continuously differentiable function $r : \mathbb{T} \to \mathrm{M}$ such that $\dot{r}(\boldsymbol{t}) \in \mathrm{V}(1)$ for all $\boldsymbol{t}$ in the domain of r.

A world line function determines a unique world line: its range. A world line determines several world line functions but those differ only in translations of their domains, i. e., if r_1 and r_2 have the same range then there is an $\boldsymbol{s}$ such that $r_2(\boldsymbol{t}) = r_1(\boldsymbol{t} + \boldsymbol{s})$ for all $\boldsymbol{t}$ in the domain of r_2.

2.4 Observers

2.4.1 Physical meaning of an observer

It is highly important to clarify the notions of observer, reference frame, coordinate system because they are used in the literature with several intuitive meanings, and different terminologies may refer to the same object or the same terminology refers to different objects as it is detailed in Section 16.1.

Observer often means a human being who makes experiments [1, 18] and is considered point-like. This is not satisfactory, however. A single material point cannot observe anything. A set of material points is necessary for making observations (experiments). For instance, if the cloud-chamber were a single material point then an ionization path could not come into being. An experimental device consists of a lot of material points.

An **observer**, physically, is an experimental device; in a wider sense, observers are the room and the car treated in Section 1.1. It is important that these objects are extended and not point-like.

Room and room-space, car and car-space: observer and its space are strictly related notions as it turns out from the definitions below.

A material point of an observer (a corner of the room, a button on the dashboard of the car) is conceived as a **space point of the observer**.

The history of a material point is a world line in spacetime, so:

A space point of an observer is a world line in spacetime.

Thinking about it a while, we do not find it so peculiar. The corner of the room existed yesterday, exists today and will exist (hopefully) tomorrow. The corner of the room is the history of the corresponding material point.

2.4.2 General observers

The tangents vectors of a world line are time-like vectors. All the time-like vectors can be obtained as nonzero multiples of absolute velocities. That is why we find it convenient to consider from now on tangents of world lines to be elements of V(1), which is automatically satisfied if the world lines are parameterized by their proper times.

In order that an observer be suitable for observations, its space points must be "side by side," i. e., the corresponding world lines must fill "continuously" a domain of spacetime. This can be best formulated with the aid of absolute velocities. Namely, attaching the tangent vector to all world points of all world lines representing the space points of the observer, we get an absolute velocity field, i. e., a function assigning an absolute velocity to world points (see Figure 2.10).

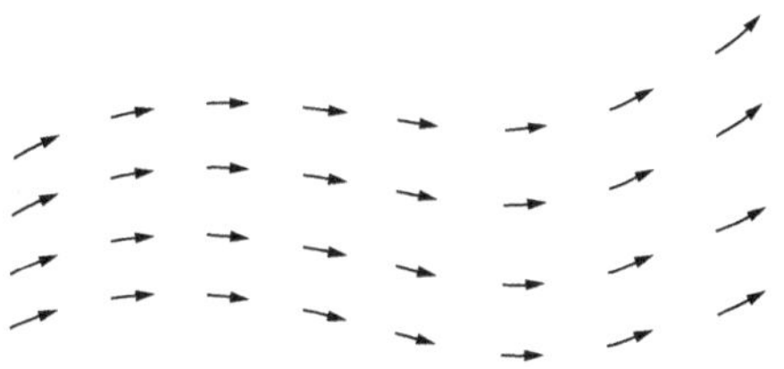

Figure 2.10: An observer.

That is why we find it convenient to accept:

An **observer** in a spacetime model is a smooth function $\boldsymbol{U} : \mathrm{M} \to \mathrm{V}(1)$ defined on a connected open subset.

It is well known that the vector field $\boldsymbol{U}$ determines uniquely world lines whose tangents are the prescribed absolute velocities. These world lines (see Figure 2.11), called $\boldsymbol{U}$**-lines**, are the solutions of the differential equation

$$(x : \mathbb{T} \to \mathrm{M})? \quad \dot{x} = \boldsymbol{U}(x).$$

U*-lines** are the **space points** of the observer ***U and ***U*-space**, denoted by $S_{\boldsymbol{U}}$, is the set of ***U*-lines**.

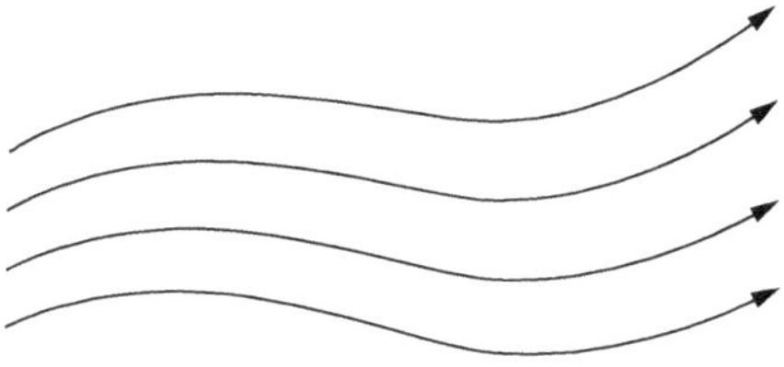

Figure 2.11: Space points of an observer.

Let us repeat in order to fix in our mind:

> *A space point of an observer is a subset of spacetime: A one-dimensional submanifold, a world line. The collection of these world lines is the space of the observer.*
>
> *The space of an observer, the collection of its space points, is not a subset of spacetime; it is a set whose elements are subsets of spacetime.*

We do not need it, therefore, we mention only that the space of an observer is a three-dimensional smooth manifold.

2.4.3 Inertial observers

The mathematical notions included in this subsection are treated in Subsection 18.5.2 and Subsection 18.1.4 of the mathematical supplement.

Heuristically, an inertial observer consists of inertial material points being "at rest relative to each other." This can be expressed by the fact that the space points of an inertial observer are inertial world lines having the same absolute velocity. That is why we accept:

An **observer is inertial** if the prescribed absolute velocity in every world point is the same: the absolute velocity field is a constant map.

From now on, we refer to an inertial observer by its constant velocity value, i. e., an inertial observer $\boldsymbol{u}$ means the observer $\boldsymbol{U}(x) = \boldsymbol{u}$ $(x \in \mathrm{M})$.

The space points of the inertial observer $\boldsymbol{u}$ are the parallel straight lines directed by $\boldsymbol{u}$. Such a line passing through the world point is $x + \mathbb{T}\boldsymbol{u}$ where

$$\mathbb{T}\boldsymbol{u} := \{t\boldsymbol{u} \mid t \in \mathbb{T}\}$$

is the one-dimensional subspace in $\mathbf{M}$, determined by $\boldsymbol{u}$. It is evident that

$$y + \mathbb{T}\boldsymbol{u} = x + \mathbb{T}\boldsymbol{u} \quad \text{if and only if} \quad y - x = \boldsymbol{t}\boldsymbol{u} \tag{2.5}$$

for some (by the way, uniquely determined) $\boldsymbol{t} \in \mathbb{T}$.

$S_{\boldsymbol{u}}$, the space of the inertial observer $\boldsymbol{u}$, is the set of the straight lines in M, directed by $\boldsymbol{u}$, illustrated by Figure 2.12.

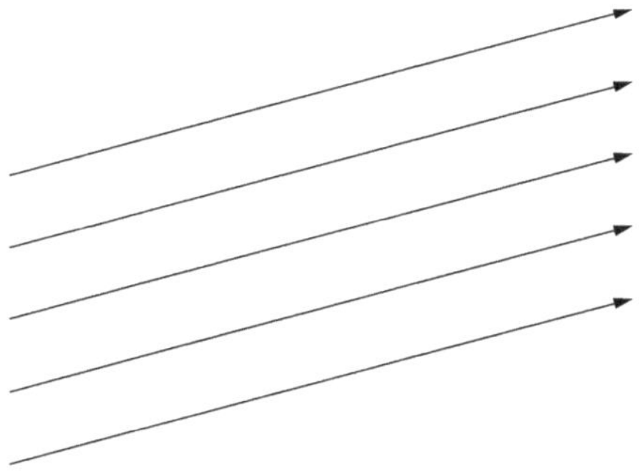

Figure 2.12: Space points of an inertial observer.

The experimental facts ($\mathcal{S}1$) and ($\mathcal{S}2$) say that the space of an inertial observer is a three-dimensional affine space. Accordingly, we have now the task to define the affine structure of $S_{\boldsymbol{u}}$; this is well known in mathematics: $S_{\boldsymbol{u}}$ is the factor space $M/\mathbb{T}\boldsymbol{u}$, an affine space over the factor space $\mathbf{M}/\mathbb{T}\boldsymbol{u}$ defined by the subtraction

$$(y + \mathbb{T}\boldsymbol{u}) - (x + \mathbb{T}\boldsymbol{u}) := (y - x) + \mathbb{T}\boldsymbol{u}.$$

The $\boldsymbol{u}$-space points in the model representing, e. g., the points of our room, can be well comprehended and shown in pictures as inertial world lines, elements of the factor space $M/\mathbb{T}\boldsymbol{u}$. On the other hand, the $\boldsymbol{u}$-space vectors in the model, elements of the factor space $\mathbf{M}/\mathbb{T}\boldsymbol{u}$, can hardly be seen to represent our vectors between the points of the room.

For the time being, we have to accept that we have a mathematically perfect tool for the space vectors of inertial observers without being able to visualize them.

Fortunately, this tool is not complicated, we can use it without hesitation, knowing only that $\mathbf{M}/\mathbb{T}\boldsymbol{u}$ is three-dimensional, the addition and the multiplication by numbers are defined by

$$(\boldsymbol{x} + \mathbb{T}\boldsymbol{u}) + (\boldsymbol{y} + \mathbb{T}\boldsymbol{u}) := (\boldsymbol{x} + \boldsymbol{y}) + \mathbb{T}\boldsymbol{u}, \quad \alpha(\boldsymbol{x} + \mathbb{T}\boldsymbol{u}) := \alpha\boldsymbol{x} + \mathbb{T}\boldsymbol{u},$$

and relation (2.5) holds for spacetime vectors instead of spacetime points, too.

As concerns orientation: our experiences that proper times are oriented and our space is oriented suggested accepting spacetime to be oriented. Now, in the model, spacetime orientation and proper time orientation (expressed by the set of future-like vectors) give the orientation of inertial spaces: let the ordered basis $(\boldsymbol{x}_1 + \mathbb{T}\boldsymbol{u}, \boldsymbol{x}_2 + \mathbb{T}\boldsymbol{u}, \boldsymbol{x}_3 + \mathbb{T}\boldsymbol{u})$ in $\frac{\mathbf{M}}{\mathbb{T}\boldsymbol{u}}$ **positively oriented** if $(\boldsymbol{t}\boldsymbol{u}, \boldsymbol{x}_1, \boldsymbol{x}_2, \boldsymbol{x}_3)$ is a positively oriented basis in $\mathbf{M}$ for an arbitrary positive element $\boldsymbol{t}$ of $\mathbb{T}$.

This definition is good, i. e., if $(\boldsymbol{t}'\boldsymbol{u}, \boldsymbol{x}_1', \boldsymbol{x}_2', \boldsymbol{x}_3')$ in $\mathbf{M}$ is equally oriented as the previous basis then $(\boldsymbol{x}_1' + \mathbb{T}\boldsymbol{u}, \boldsymbol{x}_2' + \mathbb{T}\boldsymbol{u}, \boldsymbol{x}_3' + \mathbb{T}\boldsymbol{u})$ in $\mathbf{M}/\mathbb{T}\boldsymbol{u}$ is equally oriented as the previous basis.

Indeed, introducing the notation $\boldsymbol{x}_0 := \boldsymbol{t}\boldsymbol{u}$ and $\boldsymbol{x}_0' := \boldsymbol{t}'\boldsymbol{u}$, the formula $\boldsymbol{x}_i' = \sum_{k=0}^{3} A_{ki}\boldsymbol{x}_k$ gives the 4×4 matrix $\{A_{ki} \mid i, k = 0, 1, 2, 3\}$ of the transition from the "unprimed" basis to the "primed" basis of $\mathbf{M}$. The transition from the "unprimed" basis to the "primed" basis of $\mathbf{M}/\mathbb{T}\boldsymbol{u}$ is given by the 3×3 matrix $\{A_{ki} \mid i, k = 1, 2, 3\}$. Since $A_{00} = \frac{t'}{t}$, $A_{k0} = 0$ for $k = 1, 2, 3$, the determinant of the 4×4 matrix equals the determinant of the 3×3 matrix multiplied by $\frac{t'}{t} > 0$. Consequently, if the determinant of the 4×4 matrix is positive then the determinant of the 3×3 matrix is positive.

In summary:

$S_u = \mathbf{M}/\mathbb{T}\boldsymbol{u}$, the space of the inertial observer $\boldsymbol{u}$ is a three-dimensional oriented affine space.

In the nonrelativistic as well as in the relativistic spacetime model, we can give nicer objects for space vectors of inertial observers.

2.4.4 Euclidean structures

The mathematical notions included in this subsection are treated in Subsection 18.3.1 of the mathematical supplement.

For measuring distances in observer spaces, first of all, we accept:

The structure of a flat spacetime model must contain the **measure line of distances**, denoted by $\mathbb{L}$.

According to ($\mathcal{S}$**4**) in Section 1.1, the space of an inertial observer has a Euclidean structure, which will be put in the model so that for all $\boldsymbol{u} \in \mathrm{V}(1)$ a **Euclidean form**, a symmetric, bilinear, positive definite map

$$\overline{\boldsymbol{b}_u} : \mathbf{M}/\mathbb{T}\boldsymbol{u} \times \mathbf{M}/\mathbb{T}\boldsymbol{u} \to \mathbb{L} \otimes \mathbb{L}$$

must be given, which determines the length of space vectors and the angle between space vectors.

As said, $\boldsymbol{u}$-space vectors appear in a nonpractical formulation, so this also holds for a Euclidean form on them. Fortunately, such a Euclidean form can be replaced with a nicer, equivalent object: a symmetric, bilinear, positive semidefinite map

$$\boldsymbol{b}_u : \mathbf{M} \times \mathbf{M} \to \mathbb{L} \otimes \mathbb{L}$$

having $\mathbb{T}\boldsymbol{u}$ as its kernel, which means that:

1. $\boldsymbol{b}_u(\boldsymbol{x},\boldsymbol{y}) = \boldsymbol{b}_u(\boldsymbol{y},\boldsymbol{x})$ for all $\boldsymbol{x},\boldsymbol{y} \in \mathbf{M}$.
2. $\boldsymbol{b}_u(\boldsymbol{x},\boldsymbol{y}) = 0$ for all $\boldsymbol{y} \in \mathbf{M}$ if and only $\boldsymbol{x}$ is parallel to $\boldsymbol{u}$.
3. $\boldsymbol{b}_u(\boldsymbol{x},\boldsymbol{x}) \geq 0$ and equality occurs if and only if $\boldsymbol{x}$ is parallel to $\boldsymbol{u}$.

Then

$$\overline{\boldsymbol{b}_u}(\boldsymbol{x} + \mathbb{T}\boldsymbol{u}, \boldsymbol{y} + \mathbb{T}\boldsymbol{u}) = \boldsymbol{b}_u(\boldsymbol{x},\boldsymbol{y}). \tag{2.6}$$

Given a Euclidean form $\overline{\boldsymbol{b}_u} : \mathbf{M}/\mathbb{T}\boldsymbol{u} \times \mathbf{M}/\mathbb{T}\boldsymbol{u} \to \mathbb{L}\otimes\mathbb{L}$, equality (2.6) with =: instead of = defines evidently a positive semidefinite, symmetric, bilinear map $\mathbf{M} \times \mathbf{M} \to \mathbb{L} \otimes \mathbb{L}$ having $\mathbb{T}\boldsymbol{u}$ as its kernel.

Conversely, given a positive semidefinite, symmetric, bilinear map $\boldsymbol{b}_u : \mathbf{M}\times\mathbf{M} \to \mathbb{L}\otimes\mathbb{L}$ having $\mathbb{T}\boldsymbol{u}$ as its kernel, equality (2.6) with := instead of = defines a well-posed Euclidean form $\overline{\boldsymbol{b}_u}$ on $\mathbf{M}/\mathbb{T}\boldsymbol{u}$.

Indeed, $\overline{\boldsymbol{b}_u}$ is well-defined: if $\boldsymbol{x}' + \mathbb{T}\boldsymbol{u} = \boldsymbol{x} + \mathbb{T}\boldsymbol{u}$ and $\boldsymbol{y}' + \mathbb{T}\boldsymbol{u} = \boldsymbol{y} + \mathbb{T}\boldsymbol{u}$, then there are $\boldsymbol{t},\boldsymbol{s} \in \mathbb{T}$ such that $\boldsymbol{x}' = \boldsymbol{x} + \boldsymbol{t}\boldsymbol{u}$ and $\boldsymbol{y}' = \boldsymbol{y} + \boldsymbol{s}\boldsymbol{u}$; thus, $\boldsymbol{b}_u(\boldsymbol{x}',\boldsymbol{y}') = \boldsymbol{b}_u(\boldsymbol{x},\boldsymbol{y})$.

It is trivial that $\overline{\boldsymbol{b}_u}$ is symmetric and bilinear.

If $\overline{\boldsymbol{b}_u}(\boldsymbol{x} + \mathbb{T}\boldsymbol{u}, \boldsymbol{x} + \mathbb{T}\boldsymbol{u}) = 0$ then $\boldsymbol{b}_u(\boldsymbol{x},\boldsymbol{x}) = 0$, implying that $\boldsymbol{x}$ is parallel to $\boldsymbol{u}$, so $\boldsymbol{x} + \mathbb{T}\boldsymbol{u}$ equals $\mathbb{T}\boldsymbol{u}$, which is the zero vector of $\mathbf{M}/\mathbb{T}\boldsymbol{u}$; thus, $\overline{\boldsymbol{b}_u}$ is positive definite.

Summing up: $\overline{\boldsymbol{b}_u}$ and $\boldsymbol{b}_u$ determine each other uniquely.

Instead of $\overline{\boldsymbol{b}_u}$, we shall consider $\boldsymbol{b}_u$:

The structure of a flat spacetime model must contain a **semi-Euclidean form** for all $\boldsymbol{u} \in V(1)$, a symmetric, bilinear, positive semidefinite map $\boldsymbol{b}_u : \mathbf{M}\times\mathbf{M} \to \mathbb{L}\otimes\mathbb{L}$ having $\mathbb{T}\boldsymbol{u}$ as its kernel, and $\boldsymbol{b}_u$ depends continuously on $\boldsymbol{u}$.

The continuity means that the mapping $\boldsymbol{u} \mapsto \boldsymbol{b}_u(\boldsymbol{x},\boldsymbol{y})$ should be continuous for all fixed $\boldsymbol{x},\boldsymbol{y} \in \mathbf{M}$.

Then the distance between the $\boldsymbol{u}$-space points $y + \mathbb{T}\boldsymbol{u}$ and $x + \mathbb{T}\boldsymbol{u}$ is the length of the vector $(y + \mathbb{T}\boldsymbol{u}) - (x + \mathbb{T}\boldsymbol{u}) = (y - x) + \mathbb{T}\boldsymbol{u}$,

$$\sqrt{\boldsymbol{b}_u(y - x, y - x)}; \tag{2.7}$$

in other words, the distance between the $\boldsymbol{u}$-space points q and p (straight lines directed by $\boldsymbol{u}$ in M) is (2.7) where y and x are arbitrary elements of q and p, respectively.

We note that $\boldsymbol{b}_u$ can be applied to vectors multiplied or divided by measure line elements, too. For instance, if $\boldsymbol{u}'$ is an absolute velocity then $\boldsymbol{b}_u(\boldsymbol{x},\boldsymbol{u}') := \frac{\boldsymbol{b}_u(\boldsymbol{x},\boldsymbol{t}\boldsymbol{u}')}{\boldsymbol{t}}$ for an arbitrary $0 \neq \boldsymbol{t} \in \mathbb{T}$. It is a simple fact that this is well-defined, i. e. $\frac{\boldsymbol{b}_u(\boldsymbol{x},\boldsymbol{s}\boldsymbol{u}')}{\boldsymbol{s}} = \frac{\boldsymbol{b}_u(\boldsymbol{x},\boldsymbol{t}\boldsymbol{u}')}{\boldsymbol{t}}$ for any nonzero $\boldsymbol{s},\boldsymbol{t} \in \mathbb{T}$.

2.5 Movements

2.5.1 Paths of movements

Let us take an observer $\boldsymbol{U}$. A material point with world line C, in general, moves with respect to the observer; the **path** of this movement consists of the $\boldsymbol{U}$-space points met by the material point:

$$\{q \in \mathrm{S}_U \mid q \cap \mathrm{C} \neq \emptyset\};$$

this is a curve (possibly a single point) in S_U. For an inertial observer $\boldsymbol{u}$, the path of the movement corresponding to the world line is

$$\{x + \mathbb{T}\boldsymbol{u} \mid x \in \mathrm{C}\}.$$

If C is also an inertial world line with absolute velocity $\boldsymbol{u}'$, then:

(1) For $\boldsymbol{u}' = \boldsymbol{u}$, C coincides with a $\boldsymbol{u}$-space point, the material point is at rest with respect to the inertial observer.
(2) For $\boldsymbol{u}' \neq \boldsymbol{u}$, the material point moves with respect to the inertial observer.

If x and y are different points of C, then $x + \mathbb{T}\boldsymbol{u}$ and $y + \mathbb{T}\boldsymbol{u}$ are different points of the path of the movement and $(y + \mathbb{T}\boldsymbol{u}) - (x + \mathbb{T}\boldsymbol{u}) = (y - x) + \mathbb{T}\boldsymbol{u}$ is the $\boldsymbol{u}$-space vector between them. There is unique nonzero element $\boldsymbol{t}'$ of $\mathbb{T}$ such that $y - x = \boldsymbol{t}'\boldsymbol{u}'$. Since every element of $\mathbb{T}$ is a real multiple of $\boldsymbol{t}'$, $\mathbb{T} = \mathbb{R}\boldsymbol{t}' = \boldsymbol{t}'\mathbb{R}$,

$$(y + \mathbb{T}\boldsymbol{u}) - (x + \mathbb{T}\boldsymbol{u}) = \boldsymbol{t}'\boldsymbol{u}' + \mathbb{T}\boldsymbol{u} = \boldsymbol{t}'(\boldsymbol{u}' + \mathbb{R}\boldsymbol{u}). \tag{2.8}$$

The $\boldsymbol{u}$-space vectors between two points of the path are proportional to $\boldsymbol{u}' + \mathbb{R}\boldsymbol{u}$, an element of the factor space $\frac{\mathbb{M}}{\mathbb{T}}/\mathbb{R}\boldsymbol{u}$. The laws of operations between vector spaces and their tensor products and quotients by measure lines are analogous to operations with real numbers, so it is straightforward to consider $\boldsymbol{u}'+\mathbb{R}\boldsymbol{u}$ the direction vector; this choice will make drafting easier.

In $\boldsymbol{u}$-space, the path of movement of an inertial material point with absolute velocity $\boldsymbol{u}' \neq \boldsymbol{u}$ is a straight line with direction vector $\boldsymbol{u}' + \mathbb{R}\boldsymbol{u}$.

Property ($\mathcal{M}$**1**) in Subsection 1.3.3 says that for every vector in $\boldsymbol{u}$-space there is an absolute velocity $\boldsymbol{u}'$ such that the vector is a multiple of $\boldsymbol{u}' + \mathbb{R}\boldsymbol{u}$.

Remark. Now the attentive reader may feel some confusion.

(1) In Section 1.1, we stated that space is constituted by some material objects, e. g., a room-space is constituted by the walls and the furnitures.
(2) In Subsection 2.4.1, an experimental device is considered an observer having its own space; in this sense, a room with convenient “measuring furnitures” is an observer.

(3) In Subsection 2.4.2, it is required that the space points of an observer be densely side-by-side.

Physically, the molecules of an observer serve to describe spatial relations. But then there is some trouble with movements: if the room is full to the brim then even a fly cannot move in it.

In practice, however, only some furniture is sufficient: customary furnishings allow us to define every room-space point, e. g., by measuring sticks (putting aside the earthquake-shaken rubber room). Therefore, in modeling we may pretend the stick is there virtually, i. e., we consider it an inessential implementation detail that we put it there and removing it all the time. This means that even though theoretically the stick is "there," it does not have any influence on a moving body.

So, in this respect, $\boldsymbol{U}$-space is considered permeable, "ethereal": a material point can go through without any interaction, any harm.

2.5.2 Faster–slower

Let us consider two inertial material points with absolute velocities $\boldsymbol{u}_1$ and $\boldsymbol{u}_2$, respectively, which met in a world point x_0; then their world lines are $x_0 + \mathbb{T}\boldsymbol{u}_1$ and $x_0 + \mathbb{T}\boldsymbol{u}_2$.

Let the material points move along the same path in the same direction in the space of the inertial observer $\boldsymbol{u}$ and let $\boldsymbol{u}_2$ be faster than $\boldsymbol{u}_1$, with respect to $\boldsymbol{u}$; then (see Figure 2.13) there are positive time periods $\boldsymbol{t}_2, \boldsymbol{t}_1, \boldsymbol{t}$ such that

$$\boldsymbol{t}_1\boldsymbol{u}_1 = \boldsymbol{t}_2\boldsymbol{u}_2 + \boldsymbol{t}\boldsymbol{u}. \tag{2.9}$$

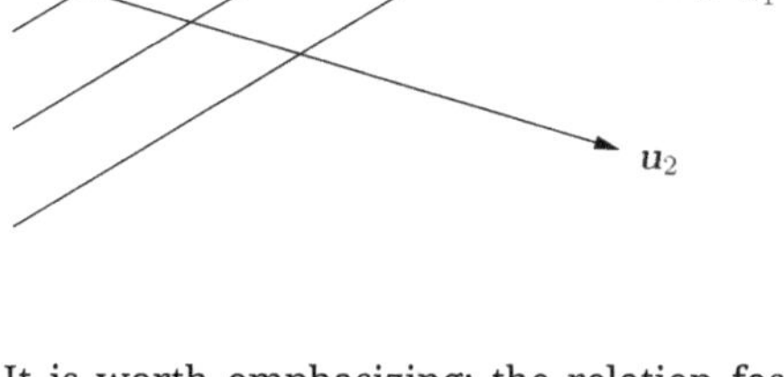

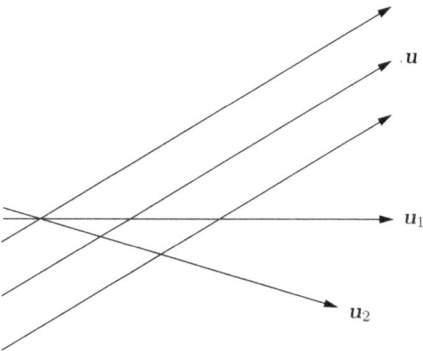

Figure 2.13: According to $\boldsymbol{u}$, $\boldsymbol{u}_2$ is faster than $\boldsymbol{u}_1$: $\boldsymbol{u}_2$ meets $\boldsymbol{u}$-space points earlier than $\boldsymbol{u}_1$.

It is worth emphasizing: the relation faster–slower is meaningful only with respect to the same path in the same direction in the space of a given inertial observer. In general, if two inertial material points—called 1 and 2—meet in a world point, then there are inertial observers

- For which the material point 2 is faster than the material point 1
- For which the material point 1 is faster than the material point 2
- Which cannot decide which of 1 and 2 is faster (because the paths or the directions are different in the observer space)

Then the properties of movements stated in Subsection 1.3.3 can be formulated as follows:

($\mathcal{M}$3):

- For all $\boldsymbol{u} \in \mathrm{V}(1)$
- For all $\boldsymbol{u}_1 \in \mathrm{V}(1)$ and for all positive $\boldsymbol{t}_1 \in \mathbb{T}$
- There are a $\boldsymbol{u}_2 \in \mathrm{V}(1)$ and a positive $\boldsymbol{t}_2 \in \mathbb{T}$
- There is a positive $\boldsymbol{t} \in \mathbb{T}$

such that (2.9) holds.

($\mathcal{M}$4):

- For all $\boldsymbol{u} \in \mathrm{V}(1)$ and for all positive $\boldsymbol{t} \in \mathbb{T}$
- For all $\boldsymbol{u}_2 \in \mathrm{V}(1)$ and for all positive $\boldsymbol{t}_2 \in \mathbb{T}$
- There is a $\boldsymbol{u}_1 \in \mathrm{V}(1)$ and a positive $\boldsymbol{t}_1 \in \mathbb{T}$

such that (2.9) holds.

2.6 About the future-like vectors

Formulas of the preceding subsection allow us to prove the fundamental properties of $\mathrm{T}^{\rightarrow}$.

The set of future-like vectors is a convex cone with apex at zero.

($\mathcal{M}$4) can be rewritten as follows:

- For all $\boldsymbol{y} \in \mathrm{T}^{\rightarrow}$ $(\boldsymbol{y} = \boldsymbol{t}\boldsymbol{u})$
- For all $\boldsymbol{x} \in \mathrm{T}^{\rightarrow}$ $(\boldsymbol{x} = \boldsymbol{t}_2\boldsymbol{u}_2)$

$$\boldsymbol{x} + \boldsymbol{y} \in \mathrm{T}^{\rightarrow} \quad (\boldsymbol{x} + \boldsymbol{y} = \boldsymbol{t}_1\boldsymbol{u}_1). \tag{2.10}$$

Since positive multiples of future-like vectors are future-like, (2.10) implies that for all $\boldsymbol{x}, \boldsymbol{y} \in \mathrm{T}^{\rightarrow}$ and for all positive real numbers α, β, $\alpha\boldsymbol{x} + \beta\boldsymbol{y} \in \mathrm{T}^{\rightarrow}$.

The set of future-like vectors is open.

($\mathcal{M}$3) can be rewritten as follows:

- For all $\boldsymbol{y} \in \mathrm{T}^{\rightarrow}$ $(\boldsymbol{u} = \frac{\boldsymbol{y}}{P(\boldsymbol{y})})$
- For all $\boldsymbol{x} \in \mathrm{T}^{\rightarrow}$ $(\boldsymbol{x} = \boldsymbol{t}_1\boldsymbol{u}_1)$.

– There is a positive real number $\lambda(\boldsymbol{x},\boldsymbol{y})$ ($\lambda(x,y) = \frac{t}{P(y)}$) such that

$$\boldsymbol{x} - \lambda(\boldsymbol{x},\boldsymbol{y})\boldsymbol{y} \in \mathrm{T}^{\rightarrow} \quad \big(\boldsymbol{x} - \lambda(\boldsymbol{x},\boldsymbol{y})\boldsymbol{y} = \boldsymbol{t}_2\boldsymbol{u}_2\big). \tag{2.11}$$

According to our basic requirement, $(\boldsymbol{x},\boldsymbol{y}) \mapsto \lambda(\boldsymbol{x},\boldsymbol{y})$ is continuous.

According to ($\mathcal{M}$**1**), for all $\boldsymbol{u} \in \mathrm{V}(1)$ there are linearly independent future-like vectors $\boldsymbol{x}_1$, $\boldsymbol{x}_2$, $\boldsymbol{x}_3$ such that $\boldsymbol{x}_1 + \mathbb{T}\boldsymbol{u}$, $\boldsymbol{x}_2 + \mathbb{T}\boldsymbol{u}$, and $\boldsymbol{x}_3 + \mathbb{T}\boldsymbol{u}$ form a basis for $\boldsymbol{u}$-space vectors. Then for arbitrary $\boldsymbol{s} \in \mathbb{T}^+$ the future-like vectors $\boldsymbol{su}$, $\boldsymbol{x}_1$, $\boldsymbol{x}_2$, $\boldsymbol{x}_3$ form a basis in $\mathbf{M}$. Consequently, every vector $\boldsymbol{z}$ of $\mathbf{M}$, being a linear combination of that basis, can be represented as a difference $\boldsymbol{z} = \boldsymbol{z}_1 - \boldsymbol{z}_2$ where both $\boldsymbol{z}_1$ and $\boldsymbol{z}_2$ are in $\mathrm{T}^{\rightarrow}$. Indeed, this is trivial if $\boldsymbol{z} = 0$. If $\boldsymbol{z} \neq 0$ is not time-like, then $\boldsymbol{z}_1$ is the sum of the basis elements with positive coefficients and $\boldsymbol{z}_2$ is the sum with negative coefficients. If $\boldsymbol{z}$ is future-like, then $\boldsymbol{z}_1 := 2\boldsymbol{z}$, $\boldsymbol{z}_2 := \boldsymbol{z}$, and if $\boldsymbol{z}$ is past-like, then $\boldsymbol{z}_1 := -\boldsymbol{z}$, $\boldsymbol{z}_2 := -2\boldsymbol{z}$.

Let us fix a future-like vector $\boldsymbol{x}$ and some norm $\|\ \|$ on $\mathbf{M}$. Taking an arbitrary vector $\boldsymbol{z} = \boldsymbol{z}_1 - \boldsymbol{z}_2$ with $\|\boldsymbol{z}\| = 1$ and let $\boldsymbol{z}_2$ (depending continuously on $\boldsymbol{z}$) play the role of $\boldsymbol{y}$ in (2.11): $\boldsymbol{x} - \lambda(\boldsymbol{x},\boldsymbol{z}_2(\boldsymbol{z}))\boldsymbol{z}_2 \in \mathrm{T}^{\rightarrow}$. The positive valued continuous function $\boldsymbol{z} \mapsto \lambda(\boldsymbol{x},\boldsymbol{z}_2(\boldsymbol{z}))$ on the compact set of vectors of norm 1 has a minimum $\lambda_{\min}(\boldsymbol{x}) > 0$. Then for all $0 < \eta(\boldsymbol{x}) < \lambda_{\min}(\boldsymbol{x})$, $\boldsymbol{x} - \eta(\boldsymbol{x})\boldsymbol{z}_2 = \boldsymbol{x} - \lambda(\boldsymbol{x},\boldsymbol{z}_2(\boldsymbol{z}))\boldsymbol{z}_2 + (\lambda(\boldsymbol{x},\boldsymbol{z}_2(\boldsymbol{z})) - \lambda_{\min}(\boldsymbol{x}))\boldsymbol{z}_2$ is in $\mathrm{T}^{\rightarrow}$. In the same way, for all $0 < \eta(\boldsymbol{x}) < \lambda_{\min}(\boldsymbol{x})$, $\boldsymbol{x} - \eta(\boldsymbol{x})\boldsymbol{z}_2$ is future-like and, of course, $\boldsymbol{x} + \eta(\boldsymbol{x})\boldsymbol{z}_1$ is future-like; so, the half of their sum, $\boldsymbol{x} + \frac{\eta(\boldsymbol{x})}{2}\boldsymbol{z}$, is future-like, too, i. e., the neighborhood $\{\boldsymbol{x}' \in \mathrm{T}^{\rightarrow} \mid \|\boldsymbol{x}' - \boldsymbol{x}\| < \lambda_{\min}(\boldsymbol{x})/2\}$ of $\boldsymbol{x}$ is a subset of $\mathrm{T}^{\rightarrow}$.

3 Mathematical structure of flat spacetime models

3.1 Exact definition

Now we have the possibility to compose an exact mathematical definition:

A **flat spacetime model** is $(\mathrm{M}, \mathbb{T}, \mathbb{L}, \mathrm{T}^{\rightarrow}, \boldsymbol{P}, \boldsymbol{b})$ where

- M is **spacetime**, a four-dimensional oriented affine space over the vector space **M**.
- $\mathbb{T}$ is the measure line of **time periods**.
- $\mathbb{L}$ is the measure line of **distances**.
- $\mathrm{T}^{\rightarrow}$ is the set of **future-like vectors**, a proper subset of $\mathbf{M}\setminus\{0\}$ which is an open and convex cone having the zero as its apex.
- $\boldsymbol{P} : \mathrm{T}^{\rightarrow} \to \mathbb{T}^{+}$ is the **inertial time progress**, a continuous, positive homogeneous map, and the latter two objects determine

$$\mathrm{V}(1) := \left\{ \boldsymbol{u} \in \frac{\mathbf{M}}{\mathbb{T}} \;\middle|\; \boldsymbol{u} \text{ is future-like, } \boldsymbol{P}(\boldsymbol{u}) = 1 \right\},$$

  the set of **absolute velocities**.
- $\boldsymbol{b}$ is the collection of **semi-Euclidean forms**, which assigns to every absolute velocity $\boldsymbol{u}$ a symmetric, bilinear, positive semidefinite map $\boldsymbol{b}_u : \mathbf{M}\times\mathbf{M} \to \mathbb{L}\otimes\mathbb{L}$ having $\mathbb{T}\boldsymbol{u}$ as its kernel, and this assignment is continuous with respect to $\boldsymbol{u}$.

In such a spacetime model

- $\mathrm{T}^{\leftarrow} := -\mathrm{T}^{\rightarrow}$ is the set of past-like vectors and $\mathrm{T}^{\leftarrow} \cup \mathrm{T}^{\rightarrow}$ is the set of time-like vectors.
- A world line is an oriented curve whose oriented tangents are future-like, an inertial world line is a straight line with future-like direction vector.
- The proper time passed between the occurrences x and y of an inertial world line equals $\boldsymbol{P}(y-x)$ if $y-x$ is future-like and $-\boldsymbol{P}(-(y-x))$ if $y-x$ is past-like.
- The proper time passed between the occurrences x and y of a world line C is

$$\boldsymbol{t}_{\mathrm{C}}(x,y) := \int_{p^{-1}(x)}^{p^{-1}(y)} \boldsymbol{P}(\dot{p}(a))da$$

  where p is an arbitrary progressive parametrization of the world line (i. e., $\boldsymbol{P}(\dot{p}(a)) > 0$ for all a).
- An observer is a smooth absolute velocity field $\boldsymbol{U} : \mathrm{M} \to \mathrm{V}(1)$ defined on a connected open subset.
- A space point of the observer $\boldsymbol{U}$ (briefly: a $\boldsymbol{U}$-space point) is a $\boldsymbol{U}$-line, a maximal integral curve of the vector field $\boldsymbol{U}$; the space of the observer (briefly: $\boldsymbol{U}$-space) is the collection of its space points.
- An inertial observer is a constant absolute velocity field $\boldsymbol{u}$, its space points are parallel straight lines directed by $\boldsymbol{u}$, its space is $\mathrm{S}_u := \mathrm{M}/\mathbb{T}\boldsymbol{u}$, a three-dimensional oriented affine space over $\mathbf{M}/\mathbb{T}\boldsymbol{u}$.

https://doi.org/10.1515/9783112219553-004

- The inner product of the $\boldsymbol{u}$-space vectors $\boldsymbol{x} + \mathbb{T}\boldsymbol{u}$ and $\boldsymbol{y} + \mathbb{T}\boldsymbol{u}$ is $\boldsymbol{b}_{\boldsymbol{u}}(\boldsymbol{x},\boldsymbol{y})$; thus, the distance between the $\boldsymbol{u}$-space points $x + \mathbb{T}\boldsymbol{u}$ and $y + \mathbb{T}\boldsymbol{u}$ is $\sqrt{\boldsymbol{b}_{\boldsymbol{u}}(x-y, x-y)}$.

3.2 Isomorphisms

We can construct various spacetime models according to our experience. It may happen that two spacetime models, expressed in different mathematical formulas, have the same physical content. We expect that two models have the same physical content if and only if their mathematical structures are isomorphic as formulated below.

Let us consider the models $(\mathrm{M}, \mathbb{T}, \mathbb{L}, \mathrm{T}^{\to}, \boldsymbol{P}, \boldsymbol{b})$ and $(\widehat{\mathrm{M}}, \widehat{\mathbb{T}}, \widehat{\mathbb{L}}, \widehat{\mathrm{T}^{\to}}, \widehat{\boldsymbol{P}}, \widehat{\boldsymbol{b}})$. As concerns their mathematical structures, M and $\widehat{\mathrm{M}}$, $\mathbb{T}$ and $\widehat{\mathbb{T}}$, as well as $\mathbb{L}$ and $\widehat{\mathbb{L}}$ are isomorphic as a matter of course, i. e., there are (continuum many):

(i) Orientation preserving affine bijections $L : \mathrm{M} \to \widehat{\mathrm{M}}$ (over the linear bijection $\boldsymbol{L} : \mathbf{M} \to \widehat{\mathbf{M}}$).
(ii) Orientation preserving linear bijections $\boldsymbol{F} : \mathbb{T} \to \widehat{\mathbb{T}}$.
(iii) Orientation preserving linear bijections $\boldsymbol{Z} : \mathbb{L} \to \widehat{\mathbb{L}}$.

In other words, the first three components of every flat spacetime model are essentially the same, only the last three components can be different

The two spacetime models are called **isomorphic** if there are L, $\boldsymbol{F}$, and $\boldsymbol{Z}$, which transform $\mathrm{T}^{\to}$ into $\widehat{\mathrm{T}^{\to}}$, $\boldsymbol{P}$ into $\widehat{\boldsymbol{P}}$, $\boldsymbol{b}$ into $\widehat{\boldsymbol{b}}$ "conveniently" as follows:

(I) $\boldsymbol{L}[\mathrm{T}^{\to}] = \widehat{\mathrm{T}^{\to}}$.
(II) $\frac{\widehat{\boldsymbol{P}}(\boldsymbol{L}\cdot\boldsymbol{x})}{\boldsymbol{F}(\boldsymbol{t})} = \frac{\boldsymbol{P}(\boldsymbol{x})}{\boldsymbol{t}}$ for all $\boldsymbol{x} \in \mathrm{T}^{\to}$ and $0 \neq \boldsymbol{t} \in \mathbb{T}$.
(III)

$$\frac{\widehat{\boldsymbol{b}}_{\widehat{\boldsymbol{u}}}(\boldsymbol{L}\cdot\boldsymbol{x}, \boldsymbol{L}\cdot\boldsymbol{y})}{\boldsymbol{Z}(\boldsymbol{d})^2} = \frac{\boldsymbol{b}_{\boldsymbol{u}}(\boldsymbol{x},\boldsymbol{y})}{\boldsymbol{d}^2}$$

for all $\boldsymbol{u} \in \mathrm{V}(1)$, $\boldsymbol{x},\boldsymbol{y} \in \mathbf{M}$, and $0 \neq \boldsymbol{d} \in \mathbb{L}$, where $\widehat{\boldsymbol{u}} := \frac{\boldsymbol{t}}{\boldsymbol{F}(\boldsymbol{t})}\boldsymbol{L}\cdot\boldsymbol{u}$ is independent of $\boldsymbol{t}$.

Then the triplet $(L, \boldsymbol{F}, \boldsymbol{Z})$ is called an **isomorphism** between the two spacetime models.

Note that two flat spacetime models cannot be isomorphic if their sets of future-like vectors are of different types drawn in Figure 2.7 because a linear bijection cannot map the different cones onto each other.

3.3 Symmetries

An isomorphism $(L, \boldsymbol{F}, \boldsymbol{Z})$ of a flat spacetime model to itself is called an **automorphism**. A **symmetry** of the spacetime model is an automorphism if the involved linear maps $\boldsymbol{F} : \mathbb{T} \to \mathbb{T}$ and $\boldsymbol{Z} : \mathbb{L} \to \mathbb{L}$ are the identities. Treating a symmetry, we omit these identities.

Accordingly, a **symmetry** of the spacetime model $(\mathrm{M}, \mathbb{T}, \mathbb{L}, \mathrm{T}^{\to}, \boldsymbol{P}, \boldsymbol{b})$ is an orientation preserving affine bijection $L : \mathrm{M} \to \mathrm{M}$ such that the underlying linear bijection $\boldsymbol{L} : \mathbf{M} \to \mathbf{M}$—called **vectorial symmetry**—has the properties:

(I) $\boldsymbol{L}[\mathrm{T}^{\to}] = \mathrm{T}^{\to}$.

(II) $\boldsymbol{P}(\boldsymbol{L} \cdot \boldsymbol{x}) = \boldsymbol{P}(\boldsymbol{x})$ for all $\boldsymbol{x} \in \mathrm{T}^{\to}$.

(III) $\boldsymbol{b}_{\boldsymbol{L} \cdot \boldsymbol{u}}(\boldsymbol{L} \cdot \boldsymbol{x}, \boldsymbol{L} \cdot \boldsymbol{y}) = \boldsymbol{b}_{\boldsymbol{u}}(\boldsymbol{x}, \boldsymbol{y})$ for all $\boldsymbol{u} \in \mathrm{V}(1)$ and $\boldsymbol{x}, \boldsymbol{y} \in \mathbf{M}$.

Property (I) is called that $\boldsymbol{L}$ is **arrow orientation preserving**.

The **translation** by an arbitrary vector $\boldsymbol{a}$, i. e., the affine map $x \mapsto x + \boldsymbol{a}$ is a symmetry, since the underlying linear map is the identity of $\mathbf{M}$. This gives an exact meaning to the homogeneity of spacetime mentioned in Subsection 2.2.2.

The symmetries have two meanings:

1. We have an intuition when two physical objects can be considered the same from a physical point of view. This is expressed in the spacetime model in such a way that two objects are **equivalent** if there is a symmetry that maps one of them to the other.

For instance, the world lines C and C′—i. e., the histories of two material points—are equivalent if there is symmetry L such that $C' = L[C]$.

Further, the observers $\boldsymbol{U}$ and $\boldsymbol{U}'$ are equivalent if there is spacetime symmetry L (over the vectorial symmetry $\boldsymbol{L}$), which maps the domain of $\boldsymbol{U}$ onto the domain of $\boldsymbol{U}'$ and $\boldsymbol{U}'(L(x)) = \boldsymbol{L} \cdot \boldsymbol{U}(x)$ for all x in the domain of $\boldsymbol{U}$.

2. We have a simple idea what a symmetry of a physical object is (for instance, a symmetry of a cube is the rotation by an angle of $\pi/2$ degree around an axis going through the center and perpendicular to a face). This is expressed in the spacetime model that a **symmetry of an object** is a spacetime symmetry, which maps the object into itself; in other words, the object is invariant for the spacetime symmetry.

For instance, the spacetime symmetry L is the symmetry of the observer $\boldsymbol{U}$ if the domain of $\boldsymbol{U}$ is invariant for L and $\boldsymbol{U}(L(x)) = \boldsymbol{L} \cdot \boldsymbol{U}(x)$ for all x in the domain of $\boldsymbol{U}$.

3.4 Equivalence of inertial observers, boosts

We mentioned in the Introduction the generally accepted “fundamental principle” that “all systems of inertia ... are completely equivalent in respect of our description of nature.” The “system of inertia” covers several intuitive notions; one of them is already clarified by us as an observer.

“Completely equivalent” in a spacetime model is just equivalence as it is defined previously. Inertial observers are equivalent because for arbitrary two absolute velocities there is a vectorial symmetry, which maps them into each other; this is quite trivial and does not tell anything new.

We can go further by noting that usually one tries to express this principle by transformation rules between spatial coordinate systems moving with respect to each other, with the intuitive notions and tacit assumptions that:

(1) The spatial coordinate systems have parallel axes and equal scales on the axes.
(2) The respective velocities of the spatial coordinate systems are opposite to each other.

The precise definition of a spacetime model reveals that the space vectors of different inertial observers $\boldsymbol{u}$ and $\boldsymbol{u}'$ are in the different vector spaces $\mathbf{M}/\mathbb{T}\boldsymbol{u}$ and $\mathbf{M}/\mathbb{T}\boldsymbol{u}'$; consequently:

It is not self-evident:
(1) Either that a straight line in an inertial space is parallel to a straight line in another inertial space.
(2) Or that a vector in an inertial space is opposite to a vector in another inertial space.

Since velocity requires a synchronization and we avoid synchronizations in formulating physical facts, we consider directions of movements instead of velocities and we express the "fundamental principle of equivalence" in the spacetime model $(\mathrm{M}, \mathbb{T}, \mathbb{L}, \boldsymbol{P}, \boldsymbol{b})$ in the following way:

(i) A vectorial symmetry $\boldsymbol{B}_{\boldsymbol{uu}'}$ is expected to exist for all inertial observers (absolute velocities) $\boldsymbol{u}$ and $\boldsymbol{u}'$, which
(ii) Maps $\boldsymbol{u}'$ into $\boldsymbol{u}$
(iii) Maps the direction of movement of $\boldsymbol{u}$ with respect to $\boldsymbol{u}'$ into the opposite of the direction of movement of $\boldsymbol{u}'$ with respect to $\boldsymbol{u}$

In more details:

1. $\boldsymbol{B}_{\boldsymbol{uu}'}$ is an orientation and arrow orientation preserving linear bijection having the properties:

$$\boldsymbol{B}_{\boldsymbol{uu}'}[\mathrm{T}^{\rightarrow}] = \mathrm{T}^{\rightarrow}, \tag{3.1}$$

$$\boldsymbol{P}(\boldsymbol{B}_{\boldsymbol{uu}'} \cdot \boldsymbol{x}) = \boldsymbol{P}(\boldsymbol{x}) \quad (\boldsymbol{x} \in \mathrm{T}^{\rightarrow}), \tag{3.2}$$

$$\boldsymbol{b}_{\boldsymbol{u}}(\boldsymbol{B}_{\boldsymbol{uu}'} \cdot \boldsymbol{x}, \boldsymbol{B}_{\boldsymbol{uu}'} \cdot \boldsymbol{y}) = \boldsymbol{b}_{\boldsymbol{u}'}(\boldsymbol{x}, \boldsymbol{y}) \quad (\boldsymbol{x}, \boldsymbol{y} \in \mathbf{M}). \tag{3.3}$$

2. $$\boldsymbol{B}_{\boldsymbol{uu}'} \cdot \boldsymbol{u}' = \boldsymbol{u} \tag{3.4}$$

   and according to Subsection 2.5.1.

3. $$\boldsymbol{B}_{\boldsymbol{uu}'}[\boldsymbol{u} + \mathbb{R}\boldsymbol{u}'] = -(\boldsymbol{u}' + \mathbb{R}\boldsymbol{u}), \tag{3.5}$$

   which is the same as

$$(\boldsymbol{B}_{\boldsymbol{uu}'} \cdot \boldsymbol{u}) + \mathbb{R}\boldsymbol{u} = -\boldsymbol{u}' + \mathbb{R}\boldsymbol{u}. \tag{3.6}$$

   Moreover, the following natural properties are supposed, too:

4.
$$B_{u'u} = B_{uu'}^{-1}. \tag{3.7}$$

5.
$$B_{uu} = \mathbf{1} \quad \text{(the identity of } \mathbf{M}\text{)}. \tag{3.8}$$

The spacetime model is called **fair** if such a vectorial symmetry, called the **boost from u' to u** is given for every u' and u.

(3.6) can be given in a more suitable, equivalent form: there is a number $\alpha_{u,u'}$ such that

$$B_{uu'} \cdot u = -u' + 2\alpha_{uu'} u; \tag{3.9}$$

the number 2 appeared for convenience because so

$$B_{uu'} \cdot (u - \alpha_{uu'} u') = -(u' - \alpha_{uu'} u). \tag{3.10}$$

Further, it follows from property 5 that

$$\alpha_{u,u} = 1$$

and, by property 4,

$$u = B_{u'u} \cdot B_{uu'} \cdot u = B_{u'u} \cdot (-u' + 2\alpha_{uu'} u) = -(-u + 2\alpha_{u'u} u') + 2\alpha_{uu'} u'$$

implying

$$\alpha_{uu'} = \alpha_{u'u}.$$

Equalities (3.3) and (3.9) involve

$$b_{u'}(u, u) = b_u(u', u'), \tag{3.11}$$

which is a necessary condition imposed on the semi-Euclidean forms in order for the spacetime model to be fair.

3.5 Reference frames

3.5.1 Synchronizations

We treated the problem of time points, the problem of synchronization in the Introduction as well as in Subsection 1.2.3. We can give a clear and exact meaning to these notions in the framework of a spacetime model.

An observer can define simultaneity for its space points by some procedure; **synchronization** means establishing simultaneity in a continuous manner.

We accept as a fundamental property of any synchronization:

Different occurrences of any world line (different ticks of a chronometer) cannot be simultaneous, i. e., time-like separated world points cannot be simultaneous.

As a matter of course, a synchronization determines its time points (instants): simultaneous occurrences happen at the same instant.

Now a very important idea follows. Take the space of an observer according to Figure 2.11 and let us imagine it being a little thicker, i. e., let us imagine space points (world lines) in front of and behind the plane of the page, too; then the world lines corresponding to the space points of the observer form a bunch.

Let us mark the occurrences in each space point that are simultaneous according to a synchronization (say, for the sake of clarity, let us mark "midday" in each space point). These simultaneous occurrences form a cross-section of the bunch as it is seen in Figure 3.1.

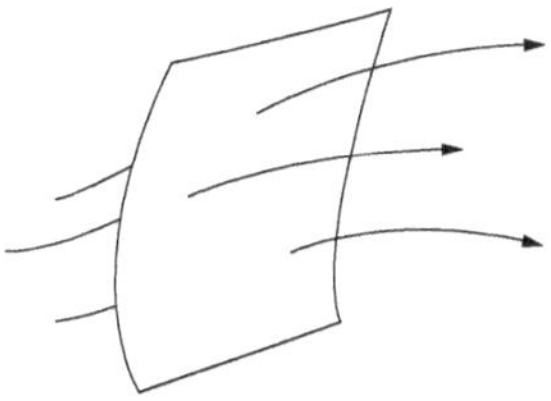

Figure 3.1: A synchronization instant.

This motivates us and it is convenient to consider a time point (instant) of a synchronization as the set of the corresponding simultaneous world points. For instance, midday as a time point is the set of middays in Greenwich, Budapest, Vienna, Prague, Paris, etc. We have to get accustomed to the following:

A time point established by a synchronization is a subset of spacetime.

It is a simple fact that a synchronization must have the following properties:
(1) Every world point x is simultaneous to itself.
(2) If x is simultaneous to y then y is simultaneous to x.
(3) If x is simultaneous to y and y is simultaneous to z, then x is simultaneous to z.

This means that a **synchronization** is an equivalence relation in spacetime; a **synchronization time point** is an equivalence class. The **synchronization time** is the set of the synchronization time points.

We require some "regularity" conditions:
- A time point should be a hypersurface (three-dimensional submanifold) in spacetime, which is called **world surface**.

- The set of time points should be continuously defined.

  Since time-like separated world points cannot be simultaneous, the tangent spaces of such a world surface are transverse to all absolute velocities. Then condition 2 in a precise setting means that there are three continuous vector fields X, Y, Z (vector valued functions) in spacetime in such a way that at any world point x the vectors:
  - $X(x)$, $Y(x)$, $Z(x)$, and $\boldsymbol{u}$ are linearly independent for all absolute velocities $\boldsymbol{u}$.
  - $X(x)$, $Y(x)$, $Z(x)$ span the tangent space of the world surface (time point) containing x.

  It is worth mentioning that the converse is not necessarily true even if the vector fields are smooth: linearly independent smooth vector fields, according to Frobenius' theorem define a synchronization only under some additional conditions.

Returning to the illustration in the plane of the page, a world surface will be drawn by a curve as it is seen in Figure 3.2.

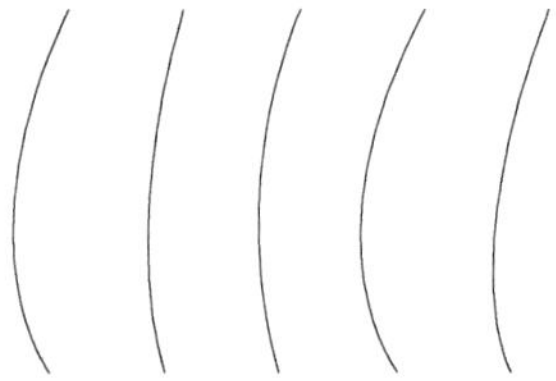

Figure 3.2: Time points of a synchronization.

Let us repeat in order to fix it in our minds:

> *An instant of a synchronization is a subset of spacetime: a three-dimensional submanifold, a world surface. The collection of these world surfaces is the synchronization time.*
>
> *The synchronization time is not a subset of spacetime; it is a set whose elements are subsets of spacetime.*

We emphasize that a synchronization is important only from a practical point of view, it is a human convention not a physical reality; that is why a synchronization is not a part of the structure of a spacetime model, it is defined in the model.

3.5.2 Splitting of spacetime

Though we illustrated a synchronization in such a way that it is established by an observer, a synchronization need not be attached to an observer. An observer can define different synchronizations and different observers can define the same synchronization. In other words, a synchronization is a notion independent of observers.

A **reference frame** is an observer and a synchronization together.

A reference frame assigns to a world point x the corresponding synchronization instant (world surface) $t(x)$ and the observer space point (world line) $q(x)$ that contain the world point in question (see Figure 3.3). We say that the reference frame **splits** spacetime into time and space.

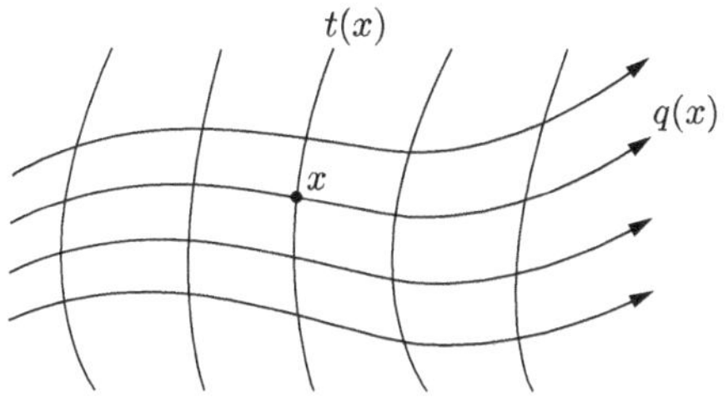

Figure 3.3: Splitting of spacetime by a reference frame.

This splitting has the physical meaning that the reference frame characterizes occurrences by "when" and "where" they happen. In other words, splitting of spacetime in the model corresponds to that an observer, having introduced a synchronization, perceives spacetime as time and space separately.

The splits due to different reference frames are different, i. e., different reference frames assign different time points and space points to the same world point.

It may happen that different proper times may elapse in different space points of an observer between two synchronization time points.

3.5.3 Description of movements

Motion of a material point with respect to an observer is meaningful without synchronization but its description also requires a synchronization: movements can be described only in a reference frame, which gives when (a time point) and where (a space point) the material point is.

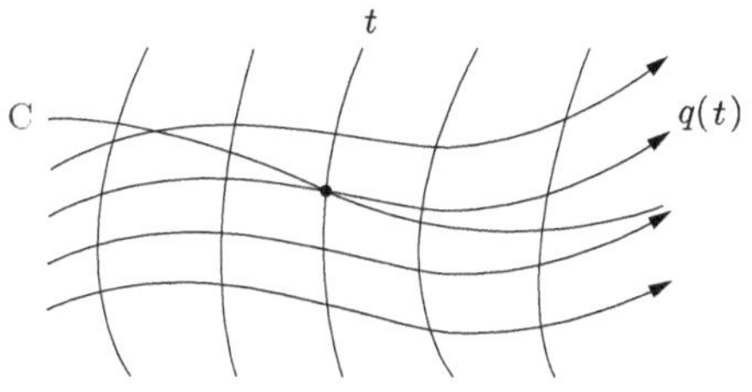

Figure 3.4: Description of how C moves with respect to a reference frame.

Let us take a world line C representing the history of a material point. Let us consider an observer and a synchronization. At the synchronization time point (world surface) t, the material point meets the observer space point (world line) $q(t)$: the occurrence of the material point at t and the occurrence of the observer space point at t correspond to the same world point: $t \cap C = t \cap q(t)$ (see Figure 3.4). The description of the movement in the reference frame is the function $t \mapsto q(t)$.

The velocity of the movement relative to the reference frame, the function $t \mapsto \frac{dq(t)}{dt}$ (whose precise definition is complicated in general) shows very well,

Velocity with respect to an observer only makes sense if a synchronization is specified. Velocity is meaningful only in a reference frame

which has been emphasized in Subsection 1.3.3. The same observer, the same movement, different synchronizations: different velocities.

3.5.4 Uniform synchronizations

The mathematical form of a synchronization, in general, is somewhat complicated. Exceptions are the **uniform synchronizations** in which the time points are parallel three-dimensional hyperplanes.

A uniform synchronization is defined by a three-dimensional linear subspace $\mathbf{S}_S$ of $\mathbf{M}$ in such a way that the world points x and y are simultaneous if and only if $y - x \in \mathbf{S}_S$. According to the basic properties of synchronizations, $\mathbf{S}_S$ must be transverse to all absolute velocities.

For the sake of simpler wording, the synchronization itself will be called $\mathbf{S}_S$ instead of the synchronization established by $\mathbf{S}_S$.

The world points $\mathbf{S}_S$-simultaneous with the world point x are the elements of $x + \mathbf{S}_S$. In other words, the time points (instants) of this synchronization are hyperplanes in M, directed by $\mathbf{S}_S$ (see Figure 3.5). The time T_S of this synchronization is the collection of the time points; it is just the factor space $\mathrm{M}/\mathbf{S}_S$, a one-dimensional affine space over $\mathbf{M}/\mathbf{S}_S$:

$$(y + \mathbf{S}_S) - (y + \mathbf{S}_S) := (y - x) + \mathbf{S}_S. \tag{3.12}$$

This is similar to what we have for the space of an inertial observer (see Subsection 2.4.3).

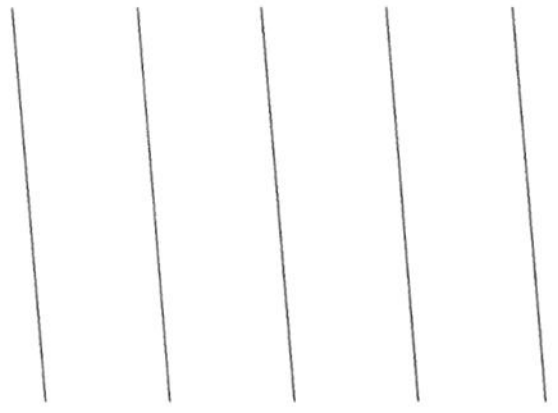

Figure 3.5: Time points of a uniform synchronization.

An orientation can be given for T_S in a natural way: the $\mathbf{S}_S$-instant t_2 is **later** or **earlier** than another $\mathbf{S}_S$-instant t_1 if for all world points x_2 in the hyperplane t_2 there is world point x_1 in the hyperplane t_1 (i. e., $t_2 = x_2 + \mathbf{S}_s$, $t_1 = x_1 + \mathbf{S}_s$) such that $x_2 - x_1$ is future-like or past-like, respectively.

3.5.5 Inertial frames

A reference frame consisting of an inertial observer and a uniform synchronization is called an **inertial frame.**

So, an inertial frame is a pair $(\boldsymbol{u}, \mathbf{S}_s)$ where $\boldsymbol{u}$ is an inertial observer (in fact, an absolute velocity) and $\mathbf{S}_s$ is a uniform synchronization (in fact, a three-dimensional linear subspace of $\mathbf{M}$ transverse to all absolute velocities).

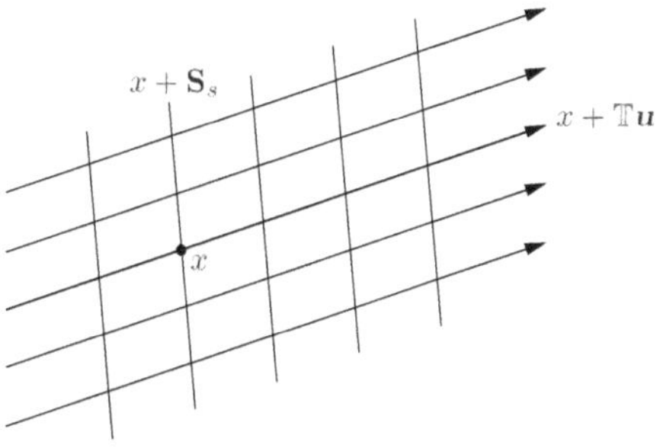

Figure 3.6: Splitting of spacetime by an inertial frame.

The splitting of spacetime according to such an inertial frame is the affine map

$$\mathrm{M} \to \mathrm{T}_s \times \mathrm{S}_{\boldsymbol{u}}, \quad x \mapsto (x + \mathbf{S}_s, x + \mathbb{T}\boldsymbol{u}).$$

Inertial frames have the following important properties:

1. *The same proper time elapses in every space point of an inertial observer between two time points of a uniform synchronization.*

Indeed, let $x + \mathbb{T}\boldsymbol{u}$ and $y + \mathbb{T}\boldsymbol{u}$ be $\boldsymbol{u}$-space points such that x and y are $\mathbf{S}_s$-simultaneous, i. e., $y - x$ is in $\mathbf{S}_s$. Suppose that $x + \boldsymbol{t}\boldsymbol{u}$ and $y + \boldsymbol{s}\boldsymbol{u}$ are $\mathbf{S}_s$-simultaneous, too, for some $\boldsymbol{t}$ and $\boldsymbol{s}$. This means that $(y + \boldsymbol{s}\boldsymbol{u}) - (x + \boldsymbol{t}\boldsymbol{u}) = y - x + (\boldsymbol{s} - \boldsymbol{t})\boldsymbol{u}$ is also in $\mathbf{S}_s$. Then $(\boldsymbol{s} - \boldsymbol{t})\boldsymbol{u}$ is an element of $\mathbf{S}_s$, which is possible if and only if $\boldsymbol{s} = \boldsymbol{t}$ because $\mathbf{S}_s$ is transverse to all absolute velocities.

2. *In an inertial frame, an inertial world line results in a uniform motion along a straight line.*

Indeed, let x and y be occurrences of the inertial line with absolute velocity $\boldsymbol{u}'$, y being later than x; then there is a positive $\boldsymbol{t}' \in \mathbb{T}$ such that $y - x = \boldsymbol{t}'\boldsymbol{u}'$.

The $\boldsymbol{u}$-space vector between the $\boldsymbol{u}$-space points $x + \mathbb{T}\boldsymbol{u}$ and $y + \mathbb{T}\boldsymbol{u}$ is $(y + \mathbb{T}\boldsymbol{u}) - (x + \mathbb{T}\boldsymbol{u}) = (y - x) + \mathbb{T}\boldsymbol{u} = \boldsymbol{t}'\boldsymbol{u}' + \mathbb{T}\boldsymbol{u} = \boldsymbol{t}'(\boldsymbol{u}' + \mathbb{R}\boldsymbol{u})$.

The $\mathbf{S}_s$-time interval between the $\mathbf{S}_s$-time points $x+\mathbf{S}_s$ and $y+\mathbf{S}_s$ is $(y+\mathbf{S}_s)-(x+\mathbf{S}_s) = (y-x)+\mathbf{S}_s = t'\boldsymbol{u}' + \mathbf{S}_s = t'(\boldsymbol{u}' + \frac{\mathbf{S}_s}{\mathbb{T}})$.

We obtained that any straight line section in $\boldsymbol{u}$-space covered by this movement is proportional to the corresponding $\mathbf{S}_s$-time period: the velocity of the movement is constant. Note that the velocity of this $\boldsymbol{u}$-movement depends on $\mathbf{S}_s$.

Let us repeat and summarize our results:

(1) In the space of an inertial observer, the path of movement of an inertial material point is a straight line (see Subsection 2.5.1).
(2) If the observer chooses an arbitrary uniform synchronization, then the movement becomes uniform, i. e., has constant velocity.
(3) For different uniform synchronizations the constant velocities of the movement are different.

Remark. "Rest frame," "moving," and "being at rest" are frequently used notions in the literature. It is important to keep in mind that it cannot be emphasized enough that
- There are no "rest frames" and "moving frames."
- "Being at rest" and "moving" are not meaningful in themselves.
- Only "being at rest with respect to an observer" and "moving with respect to an observer" (without any synchronization) are meaningful.

3.6 Coordinate systems

It is worth repeating: An observer and the time passing in the space points of an observer are physical facts but a synchronization—though in practice it is defined by a physical procedure—is a human convention. A physical fact and a human convention are mixed in a reference frame.

A further human convention results in a **coordinate system of spacetime**: The synchronization instants are represented by real numbers and observer space points are represented by triplets of real numbers; thus, world points are represented by quartets of real numbers. Coordinate systems are important only from a practical point of view, e. g., for solving concrete problems.

An inertial frame—a uniform synchronization and an inertial observer together—can give coordinates by choosing:
(i) A unit time interval (e. g., second or year).
(ii) An origin in the synchronization time (e. g., today midday or "a culturally significant event").
(iii) A unit length (e. g., meter or yard).
(iv) An origin in the observer space (e. g., the corner of our room or Piccadilly Circus in London).
(v) Three axes, straight lines in the space origin (e. g., the edges of walls in our room or edges of some buildings on Piccadilly Circus).

All these will be treated in a mathematically exact way in the nonrelativistic spacetime model as well as in the special relativistic spacetime model.

We see how many arbitrariness there is in a coordinate system, which points out that treating spacetime by coordinates may cause unnecessary complications: the formulas depend on the chosen coordinates and one must permanently control their proper physical meaning.

3.7 Exercises

1. Demonstrate that two absolute velocities are parallel if and only if they are equal.
2. Show that the paths in $\boldsymbol{u}$-space of the inertial world lines directed by $\boldsymbol{u}_1$ and $\boldsymbol{u}_2$ are parallel if and only if $\boldsymbol{u}$, $\boldsymbol{u}_1$, $\boldsymbol{u}_2$ are coplanar.
3. Redraw Figure 2.13 keeping the world lines with $\boldsymbol{u}_1$ and $\boldsymbol{u}_2$ but with another observers $\boldsymbol{u}$ in such a way that
   a) $\boldsymbol{u}_1$ be faster than $\boldsymbol{u}_2$ along the same $\boldsymbol{u}$-path in the same direction.
   b) $\boldsymbol{u}_1$ and $\boldsymbol{u}_2$ have the same $\boldsymbol{u}$-path but opposite directions.
4. Let three inertial material points with absolute velocities $\boldsymbol{u}_1$, $\boldsymbol{u}_2$, and $\boldsymbol{u}_3$ move along the same path in the same direction in the space of the observer $\boldsymbol{u}$. It makes sense (without synchronization) that $\boldsymbol{u}_2$ is three times faster than $\boldsymbol{u}_1$ as $\boldsymbol{u}_3$ is faster than $\boldsymbol{u}_2$; how? Draw a figure.
5. What are the symmetries of an inertial observer $\boldsymbol{u}$?
6. Demonstrate that the orientation of a uniform synchronization time given in Subsection 3.5.4 is well-defined, i. e., the $\mathbf{S}_s$-time point t cannot be both later and earlier than s.
7. Take the toy model in which spacetime is $\mathbb{R}^4$ as an affine space with the underlying vector space $\mathbb{R}^4$ and every measure line is $\mathbb{R}$, all these objects having the standard orientations.
   Let the world points and world vectors be denoted in the form $x = (\xi^1, \xi^2, \xi^3, \xi^4)$ and $\mathbf{x} = (x^1, x^2, x^3, x^4)$, respectively.
   Let the set of future-like vectors be
   $$\mathrm{T}^{\rightarrow} := \{\mathbf{x} \mid x^1 > 0, x^2 > 0\}$$
   and let the inertial time progress be
   $$\boldsymbol{P}(\mathbf{x}) := x^1 + x^2.$$
   (i) Show that the set $\mathrm{T}^{\rightarrow}$ is an open convex cone with apex at zero and that $\boldsymbol{P}$ is positive homogeneous.
   (ii) What is V(1)?

(iii) Let $\cdot$ denote the usual scalar product on $\mathbb{R}^4$. Show that

$$\boldsymbol{b}_{\boldsymbol{u}}(\boldsymbol{y},\boldsymbol{x}) := \big((\boldsymbol{u}\cdot\boldsymbol{u})\boldsymbol{y} - \boldsymbol{u}(\boldsymbol{u}\cdot\boldsymbol{y})\big)\cdot\big((\boldsymbol{u}\cdot\boldsymbol{u})\boldsymbol{x} - \boldsymbol{u}(\boldsymbol{u}\cdot\boldsymbol{x})\big)$$

is a semi-Euclidean form, i. e., has the properties listed in Section 2.4.4.

(iv) Is this model fair?

(v) Give the general form of world line functions and give a concrete non-inertial world line function.

(vi) Give a uniform synchronization, i. e., a three-dimensional linear subspace which is transverse to all absolute velocities.

Part II: **Absolute time progress**

In this part, we treat a fair flat spacetime, the **nonrelativistic spacetime model** arising from everyday experience on time progress and the speed of movements. The way to it is simple and instructive—it is worth going through, everything is well understandable even disregarding the proofs in small size.

https://doi.org/10.1515/9783112219553-005

4 Construction of the nonrelativistic spacetime model

4.1 Basic experiences

As previously mentioned, our simple and superficial experience indicates that the chronometer left at home and the one that accompanied me for my holidays have ticked the same number of times between their two meetings. The classical idea of spacetime—tacitly—is based on the assumption that time progresses in the same way for everything.

This is one of our (superficial) experience we put in the spacetime model:

($\mathcal{A}$1). Time progress is absolute, *i. e., if two world lines have two occurrences in common (they meet in two world points), then the same proper time passes on both world lines between their corresponding two occurrences.*

Another our (superficial) experience is that "arbitrarily fast" motion can be realized in an arbitrary inertial frame. This property is stronger than (**$\mathcal{M}$3**) in Subsection 1.3.3 and cannot be formulated in such a way that there is no limit on speeds relative to an inertial frame because this would require a synchronization, which cannot be applied in the construction of a spacetime model. Round-way speed (see Subsection 1.3.4), however, is meaningful without synchronization, so we can accept:

($\mathcal{A}$2). Arbitrarily fast round-way speed can be realized *in the space of every inertial observer.*

In what follows, time passes from left to right in our figures as previously but the orientation of world lines will not be shown in some figures.

4.2 Absolute time progress

Property (**$\mathcal{A}$1**) can be formulated as follows: if x and y are world points such that $y - x$ is future-like, then $\boldsymbol{t}_{\mathrm{C}_1}(x,y) = \boldsymbol{t}_{\mathrm{C}_2}(x,y)$ for arbitrary world lines C_1 and C_2 meeting at x and y.

This implies:

There is a unique linear map $\boldsymbol{\tau} : \mathbf{M} \to \mathbb{T}$ whose restriction onto $\mathrm{T}^{\to}$ equals $\boldsymbol{P}$.

Let C_1 be the inertial world line passing through x and y, and let C_2 consist of two inertial world line segments, one of which goes from x to z and the other goes from z to y; then we obtain by (**$\mathcal{A}$1**) that $\boldsymbol{P}(y - x) = \boldsymbol{P}(y - z) + \boldsymbol{P}(z - x)$. Since both $y - z$ and $z - x$ are future-like, and $y - x = (y - z) + (z - x)$, with the notation $\boldsymbol{y} := y - z$, $\boldsymbol{x} := z - x$, we have $\boldsymbol{P}(\boldsymbol{x} + \boldsymbol{y}) = \boldsymbol{P}(\boldsymbol{x}) + \boldsymbol{P}(\boldsymbol{y})$ and, taking into account that $\boldsymbol{P}$ is positive homogeneous,

https://doi.org/10.1515/9783112219553-006

$$\boldsymbol{P}(\alpha \boldsymbol{x} + \beta \boldsymbol{y}) = \alpha \boldsymbol{P}(\boldsymbol{x}) + \beta \boldsymbol{P}(\boldsymbol{y}) \tag{4.1}$$

for all positive real numbers α, β and future-like vectors $\boldsymbol{x}, \boldsymbol{y}$.

As a consequence, $\boldsymbol{P}$ can be uniquely extended to a linear map on $\mathbf{M}$. Indeed, according to Section 2.6 every element of $\mathbf{M}$ is of the form $\boldsymbol{y} - \boldsymbol{x}$ where $\boldsymbol{y}$ and $\boldsymbol{x}$ are in $\mathrm{T}^{\rightarrow}$. Let us define the linear map $\boldsymbol{\tau}$ by $\boldsymbol{\tau} \cdot (\boldsymbol{y} - \boldsymbol{x}) := \boldsymbol{P}(\boldsymbol{y}) - \boldsymbol{P}(\boldsymbol{x})$.

The definition is correct: if $\boldsymbol{y}_1 - \boldsymbol{x}_1 = \boldsymbol{y}_2 - \boldsymbol{x}_2$, then $\boldsymbol{y}_1 + \boldsymbol{x}_2 = \boldsymbol{y}_2 + \boldsymbol{x}_1$ and here both sides are elements of $\mathrm{T}^{\rightarrow}$, so they are in the domain of $\boldsymbol{P}$ and by (4.1) we get $\boldsymbol{P}(\boldsymbol{y}_1) + \boldsymbol{P}(\boldsymbol{x}_2) = \boldsymbol{P}(\boldsymbol{y}_2) + \boldsymbol{P}(\boldsymbol{x}_1)$ involving $\boldsymbol{P}(\boldsymbol{y}_1) - \boldsymbol{P}(\boldsymbol{x}_1) = \boldsymbol{P}(\boldsymbol{y}_2) - \boldsymbol{P}(\boldsymbol{x}_2)$.

$\boldsymbol{\tau}$ is linear:

– For arbitrary $\boldsymbol{y}_1, \boldsymbol{x}_1, \boldsymbol{y}_2, \boldsymbol{x}_2 \in \mathrm{T}^{\rightarrow}$

$$\begin{aligned}\boldsymbol{\tau} \cdot \big((\boldsymbol{y}_1 - \boldsymbol{x}_1) + (\boldsymbol{y}_2 - \boldsymbol{x}_2)\big) &= \boldsymbol{\tau} \cdot \big((\boldsymbol{y}_1 + \boldsymbol{y}_2) - (\boldsymbol{x}_1 + \boldsymbol{x}_2)\big) \\ &= \boldsymbol{P}(\boldsymbol{y}_1 + \boldsymbol{y}_2) - \boldsymbol{P}(\boldsymbol{x}_1 + \boldsymbol{x}_2) = \boldsymbol{P}(\boldsymbol{y}_1) + \boldsymbol{P}(\boldsymbol{y}_2) - \big(\boldsymbol{P}(\boldsymbol{x}_1) + \boldsymbol{P}(\boldsymbol{x}_2)\big) \\ &= \big(\boldsymbol{P}(\boldsymbol{y}_1) - \boldsymbol{P}(\boldsymbol{x}_1)\big) + \big(\boldsymbol{P}(\boldsymbol{y}_2) - \boldsymbol{P}(\boldsymbol{x}_2)\big) \\ &= \boldsymbol{\tau} \cdot (\boldsymbol{y}_1 - \boldsymbol{x}_1) + \boldsymbol{\tau} \cdot (\boldsymbol{y}_2 - \boldsymbol{x}_2).\end{aligned}$$

– For arbitrary $\boldsymbol{y}, \boldsymbol{x} \in \mathrm{T}^{\rightarrow}$ and real number α, $\alpha(\boldsymbol{y} - \boldsymbol{x}) = \alpha\boldsymbol{y} - \alpha\boldsymbol{x}$ if α is positive and $\alpha(\boldsymbol{y} - \boldsymbol{x}) = |\alpha|\boldsymbol{x} - |\alpha|\boldsymbol{y}$ if α is negative; thus, the positive homogeneity of $\boldsymbol{P}$ implies $\boldsymbol{\tau} \cdot (\alpha(\boldsymbol{y} - \boldsymbol{x})) = \alpha\boldsymbol{\tau} \cdot (\boldsymbol{y} - \boldsymbol{x})$.

For the sake of precision, we mention that the deduction above has departed from the original definition of a world line because C_2, consisting of two straight line segments, has a breakpoint. We could have defined world lines by allowing breakpoints but we avoided this somewhat cumbersome definition and we allowed here the simply understandable curve consisting of two straight line segments.

In the sequel, we write $\boldsymbol{\tau}$ instead of $\boldsymbol{P}$; then for an arbitrary world line C and $x, y \in \mathrm{C}$, we have

$$\boldsymbol{t}_{\mathrm{C}}(x, y) = \boldsymbol{\tau} \cdot (y - x).$$

$\boldsymbol{\tau}$ can be transferred to a linear map $\frac{\mathbf{M}}{\mathbb{T}} \to \mathbb{R}$ in the well-known way; then, according to the general formula (2.3),

$$\boldsymbol{\tau} \cdot \boldsymbol{u} = 1$$

for all absolute velocities.

4.3 Future-like vectors

First of all, note that if $\boldsymbol{x}$ is future-like, then $\boldsymbol{\tau} \cdot \boldsymbol{x} > 0$.

Property ($\mathcal{A}2$) allows us to prove the contrary, too: if $\boldsymbol{\tau} \cdot \boldsymbol{x} > 0$, then $\boldsymbol{x}$ is future-like.

Taking an inertial observer $\boldsymbol{u}$, let us consider the $\boldsymbol{u}$-space points $p := x + \mathbb{T}\boldsymbol{u}$ and $q := y + \mathbb{T}\boldsymbol{u}$ such that $\boldsymbol{\tau} \cdot (y - x) = 0$. For an arbitrary "small" time interval $\boldsymbol{t} > 0$, there is a two-way motion:

(i) Starting at the occurrence x of p.

(ii) Turning round at the occurrence $y + \boldsymbol{s}\boldsymbol{u}$ of q for some $\boldsymbol{s} > 0$.

(iii) Arriving back at the occurrence $x + \boldsymbol{t}\boldsymbol{u}$ of p.

Then

(1) $(y + \boldsymbol{s}u) - x = y - x + \boldsymbol{s}u$ is future-like.

(2) $(x + \boldsymbol{t}u) - (y + \boldsymbol{s}u) = x - y + (\boldsymbol{t} - \boldsymbol{s})u$ is also future-like, so $\tau \cdot (x - y + (\boldsymbol{t} - \boldsymbol{s})u) = \boldsymbol{t} - \boldsymbol{s} > 0$.

Since x and y are arbitrary, with the notation $q := x - y$ we have that $q + \boldsymbol{t}u = (q + (\boldsymbol{t} - \boldsymbol{s})u) + \boldsymbol{s}u$ (the sum of two future-like vectors) is in $\mathrm{T}^{\rightarrow}$ for all $\tau \cdot q = 0$ and $0 < \boldsymbol{s} < \boldsymbol{t} \in \mathbb{T}^{+}$.

If x is a vector for which $\tau \cdot x > 0$ holds, then $\tau \cdot (x - (\tau \cdot x)u) = 0$ and $x = (x - (\tau \cdot x)u) + (\tau \cdot x)u$; putting $\boldsymbol{t} := \tau \cdot x$ and $q := x - \boldsymbol{t}u$, we see that $x \in \mathrm{T}^{\rightarrow}$.

Thus, we have that the set of **future-like vectors** is

$\mathrm{T}^{\rightarrow} = \{x \in \mathbf{M} \mid \tau \cdot x > 0\}$.

Such a $\mathrm{T}^{\rightarrow}$ is illustrated by the first pictures in Figures 2.7 and 2.8.

The set of time-like vectors is $\mathrm{T} = \{x \in \mathbf{M} \mid \tau \cdot x \neq 0\}$.

4.4 Space-like vectors, absolute simultaneity

The set of future-like vectors is a half-space, its boundary is

$$\mathbf{S} := \{q \in \mathbf{M} \mid \tau \cdot q = 0\},$$

a three-dimensional linear subspace of $\mathbf{M}$, being the kernel of a linear map whose range is one-dimensional.

The elements of $\mathbf{S}$ are called **absolute space-like.**

Accordingly, by (2.3), the set of absolute velocities,

$$\mathrm{V}(1) = \left\{u \in \frac{\mathbf{M}}{\mathbb{T}} \;\middle|\; \tau \cdot u = 1\right\}$$

is a three-dimensional affine hyperplane over $\frac{\mathbf{S}}{\mathbb{T}}$.

$\mathbf{S}$ is the single three-dimensional linear subspace transverse to all future-like vectors.

As a consequence:

In the nonrelativistic spacetime model, there can be a single synchronization only: the one determined by **S**, called **absolute synchronization**.

The physical realization of this synchronization is based on the assumption of absolute time progress. An observer (say the Earth) starts chronometers in a space point (say in Greenwich) and then sends them to everywhere (say to Paris, Berlin, Budapest, etc.); then the equal states of the chronometers (the number of ticks after their start) in different places represent simultaneous instants.

4.5 Boosts

Absolute simultaneity suggests that boosts be realized by establishing **instantaneous prints.**

Let us consider a vector: a "rod with tip" in the space of an inertial observer $\boldsymbol{u}'$. The corresponding vector in the space of another inertial observer $\boldsymbol{u}$ is obtained by marking those $\boldsymbol{u}$-space points that meet the rod at an arbitrary instant.

To formulate this procedure, recall that the spaces of the inertial observers $\boldsymbol{u}'$ and $\boldsymbol{u}$ are the sets of lines in spacetime directed by $\boldsymbol{u}'$ and $\boldsymbol{u}$, respectively.

Let $p' := x + \mathbb{T}\boldsymbol{u}'$ and $q' := y + \mathbb{T}\boldsymbol{u}'$ be the end points of the rod in the $\boldsymbol{u}'$-space, x and y being simultaneous: $y - x \in \mathbf{S}$; then $p' - q' = (y - x) + \mathbb{T}\boldsymbol{u}'$.

The simultaneous prints of p' and q' in $\boldsymbol{u}$-space are $p := x + \mathbb{T}\boldsymbol{u}$ and $q := y + \mathbb{T}\boldsymbol{u}$ for which $q - p = (y - x) + \mathbb{T}\boldsymbol{u}$ (see Figure 4.1). The print of the rod in $\boldsymbol{u}$-space is determined by the same vector $y - x \in \mathbf{S}$ as in the $\boldsymbol{u}'$-space.

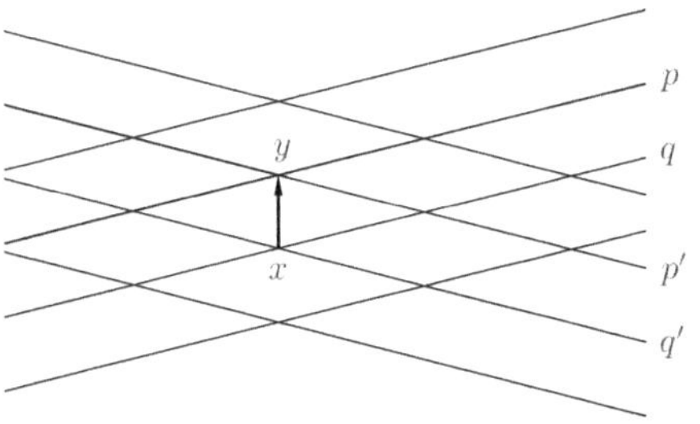

Figure 4.1: The simultaneous print of $p' - q' = y - x$ is $p - q = y - x$.

This means that the boost from $\boldsymbol{u}'$ to $\boldsymbol{u}$, restricted to $\mathbf{S}$, is the identity. That is why we define

$$\boldsymbol{B}_{\boldsymbol{u}\boldsymbol{u}'} \cdot \boldsymbol{u}' = \boldsymbol{u}, \quad \boldsymbol{B}_{\boldsymbol{u}\boldsymbol{u}'} \cdot \boldsymbol{q} = \boldsymbol{q} \quad (\boldsymbol{q} \in \mathbf{S}).$$

It is trivial that this choice satisfies requirements (3.4), (3.7), and (3.8); (3.10) is satisfied with $\alpha_{\boldsymbol{u},\boldsymbol{u}'} = 1$. Namely, $\boldsymbol{u}' - \boldsymbol{u}'$ is in $\frac{\mathbf{S}}{\mathbb{T}}$, so

$$\boldsymbol{B}_{\boldsymbol{u}\boldsymbol{u}'} \cdot \left(\boldsymbol{u} - \boldsymbol{u}'\right) = \boldsymbol{u} - \boldsymbol{u}' = -\left(\boldsymbol{u}' - \boldsymbol{u}\right).$$

Repeating, according to the sense, what has been said in Subsection 2.4.3, we see that $\boldsymbol{B}_{\boldsymbol{u}\boldsymbol{u}'}$ preserves orientation.

Further, (3.1) and (3.2) hold, too: for all $\boldsymbol{x}$, $\boldsymbol{B}_{\boldsymbol{u}\boldsymbol{u}'} \cdot \boldsymbol{x} = \boldsymbol{B}_{\boldsymbol{u}\boldsymbol{u}'} \cdot (\boldsymbol{x} - (\boldsymbol{\tau} \cdot \boldsymbol{x})\boldsymbol{u}') + \boldsymbol{B}_{\boldsymbol{u}'\boldsymbol{u}} \cdot ((\boldsymbol{\tau} \cdot \boldsymbol{x})\boldsymbol{u}') = (\boldsymbol{x} - (\boldsymbol{\tau} \cdot \boldsymbol{x})\boldsymbol{u}') + (\boldsymbol{\tau} \cdot \boldsymbol{x})\boldsymbol{u}$, so

$$\boldsymbol{\tau} \cdot (\boldsymbol{B}_{\boldsymbol{u}\boldsymbol{u}'} \cdot \boldsymbol{x}) = \boldsymbol{\tau} \cdot \boldsymbol{x},$$

which shows as well that the boost preserves arrow orientation.

As concerns (3.3), we have a very strong and nice result given in the next section.

4.6 The absolute Euclidean form

Since the restriction of $\boldsymbol{B}_{uu'}$ to $\mathbf{S}$ is the identity, it follows from equality

$$\boldsymbol{b}_u(\boldsymbol{B}_{uu'} \cdot x, \boldsymbol{B}_{uu'} \cdot y) = \boldsymbol{b}_{u'}(x, y),$$

necessary for $\boldsymbol{B}_{uu'}$ being a vectorial symmetry, that the restrictions of $\boldsymbol{b}_u$ and $\boldsymbol{b}_{u'}$ to $\mathbf{S}$ are equal, which involves:

There is a uniquely determined

$$\boldsymbol{h} : \mathbf{S} \times \mathbf{S} \to \mathbb{L} \otimes \mathbb{L}$$

bilinear, symmetric, positive definite map, the **absolute Euclidean form** such that the restriction of $\boldsymbol{b}_u$ to $\mathbf{S}$ equals $\boldsymbol{h}$ for all $\boldsymbol{u}$.

Since $\boldsymbol{x} - \boldsymbol{u}(\tau \cdot \boldsymbol{x})$ is in $\mathbf{S}$ for all $\boldsymbol{x}$ (and similarly for $\boldsymbol{y}$) and the kernel of $\boldsymbol{b}_u$ is $\mathbb{T}\boldsymbol{u}$, we can state that

$$\boldsymbol{b}_u(\boldsymbol{x}, \boldsymbol{y}) = \boldsymbol{b}_u(\boldsymbol{x} - \boldsymbol{u}(\tau \cdot \boldsymbol{x}), \boldsymbol{y} - \boldsymbol{u}(\tau \cdot \boldsymbol{y})) = \boldsymbol{h}(\boldsymbol{x} - \boldsymbol{u}(\tau \cdot \boldsymbol{x}), \boldsymbol{y} - \boldsymbol{u}(\tau \cdot \boldsymbol{y})).$$

5 Structure of the nonrelativistic spacetime model

The fair flat spacetime model constructed in the previous sections is based on (**$\mathcal{A}$1**) and (**$\mathcal{A}$2**) besides the general assumptions.

The nonrelativistic spacetime model reflects our simple ideas about time and space and this spacetime model—not in an exact setting—forms the background of the usual treatments of classical mechanics, so it is deeply imbedded in physical considerations.

In the following sections, we define and treat the nonrelativistic spacetime model without referring to the way how we obtained it, so that even those who have skipped the previous sections can perfectly understand this model. That is why some previous formulas will appear again.

5.1 Fundamental properties of the model

5.1.1 New notation

Instead of the general notation $(\mathrm{M}, \mathbb{T}, \mathbb{L}, T^{\to}, \boldsymbol{P}, \boldsymbol{b})$, a **nonrelativistic spacetime model** will be referred to by the symbol

---

$$(\mathrm{M}, \mathbb{T}, \mathbb{L}, \boldsymbol{\tau}, \boldsymbol{h})$$

where

- M is spacetime, a four-dimensional oriented affine space (over the vector space $\mathbf{M}$).
- $\mathbb{T}$ is the measure line of time periods.
- $\mathbb{L}$ is the measure line of distances.
- $\boldsymbol{\tau} : \mathbf{M} \to \mathbb{T}$ is a linear surjection giving the **absolute time progress**, its kernel

$$\mathbf{S} := \{\boldsymbol{q} \in \mathbf{M} \mid \boldsymbol{\tau} \cdot \boldsymbol{q} = 0\}, \tag{5.1}$$

  a three-dimensional linear subspace of $\mathbf{M}$ is the set of **absolute space-like vectors.**
- $\boldsymbol{h} : \mathbf{S} \times \mathbf{S} \to \mathbb{L} \otimes \mathbb{L}$ is a symmetric, positive definite bilinear map, the **absolute Euclidean form**.

Then

- The set of future-like vectors

$$T^{\to} := \{\boldsymbol{x} \in \mathbf{M} \mid \boldsymbol{\tau} \cdot \boldsymbol{x} > 0\}, \tag{5.2}$$

  is an open half-space, so it is an open convex cone with apex at zero.
- The **inertial time progress**

$$\boldsymbol{P}(\boldsymbol{x}) := \boldsymbol{\tau} \cdot \boldsymbol{x} \quad (\boldsymbol{x} \in T^{\to}) \tag{5.3}$$

  is positive homogeneous and continuous (even smooth).

https://doi.org/10.1515/9783112219553-007

- The set of **absolute velocities**

$$V(1) := \left\{ \boldsymbol{u} \in \frac{\mathbf{M}}{\mathbb{T}} \;\middle|\; \tau \cdot \boldsymbol{u} = 1 \right\} \tag{5.4}$$

  is a three-dimensional affine space over $\frac{\mathrm{S}}{\mathbb{T}}$.
- The semi-Euclidean forms of inertial observers are given by

$$\boldsymbol{b}_u(\boldsymbol{x},\boldsymbol{y}) = \boldsymbol{h}(\boldsymbol{x} - \boldsymbol{u}(\tau \cdot \boldsymbol{x}), \boldsymbol{y} - \boldsymbol{u}(\tau \cdot \boldsymbol{y})) \quad (\boldsymbol{u} \in V(1),\ \boldsymbol{x},\boldsymbol{y} \in \mathbf{M}), \tag{5.5}$$

  which depends continuously (even smoothly) on $\boldsymbol{u}$, positive semidefinite and its kernel is $\mathbb{T}\boldsymbol{u}$; in particular,

$$\boldsymbol{b}_u(\boldsymbol{q},\boldsymbol{p}) = \boldsymbol{h}(\boldsymbol{q},\boldsymbol{p}) \quad (\boldsymbol{q},\boldsymbol{p} \in \mathbf{S}, \boldsymbol{u} \in V(1)).$$

- The set of **past-like vectors** and **time-like vectors** are

$$\mathrm{T}^{\leftarrow} = \{\boldsymbol{x} \in \mathbf{M} \mid \tau \cdot \boldsymbol{x} < 0\}, \quad \mathrm{T}^{\leftarrow} \cup \mathrm{T}^{\rightarrow} = \{\boldsymbol{x} \in \mathbf{M} \mid \tau \cdot \boldsymbol{x} \neq 0\}.$$

5.1.2 Space vectors of inertial observers

Let us recall (see Subsection 2.4.3) that the space $\mathrm{S}_u = \mathrm{M}/\mathbb{T}\boldsymbol{u}$ of the inertial observer $\boldsymbol{u}$, the set of straight lines directed by $\boldsymbol{u}$ in spacetime, is a three-dimensional affine space in a mathematically perfect way but the underlying vector space $\mathbf{M}/\mathbb{T}\boldsymbol{u}$ cannot be illustrated in an eye-catching way.

Now we have a possibility to improve the situation with the aid of the absolute space-like vectors.

First of all, we note that the orientation of $\mathbf{M}$ determines an orientation of $\mathbf{S}$ in a natural way: let an ordered basis $(\boldsymbol{q}_1, \boldsymbol{q}_2, \boldsymbol{q}_3)$ of $\mathbf{S}$ be positively oriented if $(\boldsymbol{x}, \boldsymbol{q}_1, \boldsymbol{q}_2, \boldsymbol{q}_3)$ is a positively oriented basis of $\mathbf{M}$ for an arbitrary future-like vector $\boldsymbol{x}$. It is not hard to demonstrate that this orientation is well-defined.

Elements of $\mathbf{M}/\mathbb{T}\boldsymbol{u}$ are straight lines in $\mathbf{M}$, directed by $\boldsymbol{u}$. Every such a line has a unique point in $\mathbf{S}$. The point $\boldsymbol{x} + t\boldsymbol{u}$ of $\boldsymbol{x} + \mathbb{T}\boldsymbol{u}$ is in $\mathbf{S}$ if $0 = \tau \cdot (\boldsymbol{x} + t\boldsymbol{u}) = \tau \cdot \boldsymbol{x} + t$, i. e., $t = -\tau \cdot \boldsymbol{x}$. That is why it is convenient to represent the $\boldsymbol{u}$-space vector $\boldsymbol{x} + \mathbb{T}\boldsymbol{u}$ by the element $\boldsymbol{x} - (\tau \cdot \boldsymbol{x})\boldsymbol{u}$ of $\mathbf{S}$.

In other words, introducing the projection onto $\mathbf{S}$ along $\mathbb{T}\boldsymbol{u}$,

$$\pi_u := \mathbf{1} - \boldsymbol{u} \otimes \tau : \mathbf{M} \to \mathbf{S}, \quad \boldsymbol{x} \mapsto \boldsymbol{x} - \boldsymbol{u}(\tau \cdot \boldsymbol{x}), \tag{5.6}$$

we make the identification

$$\mathbf{M}/\mathbb{T}\boldsymbol{u} \equiv \mathbf{S}, \quad \boldsymbol{x} + \mathbb{T}\boldsymbol{u} \equiv \pi_u \cdot \boldsymbol{x}.$$

This identification is correct, i. e., it is an orientation preserving linear bijection.

(Identification) It must be shown that the map $\frac{\mathbf{M}}{\mathbb{T}\boldsymbol{u}} \to \mathbf{S}$, $\boldsymbol{x} + \mathbb{T}\boldsymbol{u} \mapsto \pi_{\boldsymbol{u}} \cdot \boldsymbol{x}$ is a linear bijection.
Injective: if $\boldsymbol{x} + \mathbb{T}\boldsymbol{u} = \boldsymbol{y} + \mathbb{T}\boldsymbol{u}$, then there is a $\boldsymbol{t} \in \mathbb{T}$ such that $\boldsymbol{x} - \boldsymbol{y} = \boldsymbol{t}\boldsymbol{u}$, so $\pi_{\boldsymbol{u}} \cdot (\boldsymbol{x} - \boldsymbol{y}) = 0$ that is $\pi_{\boldsymbol{u}} \cdot \boldsymbol{x} = \pi_{\boldsymbol{u}} \cdot \boldsymbol{y}$.
Surjective: if $\boldsymbol{q} \in \mathbf{S}$, then $\boldsymbol{q} + \mathbb{T}\boldsymbol{u} \mapsto \pi_{\boldsymbol{u}} \cdot \boldsymbol{q} = \boldsymbol{q}$.
Linear: $\alpha(\boldsymbol{x} + \mathbb{T}\boldsymbol{u}) + \beta(\boldsymbol{y} + \mathbb{T}\boldsymbol{u}) = (\alpha\boldsymbol{x} + \beta\boldsymbol{y}) + \mathbb{T}\boldsymbol{u} \mapsto \pi_{\boldsymbol{u}} \cdot (\alpha\boldsymbol{x} + \beta\boldsymbol{y}) = \alpha\pi_{\boldsymbol{u}} \cdot \boldsymbol{x} + \beta\pi_{\boldsymbol{u}} \cdot \boldsymbol{y}$.
(Orientation) According to Subsection 2.4.3, the basis $(\boldsymbol{x}_1 + \mathbb{T}\boldsymbol{u}, \boldsymbol{x}_2 + \mathbb{T}\boldsymbol{u}, \boldsymbol{x}_3 + \mathbb{T}\boldsymbol{u})$ in $\frac{\mathbf{M}}{\mathbb{T}\boldsymbol{u}}$ is positively oriented if $(\boldsymbol{t}\boldsymbol{u}, \boldsymbol{x}_1, \boldsymbol{x}_2, \boldsymbol{x}_3)$ is a positively oriented basis in $\mathbf{M}$ for an arbitrary positive element $\boldsymbol{t}$ of $\mathbb{T}$. $\boldsymbol{x} + \mathbb{T}\boldsymbol{u} = \boldsymbol{x} - (\boldsymbol{\tau} \cdot \boldsymbol{x})\boldsymbol{u} + \mathbb{T}\boldsymbol{u}$, so we can take space-like vectors $\boldsymbol{q}_1$, $\boldsymbol{q}_2$, $\boldsymbol{q}_3$ instead of $\boldsymbol{x}_1$, $\boldsymbol{x}_2$, $\boldsymbol{x}_3$ in the previous formulas. This shows that the identification preserves orientation.

With this identification, the subtraction in $\boldsymbol{u}$-space $\mathrm{S}_{\boldsymbol{u}}$ becomes

$$(y + \mathbb{T}\boldsymbol{u}) - (x + \mathbb{T}\boldsymbol{u}) := \pi_{\boldsymbol{u}} \cdot (y - x) \in \mathbf{S}. \tag{5.7}$$

We have nice formulas by introducing the notation

$$\pi_{\boldsymbol{u}} : \mathrm{M} \to \mathrm{S}_{\boldsymbol{u}}, \quad \pi_{\boldsymbol{u}}(x) := x + \mathbb{T}\boldsymbol{u}; \tag{5.8}$$

then $\pi_{\boldsymbol{u}}$ is an affine map over $\pi_{\boldsymbol{u}}$:

$$\pi_{\boldsymbol{u}}(y) - \pi_{\boldsymbol{u}}(x) := \pi_{\boldsymbol{u}} \cdot (y - x). \tag{5.9}$$

Lastly, we can assert:

The spaces $\mathrm{S}_{\boldsymbol{u}}$ and $\mathrm{S}_{\boldsymbol{u}'}$ of different inertial observers $\boldsymbol{u}$ and $\boldsymbol{u}'$ are **different** three-dimensional affine spaces over the **same** vector space **S**.

5.1.3 Illustrations

The arithmetic spacetime model (see Section 5.2) suggests the following illustrations:
1. The world vectors are shown in Figure 5.1.

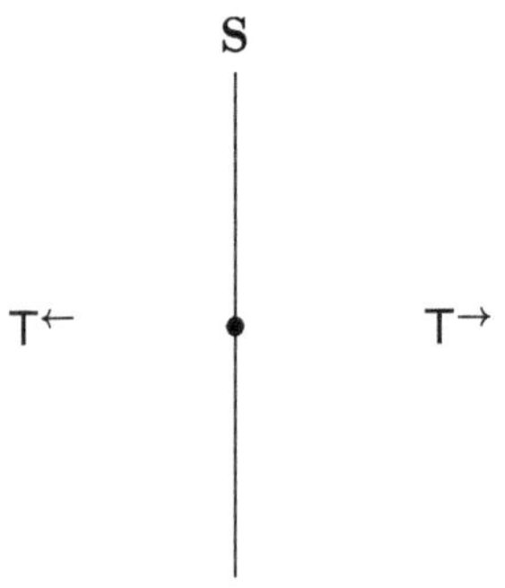

Figure 5.1: The set of world vectors.

2. V(1) is a three-dimensional affine subspace in $\frac{\mathbb{M}}{\mathbb{T}}$ over $\frac{\mathbb{S}}{\mathbb{T}}$, so it is depicted as it is seen in Figure 5.2.

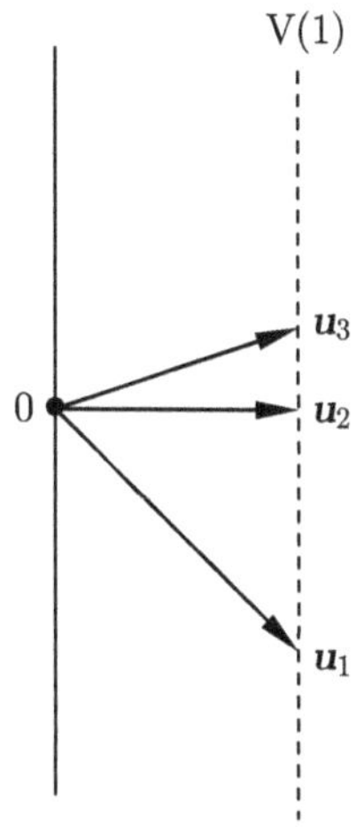

Figure 5.2: Absolute velocities.

Now we call attention again to that

- There is no zero absolute velocity
- An absolute velocity has no magnitude; in particular, it makes no sense that an absolute velocity is higher or lower than another one
- The angle between two absolute velocities is not meaningful

the illustration must not mislead us: $\boldsymbol{u}_1$ in Figure 5.2 is not higher than $\boldsymbol{u}_2$, the angle between $\boldsymbol{u}_1$ and $\boldsymbol{u}_2$ does not exist, $\boldsymbol{u}_2$ is not orthogonal to $\frac{\mathbb{S}}{\mathbb{T}}$.

Important note:
The equal time passed on straight lines with different absolute velocities are represented, in general, by sections of different lengths in our figures. The larger the angles between the horizontal line and the lines in question, the larger sections correspond to equal time periods; equal time periods correspond to equal sections in case of two absolute velocities that have the same angles—over and under—with the horizontal one. That is why when treating relations of two absolute velocities, they will be illustrated in this way. We emphasize that *"horizontal" and "angle" are not meaningful in the model; they are properties of the illustrations.*

3. The space vector between two space points of an inertial observer is illustrated by a vertical vector between the corresponding world lines. The parallel sloping lines in Figure 5.3 are space points of an inertial observer and the arrows are the space vectors between the corresponding space points.

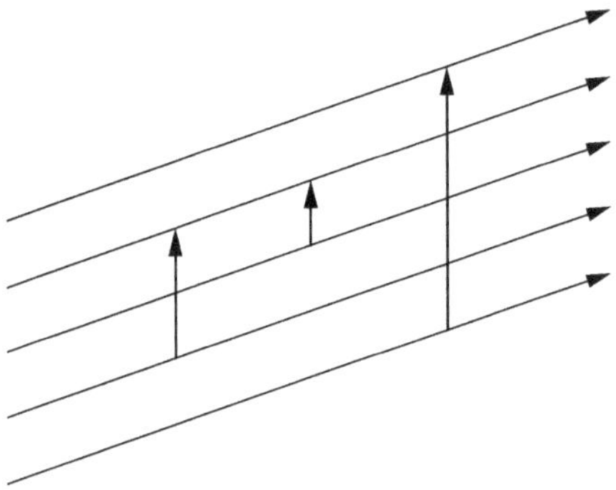

Figure 5.3: Space vectors of an inertial observer.

5.1.4 Boosts

The nonrelativistic spacetime model is fair. The boost from $\boldsymbol{u}'$ to $\boldsymbol{u}$ is

$$\boldsymbol{B}_{\boldsymbol{u}\boldsymbol{u}'} := \mathbf{1} + (\boldsymbol{u} - \boldsymbol{u}') \otimes \boldsymbol{\tau}, \tag{5.10}$$

which depends continuously (even smoothly) on $\boldsymbol{u}$ and $\boldsymbol{u}'$ and

$$\boldsymbol{B}_{\boldsymbol{u}\boldsymbol{u}'} \cdot \boldsymbol{u}' = \boldsymbol{u}, \quad \boldsymbol{B}_{\boldsymbol{u}\boldsymbol{u}'} \cdot \boldsymbol{q} = \boldsymbol{q} \quad (\boldsymbol{q} \in \mathbf{S}).$$

It is a simple fact that $\boldsymbol{B}_{\boldsymbol{u}\boldsymbol{u}'}$ preserves orientation and satisfies:

1. (3.4), (3.7), and (3.8).
2. (3.10) with $\alpha_{\boldsymbol{u},\boldsymbol{u}'} = 1$ because $\boldsymbol{u} - \boldsymbol{u}'$ is in $\frac{\mathbf{S}}{\mathbb{T}}$, so

$$\boldsymbol{B}_{\boldsymbol{u}\boldsymbol{u}'} \cdot (\boldsymbol{u} - \boldsymbol{u}') = \boldsymbol{u} - \boldsymbol{u}' = -(\boldsymbol{u}' - \boldsymbol{u}). \tag{5.11}$$

3. (3.1) and (3.2): $\boldsymbol{x} - \boldsymbol{u}'(\boldsymbol{\tau} \cdot \boldsymbol{x})$ is in $\mathbf{S}$ for all vectors $\boldsymbol{x}$, therefore, $\boldsymbol{B}_{\boldsymbol{u}\boldsymbol{u}'} \cdot \boldsymbol{x} = \boldsymbol{B}_{\boldsymbol{u}\boldsymbol{u}'} \cdot (\boldsymbol{x} - \boldsymbol{u}'(\boldsymbol{\tau} \cdot \boldsymbol{x})) + \boldsymbol{B}_{\boldsymbol{u}\boldsymbol{u}'} \cdot ((\boldsymbol{\tau} \cdot \boldsymbol{x})\boldsymbol{u}') = (\boldsymbol{x} - \boldsymbol{u}'(\boldsymbol{\tau} \cdot \boldsymbol{x})) + \boldsymbol{u}(\boldsymbol{\tau} \cdot \boldsymbol{x})$, so

$$\boldsymbol{\tau} \cdot (\boldsymbol{B}_{\boldsymbol{u}'\boldsymbol{u}} \cdot \boldsymbol{x}) = \boldsymbol{\tau} \cdot \boldsymbol{x}.$$

4. (3.3)): the kernel of $\boldsymbol{b}_{\boldsymbol{u}}$ is spanned by $\boldsymbol{u}$; thus,

$$\begin{aligned}\boldsymbol{b}_{\boldsymbol{u}}(\boldsymbol{B}_{\boldsymbol{u}\boldsymbol{u}'} \cdot \boldsymbol{x}, \boldsymbol{B}_{\boldsymbol{u}\boldsymbol{u}'} \cdot \boldsymbol{y}) &= \boldsymbol{b}_{\boldsymbol{u}}(\boldsymbol{x} - (\boldsymbol{u}' - \boldsymbol{u})(\boldsymbol{\tau} \cdot \boldsymbol{x}), \boldsymbol{y} - (\boldsymbol{u}' - \boldsymbol{u})(\boldsymbol{\tau} \cdot \boldsymbol{x})) \\ &= \boldsymbol{b}_{\boldsymbol{u}}(\boldsymbol{x} - \boldsymbol{u}'(\boldsymbol{\tau} \cdot \boldsymbol{x}), \boldsymbol{y} - \boldsymbol{u}'(\boldsymbol{\tau} \cdot \boldsymbol{x})) \\ &= \boldsymbol{h}(\boldsymbol{x} - \boldsymbol{u}'(\boldsymbol{\tau} \cdot \boldsymbol{x}), \boldsymbol{y} - \boldsymbol{u}'(\boldsymbol{\tau} \cdot \boldsymbol{x})) \\ &= \boldsymbol{b}_{\boldsymbol{u}'}(\boldsymbol{x} - \boldsymbol{u}'(\boldsymbol{\tau} \cdot \boldsymbol{x}), \boldsymbol{y} - \boldsymbol{u}'(\boldsymbol{\tau} \cdot \boldsymbol{x})) = \boldsymbol{b}_{\boldsymbol{u}'}(\boldsymbol{x}, \boldsymbol{y}).\end{aligned}$$

Furthermore, the important property of **transitivity**,

$$\boldsymbol{B}_{\boldsymbol{u}''\boldsymbol{u}} \cdot \boldsymbol{B}_{\boldsymbol{u}\boldsymbol{u}'} = \boldsymbol{B}_{\boldsymbol{u}''\boldsymbol{u}'}$$

is valid for all absolute velocities $\boldsymbol{u}$, $\boldsymbol{u}'$, and $\boldsymbol{u}''$. This has the following physical meaning: if a $\boldsymbol{u}'$-space vector $\boldsymbol{q}'$ corresponds to a $\boldsymbol{u}$-space vector $\boldsymbol{q}$ and $\boldsymbol{q}''$ corresponds to $\boldsymbol{q}'$, then $\boldsymbol{q}''$ corresponds to $\boldsymbol{q}'$.

5.1.5 Duals

The dual of the vector space $\mathbf{S}$ is $\mathbf{S}^*$, the vector space of linear maps $\mathbf{S} \to \mathbb{R}$; the absolute Euclidean form $\boldsymbol{h}$ establishes the linear bijection:

$$\frac{\mathbf{S}}{\mathbb{L} \otimes \mathbb{L}} \to \mathbf{S}^*, \quad \frac{\boldsymbol{q}}{\mathrm{m}^2} \mapsto \frac{\boldsymbol{h}(\boldsymbol{q}, \cdot)}{\mathrm{m}^2},$$

which is used for the **identification** $\frac{\mathbf{S}}{\mathbb{L}\otimes\mathbb{L}} \equiv \mathbf{S}^*$.

It is very important that no similar identification can be made for $\mathbf{M}^*$.

The kernel of the linear surjection $\boldsymbol{\tau} : \mathbf{M} \to \mathbb{T}$ is a "distinguished" linear subspace in $\mathbf{M}$, the subspace of the **absolute space-like vectors.**

The transpose of $\boldsymbol{\tau}$, $\boldsymbol{\tau}^* : \mathbb{T}^* \to \mathbf{M}^*$ is a linear injection; its range is a "distinguished" one-dimensional subspace in $\mathbf{M}^*$, the subspace of the **absolute time-like covectors.** The covector $\boldsymbol{k}$ is absolute time-like if and only if there is a (unique) $\boldsymbol{e} \in \mathbb{T}^*$ such that $\boldsymbol{k} = \boldsymbol{\tau}^*\boldsymbol{e} = \boldsymbol{e}\boldsymbol{\tau}$; in other words, $\boldsymbol{k}$ is absolute time-like if and only if maps $\mathbf{S}$ in zero.

It will be convenient to introduce the imbedding map $\boldsymbol{i} : \mathbf{S} \to \mathbf{M}$:

$$\boldsymbol{i} \cdot \boldsymbol{q} = \boldsymbol{q} \quad (\boldsymbol{q} \in \mathbf{S},\ \boldsymbol{i} \cdot \boldsymbol{q} \in \mathbf{M}). \tag{5.12}$$

$\boldsymbol{i}$ is evidently a linear injection, its transpose

$$\boldsymbol{i}^* : \mathbf{M}^* \to \mathbf{S}^*, \quad \boldsymbol{k} \mapsto \boldsymbol{i}^* \cdot \boldsymbol{k} = \boldsymbol{k} \cdot \boldsymbol{i} = \boldsymbol{k}|_{\mathbf{S}}$$

is a liner surjection where $|_{\mathbf{S}}$ denotes the restriction to $\mathbf{S}$.

5.1.6 Absolute synchronization

Since the set of future-like vectors is a half-space having $\mathbf{S}$ as its boundary, $\mathbf{S}$ is the only three-dimensional subspace transverse to all future-like vectors. That is why in the non-relativistic spacetime model only one synchronization is possible, called **absolute synchronization**, the uniform synchronization determined by $\mathbf{S}$.

The world points x and y are simultaneous if and only if $y - x \in \mathbf{S}$. In other words, the collection of world points simultaneous with x is $x + \mathbf{S}$. The **absolute time points (instants)** are hyperplanes directed by $\mathbf{S}$ (see Figure 5.4); their collection is **absolute time**, denoted by T_a.

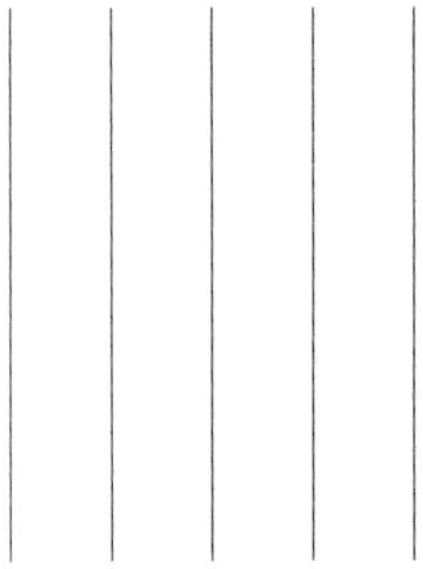

Figure 5.4: Absolute time points.

Let us introduce the map

$$\tau : \mathrm{M} \to \mathrm{T}_a, \quad x \mapsto x + \mathbf{S}, \tag{5.13}$$

called **time evaluation**. Time evaluation assigns to every world point the corresponding absolute time point.

T_a is in fact the factor space $\mathrm{M}/\mathbf{S}$, but now it can be endowed with a nicer affine structure than in the general case given in Subsection 3.5.4.

The difference of two absolute instants t and s (three-dimensional affine subspaces) is defined as the absolute time duration between any of their world points:

$$t - s := \boldsymbol{\tau} \cdot (y - x) \quad (y \in t, x \in s). \tag{5.14}$$

In other words,

$$(y + \mathbf{S}) - (x + \mathbf{S}) := \tau(y) - \tau(x) = \boldsymbol{\tau} \cdot (y - x). \tag{5.15}$$

The subtraction above turns T_a into a one-dimensional affine space over $\mathbb{T}$ and τ becomes an affine map over the linear map $\boldsymbol{\tau}$.

The world points x and y are simultaneous if and only if $\boldsymbol{\tau} \cdot (y - x) = 0$. If x and y have the relation $\tau(y) - \tau(x) = \boldsymbol{\tau} \cdot (y - x) > 0$, i. e., y is (absolutely) future-like with respect to x, then we write $\tau(y) > \tau(x)$, which means that y is later than x.

5.1.7 Exercises

1. Prove that the orientation of $\mathbf{S}$ given in Subsection 5.1.2 is well-defined.
2. Prove the transitivity of boosts.
3. Take the nonrelativistic spacetime model in which
   (1) Spacetime is $\mathbb{R}^4$ as an affine space over itself as a vector space with the usual orientation; let the world points and world vectors be denoted in the form $x = (\xi^1, \xi^2, \xi^3, \xi^4)$ and $\mathbf{x} = (x^1, x^2, x^3, x^4)$, respectively.
   (2) Every measure line is $\mathbb{R}$ with the usual orientation.

(3) The inertial time progress is $\boldsymbol{\tau} \cdot \boldsymbol{x} := x^1 + x^2$, and

give the:

(i) Future-like vectors.

(ii) Absolute velocities.

(iii) Absolute space-like vectors.

(iv) Absolute time-like covectors.

(v) Boosts.

(vi) Absolute time points.

Define an absolute Euclidean form.

4. Take the nonrelativistic spacetime model in which
   (1) Spacetime is $\mathbb{R}^2 \otimes \mathbb{R}^2$, the set of 2×2 real matrices as an affine space over itself as a vector space; the orientation is given by the basis

$$\begin{pmatrix} 1 & 0 \\ 0 & 0 \end{pmatrix}, \begin{pmatrix} 0 & 1 \\ 0 & 0 \end{pmatrix}, \begin{pmatrix} 0 & 0 \\ 1 & 0 \end{pmatrix}, \begin{pmatrix} 0 & 0 \\ 0 & 1 \end{pmatrix};$$

   let the world points and world vectors be denoted in the form $x = \begin{pmatrix} \xi^1 & \xi^2 \\ \xi^3 & \xi^4 \end{pmatrix}$ and $\boldsymbol{x} = \begin{pmatrix} x^1 & x^2 \\ x^3 & x^4 \end{pmatrix}$, respectively.

   (2) Every measure line is $\mathbb{R}$ with the standard orientation.

   (3) The inertial time progress is

$$\boldsymbol{\tau} \cdot \boldsymbol{x} := x^4.$$

   Consequently, $\boldsymbol{q}$ is an absolute space-like vector if and only if $q^4 = 0$.

   (4) The absolute Euclidean form is

$$\boldsymbol{h}(\boldsymbol{q}, \boldsymbol{p}) := \mathrm{Tr}(\boldsymbol{q}^* \boldsymbol{p})$$

   (the trace of the matrix $\boldsymbol{q}^* \boldsymbol{p}$, $\boldsymbol{q}^*$ being the transpose of $\boldsymbol{q}$).

   Give the

   (i) Absolute velocities.

   (ii) Absolute time-like covectors.

   (iii) Boosts.

   (iv) Absolute time points.

   (v) Inertial world lines.

5.2 The arithmetic spacetime model

The following nonrelativistic spacetime model constructed with the aid of real numbers is called the **arithmetic** one:

- $\mathrm{M} = \mathbb{R}^4$ with the standard orientation; then $\mathbf{M} = \mathbb{R}^4$ as well. To distinguish between the world points and world vectors, they are denoted in the form $x = (\xi^0, \xi^1, \xi^2, \xi^3)$ and $\boldsymbol{x} = (x^0, x^1, x^2, x^3)$, respectively.
- $\mathbb{T} = \mathbb{R}$ with the standard orientation.
- $\mathbb{L} = \mathbb{R}$ with the standard orientation.
- $\boldsymbol{\tau} \cdot (x^0, x^1, x^2, x^3) := x^0$,
  then the subspace of absolute space-like vectors is

$$\mathbf{S} = \{(x^0, x^1, x^2, x^3) \mid x^0 = 0\}; \tag{5.16}$$

  it is convenient to take absolute space-like vectors in the form $(0, q^1, q^2, q^3)$.
- $\boldsymbol{h}$ is the usual inner product on $\mathbb{R}^3$, $\boldsymbol{h}(\boldsymbol{q}, \boldsymbol{p}) := q^1p^1 + q^2p^2 + q^3p^3$.

Further,

$$\mathrm{T}^{\rightarrow} = \{(x^0, x^1, x^2, x^3) \mid x^0 > 0\},$$
$$\mathrm{V}(1) = \{(u^0, u^1, u^2, u^3) \mid u^0 = 1\};$$

it is convenient to take absolute velocities in the form $(1, v^1, v^2, v^3)$.

The world points $(\xi^0, \xi^1, \xi^2, \xi^3)$ and $(\eta^0, \eta^1, \eta^2, \eta^3)$ are simultaneous if and only if $\xi^0 = \eta^0$. As a consequence, the time evaluation is

$$\tau(\xi^0, \xi^1, \xi^2, \xi^3) = \xi^0.$$

Accordingly, the instants of absolute time T_a (hyperplanes directed by $\mathbf{S}$) can be identified with real numbers: the instant corresponding to the real number t is $\{(t, q_1.q_2, q_3) \mid (q_1.q_2, q_3) \in \mathbb{R}^3\}$.

Now both $\mathbf{M}$ and $\mathbf{M}^*$ are $\mathbb{R}^4$ but in reality there is no correspondence between the elements of $\mathbf{M}$ and $\mathbf{M}^*$; in order to make distinction, the elements of covectors are written with lower indexes. The covector $\boldsymbol{k} := (k_0, k_1, k_2, k_3)$ acting on the vector $\boldsymbol{x} := (x^0, x^1, x^2, x^3)$ results in $\boldsymbol{k} \cdot \boldsymbol{x} = k_0x^0 + k_1x^1 + k_2x^2 + k_3x^3 = k_ix^i$ where Einstein's rule is applied, i. e., the symbol of summation from 0 to 3 is omitted.

A clearer distinction between vectors and covectors can be made in formulas if the vectors are represented by columns and covectors are represented by rows and the action of a covector on a vector is calculated by the usual role of matrix multiplication:

$$(k_0\ k_1\ k_2\ k_3) \begin{pmatrix} x^0 \\ x^1 \\ x^2 \\ x^3 \end{pmatrix}$$

Then $\boldsymbol{\tau}$ is the row vector $(1\ 0\ 0\ 0)$.

The vector-vector linear maps are represented by 4×4 matrices. In particular,

$$\text{the boost from} \begin{pmatrix} 1 \\ 0 \\ 0 \\ 0 \end{pmatrix} \text{ to } \begin{pmatrix} 1 \\ v^1 \\ v^2 \\ v^3 \end{pmatrix} \text{ is } \begin{pmatrix} 1 & 0 & 0 & 0 \\ v^1 & 1 & 0 & 0 \\ v^2 & 0 & 1 & 0 \\ v^3 & 0 & 0 & 1 \end{pmatrix}; \tag{5.17}$$

its inverse,

$$\text{the boost from} \begin{pmatrix} 1 \\ v^1 \\ v^2 \\ v^3 \end{pmatrix} \text{ to } \begin{pmatrix} 1 \\ 0 \\ 0 \\ 0 \end{pmatrix} \text{ is } \begin{pmatrix} 1 & 0 & 0 & 0 \\ -v^1 & 1 & 0 & 0 \\ -v^2 & 0 & 1 & 0 \\ -v^3 & 0 & 0 & 1 \end{pmatrix}. \tag{5.18}$$

Exercise. Give the boost from $(1, (v')^1, (v')^2, (v')^3)$ to $(1, v^1, v^2, v^3)$.

5.3 Isomorphisms

Isomorphism of models is an important notion: two models, which are formally different, have the same physical content if and only if they are isomorphic.

5.3.1 The form of isomorphisms

Recall that $(L, \boldsymbol{F}, \boldsymbol{Z})$ is an isomorphism between the nonrelativistic spacetime models $(\mathrm{M}, \mathbb{T}, \mathbb{L}, \boldsymbol{\tau}, \boldsymbol{h})$ and $(\widehat{\mathrm{M}}, \widehat{\mathbb{T}}, \widehat{\mathbb{L}}, \hat{\boldsymbol{\tau}}, \hat{\boldsymbol{h}})$ if (see Section 3.2):

(i) $L : \mathrm{M} \to \widehat{\mathrm{M}}$ is an orientation and arrow orientation preserving affine bijection (over the linear bijection $\boldsymbol{L} : \mathbf{M} \to \widehat{\mathbf{M}}$),

(ii) $\boldsymbol{F} : \mathbb{T} \to \widehat{\mathbb{T}}$ is an orientation preserving linear bijection,

(iii) $\boldsymbol{Z} : \mathbb{L} \to \widehat{\mathbb{L}}$ is an orientation preserving linear bijection,

which transforms "conveniently" $\mathrm{T}^{\to}$ into $\widehat{\mathrm{T}^{\to}}$, $\boldsymbol{\tau}$ into $\hat{\boldsymbol{\tau}}$, $\boldsymbol{h}$ into $\hat{\boldsymbol{h}}$.

Since $\mathrm{T}^{\to}$ is determined by $\boldsymbol{\tau}$, two of the general requirements can be collected:

(I)–(II)

$$\frac{\hat{\boldsymbol{\tau}} \cdot (\boldsymbol{L} \cdot \boldsymbol{x})}{\boldsymbol{F}(\boldsymbol{t})} = \frac{\boldsymbol{\tau} \cdot \boldsymbol{x}}{\boldsymbol{t}}$$

for all $\boldsymbol{x} \in \mathbf{M}$ and $0 \neq \boldsymbol{t} \in \mathbb{T}$.

Then $\boldsymbol{L}$ maps $\mathbf{S} := \mathrm{Ker}(\boldsymbol{\tau})$ onto $\hat{\mathbf{S}} := \mathrm{Ker}(\hat{\boldsymbol{\tau}})$, so the requirement on the Euclidean forms will be simpler:

(III)

$$\frac{\widehat{h}(L \cdot q, L \cdot p)}{Z(d)^2} = \frac{h(q,p)}{d^2}$$

for all $q, p \in \mathbf{S}$, and $0 \neq d \in \mathbb{L}$.

5.3.2 Nonrelativistic spacetime models are isomorphic

We can easily demonstrate:

The arbitrary nonrelativistic spacetime model is isomorphic to the arithmetic one, which involves that arbitrary two nonrelativistic spacetime models are isomorphic.

Indeed, let $(\mathrm{M}, \mathbb{T}, \mathbb{L}, \tau, h)$ be a nonrelativistic spacetime model and let us take:
(i) A time unit $\mathrm{s} \in \mathbb{T}^+$,
(ii) A distance unit $\mathrm{m} \in \mathbb{L}^+$,
(iii) An “origin” o in M,
(iv) A future-like vector e_0 for which $\tau \cdot e_0 = \mathrm{s}$ holds,
(v) A positively oriented basis (e_0, e_1, e_2, e_3) in $\mathbf{M}$ such that e_1, e_2, e_3 is in $\mathbf{S}$ and $h(e_i, e_k) = \mathrm{m}^2\delta_{ik}$ (Kronecker-delta);

and let

$$L : \mathrm{M} \to \mathbb{R}^4, \quad x \mapsto \text{coordinates of } x - o \text{ in the basis } (e_0, e_1, e_2, e_3),$$

$$F : \mathbb{T} \to \mathbb{R}, \quad t \mapsto \frac{t}{\mathrm{s}},$$

$$Z : \mathbb{L} \to \mathbb{R}, \quad d \mapsto \frac{d}{\mathrm{m}}.$$

It is simple that in this way an isomorphism is established: if
- $x = \sum_{i=0}^{3} x^i e_i$, i. e., $L \cdot x = (x^0, x^1, x^2, x^3)$, then $x^0 = \frac{\tau \cdot x}{\mathrm{s}}$.
- $q = \sum_{i=1}^{3} q^i e_i$, $p = \sum_{i=1}^{3} p^i e_i$ are elements of $\mathbf{S}$, then $\sum_{i=1}^{3} q^i p^i = \frac{h(q,p)}{\mathrm{m}^2}$.

Accordingly, the physical content of any nonrelativistic spacetime model is the same, and we can use an arbitrary one. A special model, however, may contain extra properties, which
- Have nothing to do with the structure of the model.
- Hide the essential features of the model.

For instance, in the model given in Exercise 4 of Subsection 5.1.7, it seems that the product of world vectors (matrices) is meaningful as a world vector.

In the arithmetic model:

(1) Spacetime M and the set of spacetime vectors **M** appear in the same form, they are $\mathbb{R}^4$, so they may be confused.
(2) All measure lines are the same $\mathbb{R}$, so different types of vectors and tensors are not distinguished.
(3) Absolute time T_a and time periods $\mathbb{T}$ appear in the same form, they are $\mathbb{R}$, so they can be confused.
(4) Time evaluation τ and the time progress $\boldsymbol{\tau}$ seem to be the same covector $(1, 0, 0, 0)$.

From a practical point of view, to solve actual problems, the arithmetic model by convenient coordinates works well but from a theoretical point of view it is not advisable to use it.

If, using the arithmetic model, we do not want to make errors, we have to check permanently whether the formulas in question have a real physical meaning; this is rather tiresome and something can escape our attention. Later, in Section 6.8 we show a convincing example how the use of coordinates is misleading.

5.3.3 Exercises

1. Give an isomorphism from the spacetime model of Exercise 3. in 5.1.7 to the arithmetic model.
2. Give an isomorphism from the spacetime model of Exercise 4. in 5.1.7 to the arithmetic model.

5.4 Spacetime symmetries and reversals

5.4.1 Galilei transformations

In the nonrelativistic spacetime model, the linear bijections $\boldsymbol{L} : \mathbf{M} \to \mathbf{M}$ satisfying:

(I) $\boldsymbol{\tau} \cdot (\boldsymbol{L} \cdot \boldsymbol{x}) = \pm\boldsymbol{\tau} \cdot \boldsymbol{x}$ for all $\boldsymbol{x} \in \mathbf{M}$
(II) $\boldsymbol{h}(\boldsymbol{L} \cdot \boldsymbol{q}, \boldsymbol{L} \cdot \boldsymbol{p}) = \boldsymbol{h}(\boldsymbol{q}, \boldsymbol{p})$ for all $\boldsymbol{q}, \boldsymbol{p} \in \mathbf{S}$

are called **Galilei transformations.**

Note that according to (I), **S** is invariant for $\boldsymbol{L}$, so (II) is a meaningful requirement and says that the restriction of $\boldsymbol{L}$ is an orthogonal map in the Euclidean space **S**.

The compositions of Galilei transformations as well as the inverse of a Galilei transformation are Galilei transformations; thus, they form a group, the **Galilei group** $\mathcal{G}$.

A Galilei transformation for which the positive sign holds in (I) is called an **arrow orientation preserving.**

The orientation and arrow orientation preserving Galilei transformations are called **proper Galilei transformations.**

According to the general definition in Section 3.3 and the formulas in Subsection 5.3.1:

The proper Galilei transformations are the vectorial symmetries of the nonrelativistic spacetime model.

The proper Galilei transformations whose restriction to $\mathbf{S}$ is the identity are the **special Galilei transformations.**

If $\boldsymbol{L}$ is a special Galilei transformation, then $\boldsymbol{L}\cdot\boldsymbol{u}' - \boldsymbol{L}\cdot\boldsymbol{u} = \boldsymbol{L}\cdot(\boldsymbol{u}'-\boldsymbol{u}) = \boldsymbol{u}'-\boldsymbol{u}$ for all absolute velocities $\boldsymbol{u}$ and $\boldsymbol{u}'$, therefore, $\boldsymbol{L}\cdot\boldsymbol{u}'-\boldsymbol{u}' = \boldsymbol{L}\cdot\boldsymbol{u}-\boldsymbol{u} =: \boldsymbol{v}_L$ is independent of the absolute velocities. So, there is a unique $\boldsymbol{v}_L \in \frac{\mathbf{S}}{\mathbb{T}}$ for each special Galilei transformation $\boldsymbol{L}$ in such a way that $\boldsymbol{L}\cdot\boldsymbol{u} = \boldsymbol{u}+\boldsymbol{v}_L$ for all absolute velocities $\boldsymbol{u}$. Since $\boldsymbol{x}-\boldsymbol{u}(\boldsymbol{\tau}\cdot\boldsymbol{x})$ is in $\mathbf{S}$, we have that $\boldsymbol{L}\cdot\boldsymbol{x} = \boldsymbol{x}+\boldsymbol{v}_L(\boldsymbol{\tau}\cdot\boldsymbol{x})$ for all world vectors $\boldsymbol{x}$, i. e.,

$$\boldsymbol{L} = \mathbf{1} + \boldsymbol{v}_L \otimes \boldsymbol{\tau}. \tag{5.19}$$

The special Galilei transformations form a subgroup and $\boldsymbol{v}_{L\cdot K} = \boldsymbol{v}_L + \boldsymbol{v}_K$.

It is trivial that the boosts (see (5.10)) are special Galilei transformations, and the special Galilei transformation (5.19) is the boost from an arbitrary $\boldsymbol{u}'$ to $\boldsymbol{u}'+\boldsymbol{v}_L$.

We emphasize for later reasons (see the arithmetic formula below and Remark at the end of Subsection 5.7.1) that

the orthogonal group of $\mathbf{S}$*, "the three-dimensional orthogonal group"* $\mathcal{O}(\mathbf{S})$ *is not a subgroup of the Galilei group.*

For all absolute velocities $\boldsymbol{u}$,

$$\mathcal{O}_{\boldsymbol{u}} := \{\boldsymbol{L} \in \mathcal{G} \mid \boldsymbol{L}\cdot\boldsymbol{u} = \boldsymbol{u}\}$$

is a subgroup of the Galilei group, isomorphic to the "three-dimensional orthogonal group." These groups can be called **Wigner's nonrelativistic little groups.**

It is worth saying a few words about "time reversals" and "space reversals" in the Galilei group.

For every absolute velocity $\boldsymbol{u}$, there is a **$\boldsymbol{u}$-time reversal $\boldsymbol{T}_{\boldsymbol{u}}$** defined by

$$\boldsymbol{T}_{\boldsymbol{u}}\cdot\boldsymbol{u} = -\boldsymbol{u}, \quad \boldsymbol{T}_{\boldsymbol{u}}\cdot\boldsymbol{q} = \boldsymbol{q} \quad (\boldsymbol{q}\in\mathbf{S})$$

and there is a **$\boldsymbol{u}$-space reversal $\boldsymbol{P}_{\boldsymbol{u}}$** defined by

$$\boldsymbol{P}_{\boldsymbol{u}}\cdot\boldsymbol{u} = \boldsymbol{u}, \quad \boldsymbol{P}_{\boldsymbol{u}}\cdot\boldsymbol{q} = -\boldsymbol{q} \quad (\boldsymbol{q}\in\mathbf{S});$$

it is trivial that $\boldsymbol{P}_{\boldsymbol{u}} = -\boldsymbol{T}_{\boldsymbol{u}}$ and $\boldsymbol{T}_{\boldsymbol{u}}^{-1} = \boldsymbol{T}_{\boldsymbol{u}}$.

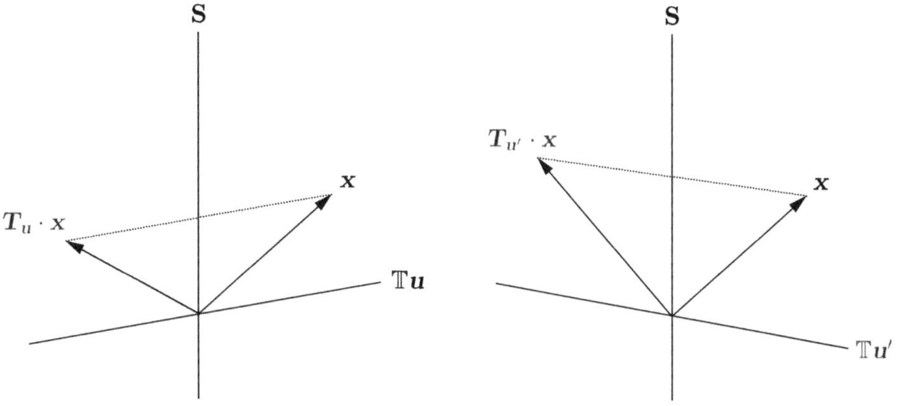

Figure 5.5: Different time reversals.

For a world vector $\boldsymbol{x}$, $\boldsymbol{x} - \boldsymbol{u}(\boldsymbol{\tau} \cdot \boldsymbol{x})$ is in $\mathbf{S}$; thus, $\boldsymbol{T}_u \cdot \boldsymbol{x} = \boldsymbol{x} - 2\boldsymbol{u}(\boldsymbol{\tau} \cdot \boldsymbol{x})$; briefly,

$$\boldsymbol{T}_u = \mathbf{1} - 2\boldsymbol{u} \otimes \boldsymbol{\tau}, \tag{5.20}$$

which shows evidently that the $\boldsymbol{u}$-time reversal (and the $\boldsymbol{u}$-space reversal) is a Galilei transformation. Two time reflections of the same vector are illustrated by Figure 5.5.

Time reversals due to different absolute velocities are different: if $\boldsymbol{u} \neq \boldsymbol{u}'$ then $\boldsymbol{T}_u \neq \boldsymbol{T}_{u'}$.

The composition of two different time reversals,

$$\boldsymbol{T}_{u'} \cdot \boldsymbol{T}_u = \mathbf{1} - 2(\boldsymbol{u}' - \boldsymbol{u}) \otimes \boldsymbol{\tau}$$

is a special Galilei transformation. Further,

$$\boldsymbol{T}_{u''} \cdot \boldsymbol{T}_{u'} \cdot \boldsymbol{T}_u = \mathbf{1} - 2(\boldsymbol{u}'' - (\boldsymbol{u}' - \boldsymbol{u})) \otimes \boldsymbol{\tau},$$

which is a time reversal.

As a consequence, all the $\boldsymbol{u}$-time (space) reversals generate a subgroup of the Galilei group consisting of the special Galilei transformations and all the $\boldsymbol{u}$-time (space) reversals.

Arithmetic formulas

The Galilei transformations in the arithmetic spacetime model are linear bijections $\mathbb{R}^4 \to \mathbb{R}^4$, given by 4×4 matrices of the form

$$\begin{pmatrix} \pm 1 & 0 & 0 & 0 \\ v^1 & R^1{}_1 & R^1{}_2 & R^1{}_3 \\ v^2 & R^2{}_1 & R^2{}_2 & R^2{}_3 \\ v^3 & R^3{}_1 & R^3{}_2 & R^3{}_3 \end{pmatrix} \tag{5.21}$$

where v^1, v^2, v^3 are arbitrary real numbers and $R^1{}_1 \dots R^3{}_3$ constitute a usual 3×3 orthogonal matrix.

Indeed, property (I) is satisfied because the row vector $\tau = (1\ 0\ 0\ 0)$ applied from the left of this matrix results in the row vector $\pm(1\ 0\ 0\ 0)$ and property (II) is satisfied because the matrix applied to a column vector whose zeroth coordinate is zero results in a column vector whose zeroth coordinate is zero and the other ones are orthogonally transformed.

The special Galilei transformations have the form

$$\begin{pmatrix} 1 & 0 & 0 & 0 \\ v^1 & 1 & 0 & 0 \\ v^2 & 0 & 1 & 0 \\ v_3 & 0 & 0 & 1 \end{pmatrix}: \tag{5.22}$$

the action of such a transformation is

$$\begin{pmatrix} x^0 \\ x^1 \\ x^2 \\ x^3 \end{pmatrix} \mapsto \begin{pmatrix} x^0 \\ x^1 + v^1 x^0 \\ x^2 + v^2 x^0 \\ x^3 + v^3 x^0 \end{pmatrix}.$$

The Galilei transformations of the shape

$$\begin{pmatrix} 1 & 0 & 0 & 0 \\ 0 & R^1{}_1 & R^1{}_2 & R^1{}_3 \\ 0 & R^2{}_1 & R^2{}_2 & R^2{}_3 \\ 0 & R^3{}_1 & R^3{}_2 & R^3{}_3 \end{pmatrix} \tag{5.23}$$

also form a subgroup; that is why it seems “the three-dimensional orthogonal group” is a subgroup of the Galilei group but this is only one of Wigner’s little group: the one corresponding to the absolute velocity $(1, 0, 0, 0)$.

One usually considers the time reversal $x^0 \mapsto -x^0$; this is

$$\text{The } \begin{pmatrix} 1 \\ 0 \\ 0 \\ 0 \end{pmatrix}\text{-time reversal,} \quad \begin{pmatrix} -1 & 0 & 0 & 0 \\ 0 & 1 & 0 & 0 \\ 0 & 0 & 1 & 0 \\ 0 & 0 & 0 & 1 \end{pmatrix}.$$

$$\text{The } \begin{pmatrix} 1 \\ v \\ 0 \\ 0 \end{pmatrix}\text{-time reversal is} \quad \begin{pmatrix} -1 & 0 & 0 & 0 \\ -2v & 1 & 0 & 0 \\ 0 & 0 & 1 & 0 \\ 0 & 0 & 0 & 1 \end{pmatrix}.$$

5.4.2 Noether transformations

The **Noether transformations** are affine transformations $\mathrm{M} \to \mathrm{M}$ over the Galilei transformations; they form a group, the **Noether group** $\mathcal{N}$.

Noether transformations over orientation preserving, arrow orientation preserving, and proper Galilei transformations are called **orientation preserving, arrow orientation preserving and proper**, respectively.

The proper Noether transformations are the symmetries of the nonrelativistic spacetime model.

Noether transformation are affine transformations of spacetime. Galilei transformations are linear transformations of space vectors; it is trivial that

the Galilei group is not a subgroup of the Noether group.

If spacetime is vectorized by an "origin" o, then the Noether transformations are given on world vectors in the following way:

$$\boldsymbol{x} \mapsto L(o + \boldsymbol{x}) - o = L(o) - o + \boldsymbol{L} \cdot \boldsymbol{x}, \tag{5.24}$$

as a consequence, for every o,

$$\mathcal{G}_o := \{L \in \mathcal{N} \mid L(o) = o\}$$

is a subgroup of the Noether group, isomorphic to the Galilei group.

In spacetime, there is no $\boldsymbol{u}$-time reversal because a reversal requires an "origin." For every world point o and absolute velocity $\boldsymbol{u}$, there is an **o-centered $\boldsymbol{u}$-time reversal**

$$T_{u,o}(x) := o + \boldsymbol{T}_{\boldsymbol{u}} \cdot (x - o) = x - 2\boldsymbol{\tau} \cdot (x - o)\boldsymbol{u}, \tag{5.25}$$

which is a Noether transformation over $\boldsymbol{T}_{\boldsymbol{u}}$.

We similarly get the **o-centered $\boldsymbol{u}$-space reversal**.

For a Noether transformation L and for all world points x, y, the time evaluation (see (5.13)) gives

$$\tau(L(y)) - \tau(L(x)) = \boldsymbol{\tau} \cdot \boldsymbol{L} \cdot (y - x) = \pm\boldsymbol{\tau} \cdot (y - x) = \pm(\tau(y) - \tau(x));$$

as a consequence, if x and y are simultaneous, then $L(x)$ and $L(y)$ are simultaneous as well, i. e., L has constant values on every hyperplane of absolute synchronization. In this way, L defines an affine transformation L_{T_a} of absolute time T_a.

L is arrow orientation preserving if and only if + stands on the right side of the above equality; then $\tau(L(x)) - \tau(x) = \tau(L(y)) - \tau(y) =: \boldsymbol{t}_L$ is independent of x and y; in other words, there is unique $\boldsymbol{t}_L \in \mathbb{T}$ such that

$$\tau(L(x)) = \tau(x) + \boldsymbol{t}_L \quad (x \in \mathrm{M}). \tag{5.26}$$

As a consequence, $L_{\mathrm{T}_a}T$ is a translation for an arrow orientation preserving L,

$$L_{\mathbb{T}}(t) = t + \boldsymbol{t}_L \quad (t \in \mathrm{T}_a). \tag{5.27}$$

The time evaluation of (5.25) gives $\tau(T_{\boldsymbol{u},o}(x)) = \tau(x) - 2(\tau(x) - \tau(o)) = \tau(x) - (\tau(x) - \tau(o)) - (\tau(x) - \tau(o)) = \tau(o) - (\tau(x) - \tau(o))$; putting $t_o := \tau(o)$ and $t := \tau(x)$, this means that for every $t_o \in \mathrm{T}_a$ there is a t_o centered reversal of absolute time:

$$T_{t_o} : \mathrm{T}_a \to \mathrm{T}_a, \quad t \mapsto t_o - (t - t_o). \tag{5.28}$$

A thorough examination of the Galilei group and the Noether group can be found in the book *Spacetime without Reference Frames.*

Remarks. 1. If world points are not distinguished from world vectors—as in usual treatment based on coordinates—then it seems with "the" origin o that a Noether transformation consists of a Galilei transformation and a translation (see (5.24)). Consequently, in usual treatments, Noether transformations are called inhomogeneous Galilei transformations and the Galilei group is considered a subgroup (with zero translations) of the Noether group.

2. "**The**" three-dimensional orthogonal group as a subgroup of the Galilei group, "**the**" time reversal, "**the**" Galilei group as a subgroup of the Noether group are errors, which show well how the arithmetic spacetime model may cause misunderstandings.

5.4.3 Exercises

1. A uniformly rotating observer $\boldsymbol{U}$ is defined by an angular velocity $\boldsymbol{\Omega}$, an antisymmetric linear map $\mathbf{S} \to \frac{\mathbf{S}}{\mathbb{T}}$ or by an (axial) vector $\boldsymbol{\omega} \in \frac{\mathbf{S}}{\mathbb{T}}$ in such a way that $\boldsymbol{U}(x + \boldsymbol{q}) - \boldsymbol{U}(x) = \boldsymbol{\Omega} \cdot \boldsymbol{q} = -\boldsymbol{\omega} \times \boldsymbol{q}$ for all $x \in \mathrm{M}$ and $\boldsymbol{q} \in \mathbf{S}$. What are the symmetries of this observer?
2. Show that the $\boldsymbol{u}$-time reversal changes orientation and arrow orientation.
3. What is the $\boldsymbol{u}$-time reversal of the absolute velocity $\boldsymbol{u}'$?
4. Show that the $\boldsymbol{u}$-space reversal changes orientation and preserves arrow orientation.
5. What are the o-centered $\boldsymbol{u}$-space reversals?
6. Give Wigner's little group corresponding to $(1, v, 0, 0)$ in the arithmetic spacetime model.

5.5 Inertial frames, relative velocities

In the nonrelativistic spacetime model, there is only a single synchronization, the absolute one, therefore, an observer determines uniquely a reference frame. That is why we could say observer instead of reference frame but we do not do that just for a clear distinction between observer and inertial frame, which is very important in the relativistic model.

In an inertial frame, an inertial world line results in a uniform motion on a straight line, as it is stated in Subsection 1.3.2. Now we examine such a motion in the nonrelativistic spacetime model.

Let us consider an inertial frame $\boldsymbol{u}$ and the world line of an inertial material point, i. e., a straight line in spacetime, directed by an absolute velocity $\boldsymbol{u}'$ (which can be an arbitrary space point of the inertial observer $\boldsymbol{u}'$). If $\boldsymbol{u}' \neq \boldsymbol{u}$, then the material point moves with respect to the observer; the relative velocity with respect to the inertial frame is defined as follows.

Let x and y be occurrences of the material point; then $y-x$ is parallel to $\boldsymbol{u}'$: $y-x = \boldsymbol{t}'\boldsymbol{u}'$ where $\boldsymbol{t}' = \tau \cdot (y - x)$.

The material point meets the $\boldsymbol{u}$-space points $\pi_{\boldsymbol{u}}(x)$ and $\pi_{\boldsymbol{u}}(y)$ at the moments $\tau(x)$ and $\tau(y)$, respectively. Consequently, the **relative velocity** of $\boldsymbol{u}'$ with respect to $\boldsymbol{u}$ is

$$\begin{aligned} \boldsymbol{v}_{\boldsymbol{u}'\boldsymbol{u}} &:= \frac{\pi_{\boldsymbol{u}}(y) - \pi_{\boldsymbol{u}}(x)}{\tau_{\boldsymbol{u}}(y) - \tau_{\boldsymbol{u}}(x)} = \frac{\pi_{\boldsymbol{u}} \cdot (y - x)}{\tau \cdot (y - x)} = \frac{\boldsymbol{t}'(\boldsymbol{u}' - \boldsymbol{u})}{\boldsymbol{t}'} \\ &= \boldsymbol{u}' - \boldsymbol{u}. \end{aligned} \tag{5.29}$$

The result is independent of the world points: an inertial material point moves uniformly on a straight line with respect to an inertial frame.

The relative velocities between arbitrary two absolute velocities are in the same three-dimensional Euclidean space $\frac{\mathbb{S}}{\mathbb{T}}$; contrary to absolute velocities,

- There is a zero relative velocity.
- The magnitude of a relative velocity is meaningful.
- The angle between two relative velocities is meaningful.

The definition shows the **reciprocity** of relative velocities,

$$\boldsymbol{v}_{\boldsymbol{u}\boldsymbol{u}'} = -\boldsymbol{v}_{\boldsymbol{u}'\boldsymbol{u}}$$

and the **transitivity** (addition) of relative velocities,

$$\boldsymbol{v}_{\boldsymbol{u}''\boldsymbol{u}'} + \boldsymbol{v}_{\boldsymbol{u}'\boldsymbol{u}} = \boldsymbol{v}_{\boldsymbol{u}''\boldsymbol{u}}.$$

These relations are in accordance with our everyday conception.

Finally, we can state with (5.29) and (5.11) that the corresponding boosts map the relative velocities into the opposite of each other,

$$B_{u'u} \cdot v_{u'u} = -v_{uu'}, \quad B_{uu'} \cdot v_{uu'} = -v_{u'u}, \tag{5.30}$$

as it must be because the direction of relative velocity equals the direction of motion.

Arithmetic formulas

The relative velocity of

$$\begin{pmatrix} 1 \\ (v')^1 \\ (v')^2 \\ (v')^3 \end{pmatrix} \quad \text{with respect to} \quad \begin{pmatrix} 1 \\ v^1 \\ v^2 \\ v^3 \end{pmatrix} \quad \text{is} \quad \begin{pmatrix} 0 \\ (v')^1 - v^1 \\ (v')^2 - v^2 \\ (v')^3 - v^3 \end{pmatrix}.$$

The relative velocities are much more complicated in the relativistic case where we shall treat only special cases for better seeing what the matter is. That is why now we write here the corresponding special cases, too:

$$\text{the relative velocity of} \quad \begin{pmatrix} 1 \\ 0 \\ v' \\ 0 \end{pmatrix} \quad \text{with respect to} \quad \begin{pmatrix} 1 \\ v \\ 0 \\ 0 \end{pmatrix} \quad \text{is} \quad \begin{pmatrix} 0 \\ -v \\ v' \\ 0 \end{pmatrix},$$

$$\text{the relative velocity of} \quad \begin{pmatrix} 1 \\ v \\ 0 \\ 0 \end{pmatrix} \quad \text{with respect to} \quad \begin{pmatrix} 1 \\ 0 \\ v' \\ 0 \end{pmatrix} \quad \text{is} \quad \begin{pmatrix} 0 \\ v \\ -v' \\ 0 \end{pmatrix} = -\begin{pmatrix} 0 \\ -v \\ v' \\ 0 \end{pmatrix}.$$

5.6 Vectorial splitting and transformation rules

5.6.1 Preliminaries

From now on, as usual, we write a dot product instead of $\boldsymbol{h}$:

$$\boldsymbol{q} \cdot \boldsymbol{p} := \boldsymbol{h}(\boldsymbol{q}, \boldsymbol{p}) \in \mathbb{L} \otimes \mathbb{L} \quad (\boldsymbol{q}, \boldsymbol{p} \in \mathbf{S}).$$

Recall the formulas in Subsection 5.1.5:

(i) $\tau : \mathbf{M} \to \mathbb{T}$, its transpose $\tau^* : \mathbb{T}^* \to \mathbf{M}^*$ is a linear injection.
(ii) $\boldsymbol{i} : \mathbf{S} \to \mathbf{M}$ is the embedding, its transpose $\boldsymbol{i}^* : \mathbf{M}^* \to \mathbf{S}^*$ is a linear surjection.
(iii) $\pi_u : \mathbf{M} \to \mathbf{S}$ is a linear surjection, its transpose $\pi_u^* : \mathbf{S}^* \to \mathbf{M}^*$ is a linear injection.

Summing up:

τ and π_u **before** a vector is the same as τ^* and π_u^* **after** it,

$$\tau \cdot x = x \cdot \tau^*, \quad \pi_u \cdot x = x \cdot \pi_u^*,$$

$\boldsymbol{i}^*$ **before** a covector is the same as $\boldsymbol{i}$ **after** it,

$$\boldsymbol{i}^* \cdot \boldsymbol{k} = \boldsymbol{k} \cdot \boldsymbol{i} = \boldsymbol{k}|_{\mathbf{S}}.$$

In more details: if $\boldsymbol{t} \in \mathbb{T}$, $\boldsymbol{e} \in \mathbb{T}^*$, $\boldsymbol{q} \in \mathbf{S}$, $\boldsymbol{p} \in \mathbf{S}^*$, then

- $(\boldsymbol{\tau}^* \boldsymbol{e}) \cdot \boldsymbol{x} = \boldsymbol{e\tau} \cdot \boldsymbol{x} =$
- $(\boldsymbol{\pi}_{\boldsymbol{u}} \cdot \boldsymbol{x}) \cdot \boldsymbol{p} = \boldsymbol{x} \cdot (\boldsymbol{\pi}^*_{\boldsymbol{u}} \cdot \boldsymbol{p})$
- $(\boldsymbol{i}^* \cdot \boldsymbol{k}) \cdot \boldsymbol{q} = \boldsymbol{k} \cdot (\boldsymbol{i} \cdot \boldsymbol{q}) = \boldsymbol{k} \cdot \boldsymbol{q}$

and

$$\boldsymbol{\pi}_{\boldsymbol{u}} \cdot \boldsymbol{u} = 0, \quad \boldsymbol{\pi}_{\boldsymbol{u}} \cdot \boldsymbol{i} = \mathbf{1}_{\mathbf{S}}, \quad \boldsymbol{i}^* \cdot \boldsymbol{\pi}^*_{\boldsymbol{u}} = \mathbf{1}_{\mathbf{S}^*}, \quad \boldsymbol{\tau} \cdot \boldsymbol{i} = \mathbf{0}_{\mathbf{S}}, \quad \boldsymbol{i}^* \cdot \boldsymbol{\tau}^* = \mathbf{0}_{\mathbf{S}^*}. \tag{5.31}$$

Arithmetic formulas

1. $\boldsymbol{i}^* \cdot (k_0\ k_1 k_2\ k_3) = \big(0\ (k_1\ k_2\ k_3)\big)$.
2. $\boldsymbol{\tau}^* e = e\boldsymbol{\tau} = (e\ 0\ 0\ 0)$.

5.6.2 Splitting

For an absolute velocity $\boldsymbol{u}$, $\mathbb{T}\boldsymbol{u}$ and $\mathbf{S}$ are complementary subspaces, so every world vector $\boldsymbol{x}$ can be uniquely given as the sum of vectors in these subspaces: $\boldsymbol{x} = \boldsymbol{u}(\boldsymbol{\tau} \cdot \boldsymbol{x}) + \boldsymbol{\pi}_{\boldsymbol{u}} \cdot \boldsymbol{x}$ (see Figure 5.6). Accordingly, the $\boldsymbol{u}$-splitting $\mathbf{M} \to \mathbb{T} \times \mathbf{S}$, $\boldsymbol{x} \mapsto (\boldsymbol{\tau} \cdot \boldsymbol{x}, \boldsymbol{\pi}_{\boldsymbol{u}} \cdot \boldsymbol{x})$ is a linear bijection.

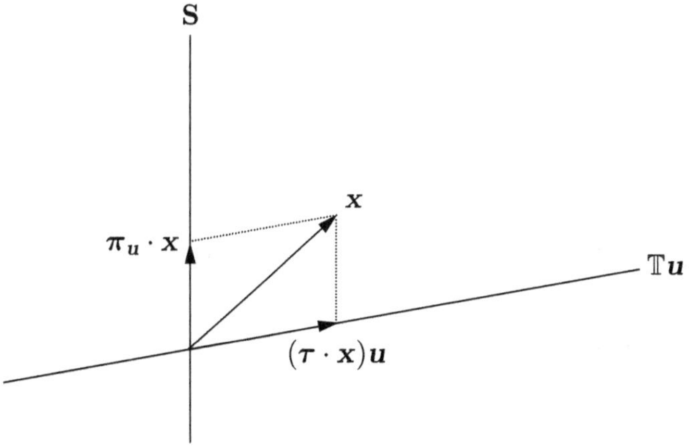

Figure 5.6: $\boldsymbol{u}$-splitting of world vectors.

For better seeing, elements of $\mathbb{T}\times\mathbf{S}$ will be written as columns in displayed formulas, and even the Cartesian product will have the form $\binom{\mathbb{T}}{\mathbf{S}}$. Then we have

$$\xi_u := \begin{pmatrix} \tau \\ \pi_u \end{pmatrix} : \mathbf{M} \to \begin{pmatrix} \mathbb{T} \\ \mathbf{S} \end{pmatrix}, \quad x \mapsto \begin{pmatrix} \tau \cdot x \\ \pi_u \cdot x \end{pmatrix}, \tag{5.32}$$

called the **splitting of world vectors** according to $\boldsymbol{u}$, or simply **vectorial $\boldsymbol{u}$-splitting**;
- $\tau \cdot x$ is the **time-like component** of x.
- $\pi_u \cdot x$ is the **$\boldsymbol{u}$-space-like component** of x.

It is trivial that

$$\xi_u^{-1} \cdot \begin{pmatrix} \boldsymbol{t} \\ \boldsymbol{q} \end{pmatrix} = \boldsymbol{t}\boldsymbol{u} + \boldsymbol{q} = \boldsymbol{u}\boldsymbol{t} + \boldsymbol{i} \cdot \boldsymbol{q} \tag{5.33}$$

$(\boldsymbol{t}, \boldsymbol{q}) \in \mathbb{T} \times \mathbf{S}$.

Of course, the tensor products and tensor quotients of $\mathbf{M}$ by measure lines such as $\frac{\mathbf{M}}{\mathbb{T}}$ and $\frac{\mathbf{M}}{\mathbb{T}\otimes\mathbb{T}}$ are split in a similar way, because the multiplication and division by measure line elements can be interchanged with linear maps. For instance,
- The time-like component of the absolute velocity $\boldsymbol{u}'$ is

$$\tau \cdot \boldsymbol{u}' = 1.$$

- The $\boldsymbol{u}$-space-like component of $\boldsymbol{u}'$,

$$\pi_u \cdot \boldsymbol{u}' = \boldsymbol{u}' - \boldsymbol{u}$$

  is just the relative velocity of $\boldsymbol{u}'$ with respect to $\boldsymbol{u}$ (see Section 5.5); thus,

$$\xi_u \cdot \boldsymbol{u}' = \begin{pmatrix} 1 \\ \boldsymbol{v}_{u'u} \end{pmatrix}. \tag{5.34}$$

The splitting of vectors determines the splitting of covectors by $(\xi_u^{-1})^*$ which maps $\mathbf{M}^*$ into $(\mathbb{T} \times \mathbf{S})^* = \mathbb{T}^* \times \mathbf{S}^*$. This Cartesian product and its elements will be written a rows in some displayed formulas.

For a covector $\boldsymbol{k}$, $(\xi_u^{-1})^* \cdot \boldsymbol{k} = \boldsymbol{k} \cdot \xi_u^{-1}$, so if $(\boldsymbol{t}, \boldsymbol{q})$ is in $\mathbb{T} \times \mathbf{S}$, then

$$\begin{aligned} ((\xi_u^{-1})^* \cdot \boldsymbol{k}) \cdot \begin{pmatrix} \boldsymbol{t} \\ \boldsymbol{q} \end{pmatrix} &= \boldsymbol{k} \cdot (\boldsymbol{t}\boldsymbol{u} + \boldsymbol{q}) = \boldsymbol{t}(\boldsymbol{k} \cdot \boldsymbol{u}) + \boldsymbol{k} \cdot \boldsymbol{q} = (\boldsymbol{k} \cdot \boldsymbol{u})\boldsymbol{t} + (\boldsymbol{k} \cdot \boldsymbol{i}) \cdot \boldsymbol{q} \\ &= (\boldsymbol{k} \cdot \boldsymbol{u} \quad \boldsymbol{k} \cdot \boldsymbol{i}) \begin{pmatrix} \boldsymbol{t} \\ \boldsymbol{q} \end{pmatrix}, \end{aligned}$$

therefore,

$$(\xi_u^{-1})^* : \mathbf{M}^* \to \mathbb{T}^* \times \mathbf{S}^*, \quad \boldsymbol{k} \mapsto (\boldsymbol{u} \cdot \boldsymbol{k}, \boldsymbol{i}^* \cdot \boldsymbol{k}) = (\boldsymbol{u} \cdot \boldsymbol{k}, \boldsymbol{i}^* \cdot \boldsymbol{k}) \tag{5.35}$$

is the **splitting of world covectors** according to $\boldsymbol{u}$, or simply **covectorial $\boldsymbol{u}$-splitting**;

- $\boldsymbol{u} \cdot \boldsymbol{k} = \boldsymbol{k} \cdot \boldsymbol{u}$ is the **$\boldsymbol{u}$-time-like component** of the covector $\boldsymbol{k}$.
- $\boldsymbol{i}^* \cdot \boldsymbol{k} = \boldsymbol{k} \cdot \boldsymbol{i} = \boldsymbol{k}|_{\mathbf{S}}$ is the **space-like component** of $\boldsymbol{k}$.

Since $(\xi_u{}^{-1})^* = (\xi_u{}^*)^{-1}$, the inverse of this splitting is $\xi_u{}^*$ and

$$\xi_u{}^* \cdot (\boldsymbol{e}, \boldsymbol{p}) = \tau^* \boldsymbol{e} + \pi_u^* \cdot \boldsymbol{p} = \boldsymbol{e}\tau + \boldsymbol{p} \cdot \pi_u$$

for $(\boldsymbol{e}, \boldsymbol{p}) \in \mathbb{T}^* \times \mathbf{S}^*$.

Let us note:

(i) *The time-like components of vectors are absolute*

(ii) *The space-like components of covectors are absolute*

where absolute means "independent of absolute velocities."

The space-like components of vectors, in general, depend on absolute velocities, except the absolute space-like vectors: the time-like component of $\boldsymbol{q} \in \mathbf{S}$ is zero, its $\boldsymbol{u}$-space-like component is $\boldsymbol{q}$ itself for all $\boldsymbol{u}$.

The time-like components of covectors, in general, depend on absolute velocities, except the absolute time-like covectors, i. e., the elements of $\tau^*[\mathbb{T}^*]$: the $\boldsymbol{u}$-time-like component of $\boldsymbol{k} := \tau^* \boldsymbol{e} = \boldsymbol{e}\tau$ is $\boldsymbol{e}$ for all $\boldsymbol{u}$, its space-like component is zero.

The formulas above can be summarized as follows:

$$\text{if} \quad \xi_u \cdot x = \begin{pmatrix} \boldsymbol{t} \\ \boldsymbol{q} \end{pmatrix} \quad \text{and} \quad (\xi_u{}^{-1})^* \cdot \boldsymbol{k} = (\boldsymbol{e}\ \boldsymbol{p}),$$

$$\text{then} \quad \boldsymbol{k} \cdot x = (\boldsymbol{e}\ \boldsymbol{p}) \begin{pmatrix} \boldsymbol{t} \\ \boldsymbol{q} \end{pmatrix} = \boldsymbol{e}\boldsymbol{t} + \boldsymbol{p} \cdot \boldsymbol{q}.$$

Splitting of vectors and covectors has the physical meaning that an inertial frame $\boldsymbol{u}$ perceives the components of vectors and covectors, not the vectors and covectors themselves; for instance, we have seen already that the $\boldsymbol{u}$-space-like component of $\boldsymbol{u}'$ is just the relative velocity of $\boldsymbol{u}'$ with respect to $\boldsymbol{u}$.

Arithmetic formulas

$\mathbb{R}^4$ is split into $\begin{pmatrix} \mathbb{R} \\ \mathbb{R}^3 \end{pmatrix}$ and $(\mathbb{R}^4)^*$ is split into $(\mathbb{R}^* \quad (\mathbb{R}^3)^*)$.

1. The vector $\begin{pmatrix} x^0 \\ x^1 \\ x^2 \\ x^3 \end{pmatrix}$ has

$$\text{the} \begin{pmatrix} 1 \\ 0 \\ 0 \\ 0 \end{pmatrix}\text{-split form} \left(\begin{pmatrix} x_0 \\ x^1 \\ x^2 \\ x^3 \end{pmatrix} \right),$$

$$\text{the } \begin{pmatrix} 1 \\ v \\ 0 \\ 0 \end{pmatrix}\text{-split form } \left(\begin{matrix} x_0 \\ \begin{pmatrix} x^1 - vx^0 \\ x^2 \\ x^3 \end{pmatrix} \end{matrix} \right).$$

$$\text{The } \begin{pmatrix} 1 \\ v \\ 0 \\ 0 \end{pmatrix}\text{-splitting of vectors has the inverse } \left(\begin{matrix} t \\ \begin{pmatrix} q^1 \\ q^2 \\ q^3 \end{pmatrix} \end{matrix} \right) \mapsto \begin{pmatrix} t \\ q^1 + tv \\ q^2 \\ q^3 \end{pmatrix}.$$

2. The covector $(k_0\ k_1\ k_2\ k_3)$ has

$$\text{the } \begin{pmatrix} 1 \\ 0 \\ 0 \\ 0 \end{pmatrix}\text{-split form } \big(k_0\ \ (k_1\ k_2\ k_3)\big),$$

$$\text{the } \begin{pmatrix} 1 \\ v \\ 0 \\ 0 \end{pmatrix}\text{-split form } \big(k_0 + vk_1\ \ (k_1\ k_2\ k_3)\big).$$

$$\text{The } \begin{pmatrix} 1 \\ v \\ 0 \\ 0 \end{pmatrix}\text{-splitting of covectors has the inverse } \big(e\ \ (p_1\ p_2\ p_3)\big) \mapsto (e - vp_1\quad p_1\quad p_2\quad p_3).$$

5.6.3 Transformation rules

Different inertial frames split world vectors differently. To see the difference, we compare them by the linear bijection

$$\xi_{u'u} := \xi_{u'} \cdot \xi_u^{-1} : \mathbb{T} \times \mathbf{S} \to \mathbb{T} \times \mathbf{S}, \tag{5.36}$$

called the **vectorial transformation rule**, which can be given explicitly as follows: let $(\boldsymbol{t}, \boldsymbol{q})$ and $(\boldsymbol{t}', \boldsymbol{q}')$ be the split forms of a vector due to $\boldsymbol{u}$ and $\boldsymbol{u}'$, respectively. Then

$$\begin{pmatrix} \boldsymbol{t}' \\ \boldsymbol{q}' \end{pmatrix} = \xi_{u'} \cdot \left(\xi_u^{-1} \cdot \begin{pmatrix} \boldsymbol{t} \\ \boldsymbol{q} \end{pmatrix} \right) = \begin{pmatrix} \tau \\ \pi_{u'} \end{pmatrix} \cdot (\boldsymbol{t}\boldsymbol{u} + \boldsymbol{q}) = \begin{pmatrix} \boldsymbol{t} \\ \boldsymbol{q} - \boldsymbol{t}\boldsymbol{v}_{u'u} \end{pmatrix},$$

that is,

$$\boldsymbol{t}' = \boldsymbol{t}, \quad \boldsymbol{q}' = \boldsymbol{q} - \boldsymbol{t}\boldsymbol{v}_{u'u} \tag{5.37}$$

which is just the **Galilei transformation rule** from $\boldsymbol{u}$ to $\boldsymbol{u}'$ (well known usually with $\mathbb{R}$ and $\mathbb{R}^3$ instead of $\mathbb{T}$ and $\mathbf{S}$, respectively).

In a block matrix form,

$$\xi_{u'u} = \begin{pmatrix} 1 & 0 \\ -\mathbf{v}_{u'u} & \mathbf{1} \end{pmatrix}. \tag{5.38}$$

Note that

a Galilei transformation is a linear map $\mathbf{M} \to \mathbf{M}$,
a Galilei transformation rule is a linear map $\mathbb{T} \times \mathbf{S} \to \mathbb{T} \times \mathbf{S}$.

The **covectorial transformation rule** is

$$((\xi_{u'}{}^{-1})^* \cdot \xi_u{}^*) \cdot (e, \boldsymbol{p}) = (\xi_{u'}{}^{-1})^* \cdot (e\tau + \boldsymbol{p} \cdot \pi_u) = (e + \boldsymbol{p} \cdot \mathbf{v}_{u'u}, \boldsymbol{p}).$$

The covectorial transformation rule and the vectorial transformation rule are different.

Arithmetic formulas

The vectorial transformation rule from

$$\begin{pmatrix} 1 \\ 0 \\ 0 \\ 0 \end{pmatrix} \text{ to } \begin{pmatrix} 1 \\ v^1 \\ v^2 \\ v^3 \end{pmatrix} \text{ is } \left(\begin{matrix} t \\ \begin{pmatrix} q^1 \\ q^2 \\ q^3 \end{pmatrix} \end{matrix} \right) \mapsto \left(\begin{matrix} t \\ \begin{pmatrix} q^1 - tv^1 \\ q^2 - tv^2 \\ q^3 - tv^3 \end{pmatrix} \end{matrix} \right),$$

so this transformation rule has the matrix

$$\left(\begin{matrix} 1 & \begin{pmatrix} 0 & 0 & 0 \end{pmatrix} \\ \begin{pmatrix} -v_1 \\ -v_2 \\ -v_3 \end{pmatrix} & \begin{pmatrix} 1 & 0 & 0 \\ 0 & 1 & 0 \\ 0 & 0 & 1 \end{pmatrix} \end{matrix} \right). \tag{5.39}$$

If the inside parentheses in the transformation rule (5.39) are omitted, then the special Galilei transformation (a boost) (5.18) is obtained. In other words, if spacetime is treated in coordinates and $\mathbb{R} \times \mathbb{R}^3$ is not distinguished from $\mathbb{R}^4$, then Galilei transformations and Galilei transformation rules have the same form, so they can be confused. We point out to such a confusion in the remark of Section 5.9.

Remark. The arithmetic vectorial transformation rule between arbitrary two absolute velocities can be obtained by a vast formula; such formulas show very well that the use of the arithmetic model may be laborious and offers possibilities for making mistakes.

Exercise

Give the arithmetic vectorial and covectorial transformation rule from the $(1, v^1, v^2, v^3)$-splitting to the $(1, (v')^1, (v')^2, (v')^3)$-splitting.

5.7 Tensorial splitting and transformation rules

5.7.1 Splitting

In a number of physical theories—e. g., in electromagnetism and continuum mechanics—not only vectors and covectors but various tensors are also included. The mathematical supplement helps the reader to be familiar with tensors.

The various tensors, i. e., elements of $\mathbf{M}\otimes\mathbf{M}$, $\mathbf{M}\otimes\mathbf{M}^*$, $\mathbf{M}^*\otimes\mathbf{M}$, and $\mathbf{M}^*\otimes\mathbf{M}^*$ can also be split by a absolute velocities. Their splitting is obtained from the splitting of $\mathbf{M}$ by $\boldsymbol{\xi}_u$ into $\mathbb{T}\times\mathbf{S}$ and the splitting of $\mathbf{M}^*$ by $(\boldsymbol{\xi}_u{}^{-1})^*$ into $\mathbb{T}^*\times\mathbf{S}^*$. The split form of these tensors can be easily calculated by considering them linear maps $\mathbf{M}^*\to\mathbf{M}$, $\mathbf{M}\to\mathbf{M}$, $\mathbf{M}^*\to\mathbf{M}^*$, and $\mathbf{M}\to\mathbf{M}^*$, respectively. Because of the identifications $(\mathbb{T}\times\mathbf{S})\otimes(\mathbb{T}\times\mathbf{S})\equiv\begin{pmatrix}\mathbb{T}\otimes\mathbb{T} & \mathbb{T}\otimes\mathbf{S}\\ \mathbf{S}\otimes\mathbb{T} & \mathbf{S}\otimes\mathbf{S}\end{pmatrix}$, etc. the split forms can be arranged in block matrices. Further, the identifications $\mathbb{T}\otimes\mathbb{T}^*\equiv\mathbb{T}^*\otimes\mathbb{T}\equiv\mathbb{R}$ and $\mathbf{S}^*\equiv\frac{\mathbf{S}}{\mathbb{T}\otimes\mathbb{T}}$, so $\mathbb{T}\otimes\mathbf{S}^*\equiv\frac{\mathbf{S}}{\mathbb{T}}$, etc. can be used in the following:

(1) **The $\boldsymbol{u}$-split form of $\boldsymbol{G}\in\mathbf{M}\otimes\mathbf{M}$**—a linear map $\mathbf{M}^*\to\mathbf{M}$—is calculated by

$$\mathbb{T}\times\mathbf{S}\xrightarrow{\boldsymbol{\xi}_u^*}\mathbf{M}\xrightarrow{\boldsymbol{G}}\mathbf{M}\xrightarrow{\boldsymbol{\xi}_u}\mathbb{T}\times\mathbf{S},$$

obtaining

$$\boldsymbol{\xi}_u\cdot\boldsymbol{G}\cdot\boldsymbol{\xi}_u{}^*=\begin{pmatrix}\boldsymbol{\tau}\cdot\boldsymbol{G}\cdot\boldsymbol{\tau}^* & \boldsymbol{\tau}\cdot\boldsymbol{G}\cdot\boldsymbol{\pi}_u^*\\ \boldsymbol{\pi}_u\cdot\boldsymbol{G}\cdot\boldsymbol{\tau}^* & \boldsymbol{\pi}_u\cdot\boldsymbol{G}\cdot\boldsymbol{\pi}_u^*\end{pmatrix}. \tag{5.40}$$

The antisymmetric elements of $\mathbf{M}\otimes\mathbf{M}$ play a fundamental role in electromagnetism. If $\boldsymbol{G}$ is an antisymmetric tensor, i. e., $\boldsymbol{G}=-\boldsymbol{G}^*$, then $\boldsymbol{\tau}\cdot\boldsymbol{G}=-\boldsymbol{G}\cdot\boldsymbol{\tau}^*$, so $\boldsymbol{\tau}\cdot\boldsymbol{G}\cdot\boldsymbol{\tau}^*=0$ and according to Formula (5.40) it has the $\boldsymbol{u}$-split form

$$\begin{pmatrix}0 & \boldsymbol{\tau}\cdot\boldsymbol{G}\\ -\boldsymbol{\tau}\cdot\boldsymbol{G} & \boldsymbol{G}-\boldsymbol{u}\wedge(\boldsymbol{\tau}\cdot\boldsymbol{G})\end{pmatrix};$$

here the two "lower" components determine the other ones, therefore, we refer to the $\boldsymbol{u}$-split form as

$$((-\boldsymbol{\tau}\cdot\boldsymbol{G},\boldsymbol{G}-\boldsymbol{u}\wedge(\boldsymbol{\tau}\cdot\boldsymbol{G})))\in(\mathbf{S}\otimes\mathbb{T})\times(\mathbf{S}\wedge\mathbf{S}). \tag{5.41}$$

The first one is called the **time-like component** of $\boldsymbol{G}$—which is independent of $\boldsymbol{u}$ –, the second one is called the **$\boldsymbol{u}$-space-like component**. It is a simple fact that if the $\boldsymbol{u}$-split form of $\boldsymbol{G}$ is $((\boldsymbol{D},\boldsymbol{H}_u))$, then

$$\boldsymbol{G}=\boldsymbol{H}_u-\boldsymbol{u}\wedge\boldsymbol{D}. \tag{5.42}$$

(2) **The $\boldsymbol{u}$-split form of $\boldsymbol{F}\in\mathbf{M}^*\otimes\mathbf{M}^*$**—a linear map $\mathbf{M}\to\mathbf{M}^*$—is obtained similarly:

$$(\boldsymbol{\xi}_u{}^{-1})^*\cdot\boldsymbol{F}\cdot\boldsymbol{\xi}_u{}^{-1}=\begin{pmatrix}\boldsymbol{u}\cdot\boldsymbol{F}\cdot\boldsymbol{u} & \boldsymbol{u}\cdot\boldsymbol{F}\cdot\boldsymbol{i}\\ \boldsymbol{i}^*\cdot\boldsymbol{F}\cdot\boldsymbol{u} & \boldsymbol{i}^*\cdot\boldsymbol{F}\cdot\boldsymbol{i}\end{pmatrix}. \tag{5.43}$$

The antisymmetric elements of $\mathbf{M}^* \otimes \mathbf{M}^*$ play a fundamental role in electromagnetism.

If $\boldsymbol{F}$ is an antisymmetric cotensor, i. e., $\boldsymbol{F} = -\boldsymbol{F}^*$, then $\boldsymbol{u}^* \cdot \boldsymbol{F} = -\boldsymbol{F} \cdot \boldsymbol{u}$ so $\boldsymbol{u}^* \cdot \boldsymbol{F} \cdot \boldsymbol{u} = 0$ and it has the $\boldsymbol{u}$-split form

$$\begin{pmatrix} 0 & -\boldsymbol{i}^* \cdot \boldsymbol{F} \cdot \boldsymbol{u} \\ \boldsymbol{i}^* \cdot \boldsymbol{F} \cdot \boldsymbol{u} & \boldsymbol{i}^* \cdot \boldsymbol{F} \cdot \boldsymbol{i} \end{pmatrix}.$$

As previously, only the "lower" components will be considered for referring to the $\boldsymbol{u}$-split form:

$$((\boldsymbol{i}^* \cdot \boldsymbol{F} \cdot \boldsymbol{u}, \boldsymbol{i}^* \cdot \boldsymbol{F} \cdot \boldsymbol{i})) \in (\mathbf{S}^* \otimes \mathbb{T}^*) \times (\mathbf{S}^* \wedge \mathbf{S}^*). \tag{5.44}$$

The first one is called the **$\boldsymbol{u}$-time-like component** of $\boldsymbol{F}$, the second one—which is independent of $\boldsymbol{u}$—is called the **space-like component.**

In view of (5.31), we find that if the $\boldsymbol{u}$-split form of $\boldsymbol{F}$ is $((\boldsymbol{E}_u, \boldsymbol{B}))$, then

$$\boldsymbol{F} = \boldsymbol{\pi}_u^* \cdot \boldsymbol{B} \cdot \boldsymbol{\pi}_u + (\boldsymbol{\pi}_u^* \cdot \boldsymbol{E}_u) \wedge \boldsymbol{\tau}. \tag{5.45}$$

(3) **The $\boldsymbol{u}$-split form of $\boldsymbol{P} \in \mathbf{M}^* \otimes \mathbf{M}$**—a linear map $\mathbf{M}^* \to \mathbf{M}^*$—is

$$(\boldsymbol{\xi}_u^{-1})^* \cdot \boldsymbol{P} \cdot \boldsymbol{\xi}_u^* = \begin{pmatrix} \boldsymbol{u} \cdot \boldsymbol{P} \cdot \boldsymbol{\tau}^* & \boldsymbol{u} \cdot \boldsymbol{P} \cdot \boldsymbol{\pi}_u^* \\ \boldsymbol{i}^* \cdot \boldsymbol{P} \cdot \boldsymbol{\tau}^* & \boldsymbol{i}^* \cdot \boldsymbol{P} \cdot \boldsymbol{\pi}_u^* \end{pmatrix}. \tag{5.46}$$

(4) **The $\boldsymbol{u}$-split form of $\boldsymbol{L} \in \mathbf{M} \otimes \mathbf{M}^*$**—a linear map $\mathbf{M} \to \mathbf{M}$—is

$$\boldsymbol{\xi}_u \cdot \boldsymbol{L} \cdot \boldsymbol{\xi}_u^{-1} = \begin{pmatrix} \boldsymbol{\tau} \cdot \boldsymbol{L} \cdot \boldsymbol{u} & \boldsymbol{\tau} \cdot \boldsymbol{L} \cdot \boldsymbol{i} \\ \boldsymbol{\pi}_u \cdot \boldsymbol{L} \cdot \boldsymbol{u} & \boldsymbol{\pi}_u \cdot \boldsymbol{L} \cdot \boldsymbol{i} \end{pmatrix}. \tag{5.47}$$

In particular, if $\boldsymbol{L}$ is a Galilei transformation then $\boldsymbol{\tau} \cdot \boldsymbol{L} = \pm\boldsymbol{\tau}$, and according to the sign $\pm\boldsymbol{L} \cdot \boldsymbol{u} \in \mathrm{V}(1)$ and the restriction of $\boldsymbol{L}$ onto $\mathbf{S}$ is an orthogonal linear map $\boldsymbol{R}_L$, we have

$$\boldsymbol{\xi}_u \cdot \boldsymbol{L} \cdot \boldsymbol{\xi}_u^{-1} = \begin{pmatrix} \pm 1 & \mathbf{0} \\ \boldsymbol{L} \cdot \boldsymbol{u} \mp \boldsymbol{u} & \boldsymbol{R}_L \end{pmatrix}. \tag{5.48}$$

If $\boldsymbol{L}$ preserves arrow orientation, then $\boldsymbol{L} \cdot \boldsymbol{u} - \boldsymbol{u} = \boldsymbol{v}_{(\boldsymbol{L} \cdot \boldsymbol{u})\boldsymbol{u}}$; in the other case, $\boldsymbol{L} \cdot \boldsymbol{u} + \boldsymbol{u} = -\boldsymbol{v}_{(-\boldsymbol{L} \cdot \boldsymbol{u}),\boldsymbol{u}}$.

The $\boldsymbol{u}$-split form of the boost from $\boldsymbol{u}$ to $\boldsymbol{u}'$ is

$$\begin{pmatrix} 1 & \mathbf{0} \\ \boldsymbol{v}_{u'u} & \mathbf{1} \end{pmatrix}. \tag{5.49}$$

The $\boldsymbol{u}'$-time reversal (5.20) has the $\boldsymbol{u}$-split form

$$\begin{pmatrix} -1 & 0 \\ -2\boldsymbol{v}_{u'u} & \mathbf{1} \end{pmatrix}, \tag{5.50}$$

i. e.,

$$(\boldsymbol{t}, \boldsymbol{q}) \mapsto (-\boldsymbol{t}, \boldsymbol{q} - 2\boldsymbol{t}\boldsymbol{v}_{u'u}). \tag{5.51}$$

Letting $\boldsymbol{L}$ run over all the Galilei transformations, $\boldsymbol{L} \cdot \boldsymbol{i}$ runs over all the orthogonal maps and $\boldsymbol{L} \cdot \boldsymbol{u} \mp \boldsymbol{u}$ runs over all the relative velocities, so we find from (5.48) that the split Galilei group is

$$\left\{ \begin{pmatrix} \pm 1 & \mathbf{0} \\ \boldsymbol{v} & \boldsymbol{R} \end{pmatrix} \,\middle|\, \boldsymbol{v} \in \frac{\mathbf{S}}{\mathbb{T}}, \boldsymbol{R} \in \mathcal{O}(\mathbf{S}) \right\}. \tag{5.52}$$

(5) Third-order tensors, $\mathbf{M} \otimes \mathbf{M} \otimes \mathbf{M}$, $\mathbf{M} \otimes \mathbf{M} \otimes \mathbf{M}^*$, $\mathbf{M}^* \otimes \mathbf{M}^* \otimes \mathbf{M}$, etc. also appear in some applications. They can be split by considering them linear maps $\mathbf{M}^* \to \mathbf{M} \otimes \mathbf{M}$, $\mathbf{M} \to \mathbf{M} \otimes \mathbf{M}$, $\mathbf{M}^* \to \mathbf{M}^* \otimes \mathbf{M}^*$, etc. In particular, we encounter antisymmetric elements of $\mathbf{M}^* \otimes \mathbf{M}^* \otimes \mathbf{M}^*$ in electromagnetism.

(6) **Tensors multiplied or divided** by some measure line also appear in practice; their splits have the same forms.

Arithmetic formulas

The arithmetic Galilei transformations have the form (5.21), the split Galilei transformations, according to (5.52), have the form

$$\left(\begin{array}{cc} \pm 1 & \begin{pmatrix} 0 & 0 & 0 \end{pmatrix} \\ \begin{pmatrix} v_1 \\ v_2 \\ v_3 \end{pmatrix} & \begin{pmatrix} R^1{}_1 & R^1{}_2 & R^1{}_3 \\ R^2{}_1 & R^2{}_2 & R^2{}_3 \\ R^3{}_1 & R^3{}_2 & R^3{}_3 \end{pmatrix} \end{array} \right). \tag{5.53}$$

If the inside parentheses in the above formula are omitted, then the special Galilei transformations are obtained. In other words, if spacetime is treated in coordinates and $\mathbb{R} \times \mathbb{R}^3$ is not distinguished from $\mathbb{R}^4$ then Galilei transformations and split Galilei transformations have the same form, so they can be—and are often—confused.

5.7.2 Transformation rules

Comparing different splits of various tensors, we get tensorial transformation rules. We treat only the formulas concerning antisymmetric tensors and cotensors.

If $((\boldsymbol{D},\boldsymbol{H}))$ and $((\boldsymbol{D}',\boldsymbol{H}'))$ are the $\boldsymbol{u}$-split form and the $\boldsymbol{u}'$-split form, respectively, of an antisymmetric tensor, then $((\boldsymbol{D}',\boldsymbol{H}')) = \xi_{u'u}((\boldsymbol{D},\boldsymbol{H}))\xi^*_{u'u}$; using the notation $\boldsymbol{v} = \boldsymbol{v}_{u'u}$, in a matrix form (see (5.38)), we have

$$\begin{pmatrix} 1 & 0 \\ -\boldsymbol{v} & 1 \end{pmatrix}\begin{pmatrix} \boldsymbol{0} & -\boldsymbol{D} \\ \boldsymbol{D} & \boldsymbol{H} \end{pmatrix}\begin{pmatrix} 1 & -\boldsymbol{v} \\ 0 & 1 \end{pmatrix} = \begin{pmatrix} 0 & -\boldsymbol{D} \\ \boldsymbol{D} & \boldsymbol{v}\wedge\boldsymbol{D}+\boldsymbol{H} \end{pmatrix};$$

thus,

$$\boldsymbol{D}' = \boldsymbol{D}, \quad \boldsymbol{H}' = \boldsymbol{v}\wedge\boldsymbol{D}+\boldsymbol{H}.$$

If $((\boldsymbol{E},\boldsymbol{B}))$ and $((\boldsymbol{E}',\boldsymbol{B}'))$ are the $\boldsymbol{u}$-split form and the $\boldsymbol{u}'$-split form, respectively, of an antisymmetric cotensor, then in a matrix form

$$\begin{pmatrix} 1 & \boldsymbol{v} \\ 0 & 1 \end{pmatrix}\begin{pmatrix} \boldsymbol{0} & -\boldsymbol{E} \\ \boldsymbol{E} & \boldsymbol{B} \end{pmatrix}\begin{pmatrix} 1 & 0 \\ \boldsymbol{v} & 1 \end{pmatrix} = \begin{pmatrix} 0 & -(\boldsymbol{E}+\boldsymbol{B}\cdot\boldsymbol{v}) \\ \boldsymbol{E}+\boldsymbol{B}\cdot\boldsymbol{v} & \boldsymbol{B} \end{pmatrix};$$

thus,

$$\boldsymbol{E}' = \boldsymbol{E}+\boldsymbol{B}\cdot\boldsymbol{v}, \quad \boldsymbol{B}' = \boldsymbol{B}.$$

5.7.3 Exercises

1. Demonstrate equality (5.50).
2. Prove equalities (5.42) and (5.45).

5.8 Spacetime splitting and transformation rules

5.8.1 Splitting

An inertial frame $\boldsymbol{u}$ characterizes the occurrences by giving when and where they happen: it **splits** spacetime into time and $\boldsymbol{u}$-space by assigning to a world point x the corresponding absolute time point $\tau(x) = x + \mathrm{S}$ and the corresponding $\boldsymbol{u}$-space point $\pi_{\boldsymbol{u}}(x) = x + \mathbb{T}\boldsymbol{u}$. The $\boldsymbol{u}$-splitting, illustrated by Figure 5.7,

$$\xi_{\boldsymbol{u}} : \mathrm{M} \to \mathrm{T}_a \times \mathrm{S}_{\boldsymbol{u}}, \quad x \mapsto (\tau(x), \pi_{\boldsymbol{u}}(x)), \tag{5.54}$$

is an affine bijection over the vectorial splitting $\boldsymbol{\xi}_{\boldsymbol{u}}$ as it is well seen from equalities (5.15), (5.9), and (5.32). The inverse of this splitting—which gives the world point corresponding to a time point and a $\boldsymbol{u}$-space point—is

$$\xi_{\boldsymbol{u}}^{-1} : \mathrm{T}_a \times \mathrm{S}_{\boldsymbol{u}} \to \mathrm{M}, \quad (t,q) \mapsto t \cap q.$$

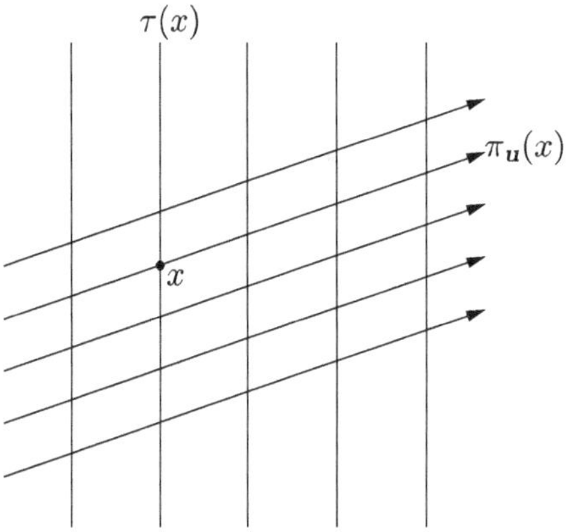

Figure 5.7: Splitting of spacetime.

Instead of affine spaces, it is often more suitable to deal with the underlying vector spaces; therefore—corresponding to the everyday usage when time points are represented by time periods that passed from a given time point, and space points are represented by vectors from an origin—an inertial frame, choosing an "initial" time point t_o and a $\boldsymbol{u}$-space "origin" q_o, vectorizes time and $\boldsymbol{u}$-space by the assignment

$$\mathrm{T}_a \times \mathrm{S}_{\boldsymbol{u}} \to \mathbb{T} \times \mathbf{S}, \quad (t, q) \mapsto (t - t_o, q - q_o). \tag{5.55}$$

Choosing a t_o and a q_o is equivalent to choosing a "spacetime origin" $o := t_o \cap q_o$, then $t_o = \tau(o) = o + \mathbf{S}$, $q_o = \pi_{\boldsymbol{u}}(o) = o + \mathbb{T}\boldsymbol{u}$. The pair $(\boldsymbol{u}, o)$ is called the **inertial frame $\boldsymbol{u}$ with origin** o.

It is a simple fact that the $\boldsymbol{u}$-splitting of spacetime followed by the vectorization of time and $\boldsymbol{u}$-space is the same as the vectorization of spacetime followed by the vectorial splitting according to $\boldsymbol{u}$; thus, the **vectorized splitting of spacetime** by $\boldsymbol{u}$ and o is

$$\xi_{\boldsymbol{u},o} : \mathrm{M} \to \mathbb{T} \times \mathbf{S}, \quad x \mapsto \boldsymbol{\xi}_{\boldsymbol{u}} \cdot (x - o) = \big(\boldsymbol{\tau} \cdot (x - o), \boldsymbol{\pi}_{\boldsymbol{u}} \cdot (x - o)\big). \tag{5.56}$$

It has the inverse

$$\mathbb{T} \times \mathbf{S} \to \mathrm{M}, \quad (\boldsymbol{t}, \boldsymbol{q}) \mapsto o + \boldsymbol{t}\boldsymbol{u} + \boldsymbol{q}, \tag{5.57}$$

briefly,

$$\xi_{\boldsymbol{u},o}^{-1} = o + \boldsymbol{\xi}_{\boldsymbol{u}}^{-1}.$$

5.8.2 Transformation rules

The transformation rule between two spacetime splits would be $\xi_{\boldsymbol{u}'} \circ \xi_{\boldsymbol{u}}^{-1}$, which is a function $\mathrm{T}_a \times \mathrm{S}_{\boldsymbol{u}} \to \mathrm{T}_a \times \mathrm{S}_{\boldsymbol{u}'}$. This is not suitable for comparison because its domain and its range are different sets. To overcome the problem, we consider vectorized splitting where the domain and the range are the same $\mathbb{T}\times\mathbf{S}$. The concrete form of the vectorized transformation rule $\xi_{u',o'}\cdot\xi_{u,o}^{-1}$ is obtained simply.

Let (t,q) and (t',q') be the vectorized split forms of the same world point, due to (u,o) and (u',o'), respectively. Then

$$\begin{aligned}(t',q') &= \xi_{u',o'}\cdot(\xi_{u,o}^{-1}(t,q))\\ &= \xi_{u',o'}\cdot(o+tu+q) = (\tau\cdot(tu+q+o-o'),\ \pi_{u'}\cdot(o-o'+tu+q))\\ &= (t+\tau\cdot(o-o'), q-tv_{u'u}+\pi_{u'}\cdot(o-o')),\end{aligned}$$

i. e., with the notation $t_0 := \tau\cdot(o'-o)$ and $q_0 = \pi_u\cdot(o'-o)$

$$t' = t - t_0,\quad q' = q - tv_{u'u} - q_0, \tag{5.58}$$

which is called the **inhomogeneous Galilei transformation rule** from u to u' (well known usually with $\mathbb{R}$ and $\mathbb{R}^3$ instead of $\mathbb{T}$ and $\mathbf{S}$).

This equals the vectorial Galilei transformation rule if the inertial frames choose the same spacetime origin ($o' = o$). In other words, the inertial frames choose the same time point for time origin ($t_o = t_{o'}$) and their space origins coincide in that instant ($t_o\cap q_o = t_o\cap q_{o'}$).

Note that

an inhomogeneous Galilei transformation rule is an affine map, which serves for comparing splits of spacetime,

a Galilei transformation rule is a linear map, which serves for comparing splits of world vectors.

5.8.3 Exercises

1. Prove that (5.54) followed by (5.55) equals (5.56).
2. Prove that $\xi_{u,o}$ is an affine map over ξ_u.
3. Show that the (u,o)-split form of the o'-centered u'-time reversal is

$$t\mapsto -t+2t_0,\quad q\mapsto q-2(t-t_0)v_{u'u}\quad (t\in\mathbb{T}, q\in\mathbf{S}) \tag{5.59}$$

where $t_0 := \tau\cdot(o'-o)$.

5.9 Transformations and transformation rules

It is worth repeating the main points of transformations and transformation rules.

1. Spacetime vectors

The $\boldsymbol{u}$-split form (5.49) of the boost from $\boldsymbol{u}$ to $\boldsymbol{u}'$ and the Galilei transformation rule (5.38) from $\boldsymbol{u}'$ to $\boldsymbol{u}$ are given by the same formula

$$\begin{pmatrix} 1 & 0 \\ \boldsymbol{v}_{u'u} & \mathbf{1} \end{pmatrix},$$

though they have different physical meaning.

Keep in mind that

Galilei transformations are linear bijections $\mathbf{M} \to \mathbf{M}$ and:
(1) Split forms of Galilei transformations are linear bijections $\mathbb{T} \times \mathbf{S} \to \mathbb{T} \times \mathbf{S}$
(2) Galilei transformation rules are also linear bijections $\mathbb{T} \times \mathbf{S} \to \mathbb{T} \times \mathbf{S}$

which may have similar mathematical forms but are completely different from a physical point of view.

2. Spacetime

Let L be a Noether transformation over the Galilei transformation $\boldsymbol{L}$. Since $L(x) = L(o) + \boldsymbol{L}(x - o)$, the $(\boldsymbol{u}, o)$-split form of L is

$$\begin{aligned} \xi_{u,o} L \xi_{u,o}^{-1} &= \xi_{u,o} L(o + \boldsymbol{\xi}_u{}^{-1}) = \xi_{u,o}(L(o) + \boldsymbol{L} \cdot \boldsymbol{\xi}_u{}^{-1}) \\ &= \boldsymbol{\xi}_u \cdot (L(o) - o + \boldsymbol{L} \cdot \boldsymbol{\xi}_u{}^{-1}) = \boldsymbol{\xi}_u \cdot (L(o) - o) + \boldsymbol{\xi}_u \cdot \boldsymbol{L} \cdot \boldsymbol{\xi}_u{}^{-1}. \end{aligned} \tag{5.60}$$

In particular, if L is a Noether transformation over the boost from $\boldsymbol{u}$ to $\boldsymbol{u}'$, putting $\boldsymbol{t}_o := \boldsymbol{\tau} \cdot (L(o) - o)$ and $\boldsymbol{q}_o := \boldsymbol{\pi}_u \cdot (L(o) - o)$, we have by (5.48) that this is the affine map

$$\mathbb{T} \times \mathbf{S} \to \mathbb{T} \times \mathbf{S}, \quad (\boldsymbol{t}, \boldsymbol{q}) \mapsto (\boldsymbol{t}', \boldsymbol{q}')$$

given by

$$\boldsymbol{t}' = \boldsymbol{t} + \boldsymbol{t}_o, \quad \boldsymbol{q}' = \boldsymbol{q} + \boldsymbol{t}\boldsymbol{v}_{u'u} + \boldsymbol{q}_o. \tag{5.61}$$

This is similar to (5.58) and enlightens why the Noether group is usually called the inhomogeneous Galilei group.

Keep in mind that

Noether transformations are affine bijections M $\to$ M and:
(1) Split forms of Noether transformations are affine bijections $\mathbb{T} \times \mathbf{S} \to \mathbb{T} \times \mathbf{S}$
(2) Inhomogeneous Galilei transformation rules are also affine bijections $\mathbb{T} \times \mathbf{S} \to \mathbb{T} \times \mathbf{S}$

which may have similar mathematical forms but are completely different from a physical point of view.

Remarks.

1. (5.51) and (5.59) show how well the treatment of spacetime in split form (and even in coordinates) may be misleading: $(\boldsymbol{t}, \boldsymbol{q}) \mapsto (-\boldsymbol{t}, \boldsymbol{q})$ is not “the” time reversal; it is only one of the time reversals.
2. If spacetime is treated in split form (and even in coordinates), then transformations (spacetime symmetries) and transformation rules may be confused because they are maps of similar form. This often causes conceptual errors, e. g., when one explains the symmetries of a physical system by showing that quantities transform in a convenient way in a change of reference frames. An excellent example for such a confusion is the following ([4]). In deducing the Lagrange function of a free material point, it is said that
   - According to the homogeneity of time and space, the properties of a closed physical system do not change when the system is translated parallel to itself (thus, spacetime translations are considered—correctly—symmetries of the physical system).
   - No essential change enters the description if one changes to another spatial coordinate system moving uniformly (thus, a transformation rule is considered—incorrectly—a symmetry of the physical system; correctly, it should be said that the properties of a closed physical system do not change when the system is put in uniform motion parallel to itself).
3. For a clear distinction, we summarize:
   (1) Split forms of transformation arise from **one** symmetry due to **one** absolute velocity.
   (2) Transformation rules compare **two** split forms of **one** symmetry that are due to **two** absolute velocities.
4. The (inhomogeneous) Galilei transformation rule is a **single** axiom in usual treatments, while it is a consequence of **two** different ways of splitting in the context of spacetime models. Among others, this exhibits the efficacy of our treatment similar to the efficacy of a magnifier glass, which shows that one smear consists of two spots.
5. Transformation rules are the cornerstone of usual treatments, while they have only practical importance in the spacetime model. On the contrary, spacetime transformations have an essential importance.

5.10 Coordinate systems

In practice, to make computations, a convenient coordinate system is introduced in which time points are represented by numbers and space points are represented by triplets of numbers; thus, world points are represented by quartets of numbers.

This is done as follows. An inertial frame $\boldsymbol{u}$ chooses:

(i) A time unit s
(ii) An “origin” t_o in time
(iii) Distance unit m

(iv) An "origin" q_o in $\boldsymbol{u}$-space
(v) Three orthogonal straight lines (coordinate axes) in $\boldsymbol{u}$-space passing through the origin, directed by the "right handed" orthogonal space vectors $(\boldsymbol{e}_1, \boldsymbol{e}_2, \boldsymbol{e}_3)$ of length m

and represents the world points by time coordinates and space coordinates, elements of $\mathbb{R} \times \mathbb{R}^3$ in such a way (in our setting) that:

1. Splits spacetime M into time T_a and $\boldsymbol{u}$-space $\mathrm{S}_{\boldsymbol{u}}$, $x \mapsto (t, q)$ where $t := \tau(x) = x + \mathbf{S}$ and $q := \pi_{\boldsymbol{u}}(x) = x + \mathbb{T}\boldsymbol{u}$ (thus, $x = t \cap q$).
2. Represents the time point t by the real number $\frac{t-t_o}{\mathrm{s}}$.
3. Represents the $\boldsymbol{u}$-space point q by the three real numbers $\frac{\boldsymbol{e}_1\cdot(q-q_o)}{\mathrm{m}^2}, \frac{\boldsymbol{e}_2\cdot(q-q_o)}{\mathrm{m}^2}, \frac{\boldsymbol{e}_3\cdot(q-q_o)}{\mathrm{m}^2}$.

This procedure having a clear physical meaning is equivalent to the following mathematical procedure: with the world "origin" $o := t_o \cap q_o$ (i. e., $t_o = o + \mathbf{S}$, $q_o = o + \mathbb{T}\boldsymbol{u}$), the world point x is represented by the four coordinates

$$\xi^0 := \frac{\boldsymbol{\tau} \cdot (x - o)}{\mathrm{s}}, \quad \xi^k := \frac{\boldsymbol{e}_k \cdot \boldsymbol{\pi}_{\boldsymbol{u}} \cdot (x - o)}{\mathrm{m}^2} \quad (k = 1, 2, 3).$$

Then, if the coordinates are known, the world point is

$$x = o + \xi^0 \mathrm{s}\boldsymbol{u} + \xi^1 \boldsymbol{e}_1 + \xi^2 \boldsymbol{e}_2 + \xi^3 \boldsymbol{e}_3.$$

It is convenient to introduce $\boldsymbol{e}_0 := \mathrm{s}\boldsymbol{u}$ and according to the formulas above, a **standard inertial coordinate system** is $(o, \mathrm{s}, \mathrm{m}, \boldsymbol{e}_0, \boldsymbol{e}_1, \boldsymbol{e}_2, \boldsymbol{e}_3)$, where o is a world point, s is a time unit, m is a distance unit, $\boldsymbol{e}_0$ is a future-like vector for which $\boldsymbol{\tau}\cdot\boldsymbol{e}_0 = \mathrm{s}$ and $(\boldsymbol{e}_1, \boldsymbol{e}_2, \boldsymbol{e}_3)$ in $\mathbf{S}$ is a positively oriented orthogonal basis, normalized to m.

Returning to Section 5.2, we can state that such a coordinate system establishes an isomorphism to the arithmetic spacetime model. We see very well how many arbitrary objects are hidden in the use of the arithmetic spacetime model: a spacetime origin, a time unit, a length unit, an inertial observer, three space-like orthogonal vectors.

In a concise form: a **coordination** of spacetime by a standard coordinate system is

$$C : \mathrm{M} \to \mathbb{R}^4, \quad x \mapsto (\xi^0, \xi^1, \xi^2, \xi^3). \tag{5.62}$$

C is an affine map, the underlying linear map $\boldsymbol{C}$ represents vectors, covectors and various tensors by coordinates, too. The coordinates $\boldsymbol{C} \cdot \boldsymbol{x}$ of a vector $\boldsymbol{x}$ are

$$x^0 := \frac{\boldsymbol{\tau} \cdot \boldsymbol{x}}{\mathrm{s}}, \quad x^i := \frac{\boldsymbol{e}_i \cdot \boldsymbol{\pi}_{\boldsymbol{u}} \cdot \boldsymbol{x}}{\mathrm{m}^2} \quad (i = 1, 2, 3);$$

the coordinates $(\boldsymbol{C}^{-1})^* \cdot \boldsymbol{k}$ of a covector $\boldsymbol{k}$ are

$$k_i := \boldsymbol{k} \cdot \boldsymbol{e}_i \quad (i = 0, 1, 2, 3).$$

The indexes of the coordinates of a covector are usually written as subscripts for distinguishing them from the coordinates of vectors.

The coordinates of a tensor $\boldsymbol{G} \in \mathbf{M} \otimes \mathbf{M}$ are $(i, k = 1, 2, 3)$:

$$G^{00} := \frac{\boldsymbol{\tau} \cdot \boldsymbol{G} \cdot \boldsymbol{\tau}^*}{\mathrm{s}^2}, \quad G^{0k} = \frac{(\boldsymbol{\tau} \cdot \boldsymbol{G} \cdot \boldsymbol{\pi}_{\boldsymbol{u}}^*) \cdot \boldsymbol{e}_k}{\mathrm{sm}}, \quad G^{k0} = \frac{\boldsymbol{e}_k \cdot (\boldsymbol{\pi}_{\boldsymbol{u}} \cdot \boldsymbol{G} \cdot \boldsymbol{\tau}^*)}{\mathrm{ms}},$$

$$G^{ik} = \frac{\boldsymbol{e}_i \cdot (\boldsymbol{\pi}_{\boldsymbol{u}} \cdot \boldsymbol{G} \cdot \boldsymbol{\pi}_{\boldsymbol{u}}^*) \cdot \boldsymbol{e}_k}{\mathrm{m}^2}.$$

The coordinates of a cotensor $\boldsymbol{F} \in \mathbf{M}^* \otimes \mathbf{M}^*$ are $(i, k = 0, 1, 2, 3)$:

$$F_{ik} = \boldsymbol{e}_i \cdot \boldsymbol{F} \cdot \boldsymbol{e}_k.$$

Remark. It is emphasized that all the formulas above are valid only for standard coordinate systems. There are other types of coordinate systems, too, e. g., in which the basis of spacetime vectors consists of four future-like vectors. In practice, spherical coordinates for space-like vectors is often applied, too.

Exercises

1. Two coordinate systems are equivalent by the general definition if there is a proper Noether transformation that maps one of them to the other. Prove that the standard coordinate systems $(o, \mathrm{s}, \mathrm{m}, \boldsymbol{e}_0, \boldsymbol{e}_1, \boldsymbol{e}_2, \boldsymbol{e}_3)$ and $(o', \mathrm{s}', \mathrm{m}', \boldsymbol{e}_0', \boldsymbol{e}_1', \boldsymbol{e}_2', \boldsymbol{e}_3')$ are equivalent if and only if $\mathrm{s}' = \mathrm{s}$, $\mathrm{m}' = \mathrm{m}$.
2. Give the coordinates of the world point x in the coordinate system defined by a world point o, and by an arbitrary basis $\boldsymbol{n}_1, \boldsymbol{n}_2, \boldsymbol{n}_3, \boldsymbol{n}_4$ of $\mathbf{M}$.
3. Define a nonstandard coordinate system in which an inertial frame splits spacetime and uses spherical coordinates in its space.

6 Elements of point mechanics in the spacetime model

Classical mechanics is a well-known and well-elaborated theory (in coordinates, as usual), and that is why it offers a good possibility to deepen our knowledge on spacetime.

6.1 World line functions

The history of a material point is a world line in spacetime. Such a world line—a curve—can be parameterized by absolute time in a natural way.

A world line C is parameterized in such a way that the common spacetime point of the world line (curve) and an absolute time point t (three-dimensional hyperplane) is assigned to the time point: $r(t) := t \cap \mathrm{C}$ (see Figure 6.1). Then

$$r : \mathrm{T}_a \to \mathrm{M}, \quad \tau(r(t)) = t. \tag{6.1}$$

and

$$\boldsymbol{\tau}(\dot{r}(t)) = 1 \quad \text{i. e.} \quad \dot{r}(t) \in \mathrm{V}(1). \tag{6.2}$$

Therefore, according to the definition in Subsection 2.3.4, $\boldsymbol{t} \mapsto r(t_0 + \boldsymbol{t})$ is a world line function for arbitrary absolute time point $t_0 \in \mathrm{T}_a$. We accept that r itself will be called a world line function.

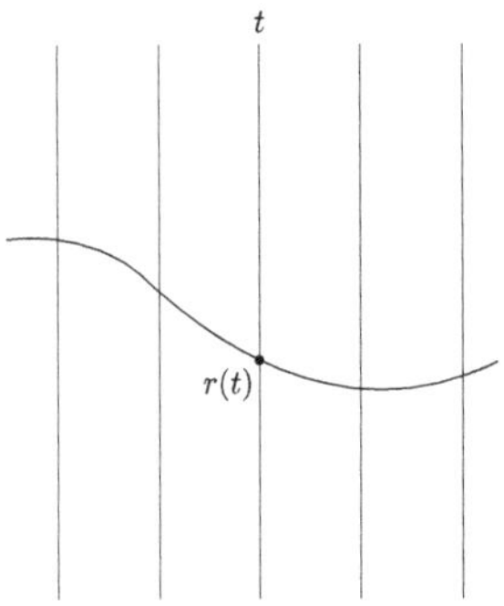

Figure 6.1: A world line function.

It is evident that if x is in the range of r, then $r^{-1}(x) = \tau(x)$. So, the definition of the proper time passed on a world line (see (2.4)) now yields that

$$\boldsymbol{t}_C(x, y) = r^{-1}(y) - r^{-1}(x) = \tau(y) - \tau(x) = \boldsymbol{\tau} \cdot (y - x):$$

the proper time passed between two occurrences of a world line is just the corresponding absolute time interval.

https://doi.org/10.1515/9783112219553-008

It is immediate that

$$\tau \cdot \ddot{r}(t) = 0, \quad \text{i. e.,} \quad \ddot{r}(t) \in \frac{\mathbf{S}}{\mathbb{T} \otimes \mathbb{T}},$$

absolute acceleration *has absolute space-like values.*

It is quite trivial that every element of $\frac{\mathbf{S}}{\mathbb{T}\otimes\mathbb{T}}$ can be an acceleration value of a world line function (e. g., $\boldsymbol{a}$ is the acceleration of $t \mapsto o + (t - \tau(o))\boldsymbol{u} + \frac{(t-\tau(o))^2}{2}\boldsymbol{a}$ for all world points o and absolute velocities $\boldsymbol{u}$); thus, the set of **absolute accelerations** is a three-dimensional Euclidean vector space. In contrast to absolute velocities,

- There is a zero absolute acceleration.
- An absolute acceleration has a magnitude.
- The angle between two absolute accelerations makes sense.

Exercises

1. Recall (5.26) and (5.27) and show that the transform of the world line function r by an arrow orientation preserving Noether transformation L,

$$t \mapsto r_L(t) := L(r(L_{\mathrm{T}_a}^{-1}(t)))$$

is a world line function.

2. Recall (5.25) and (5.28) and show that if r is a world line function then

$$t \mapsto r_{\boldsymbol{u}o}(t) := T_{\boldsymbol{u},o} r(T_{\tau(o)}^{-1}(t))$$

is a world line function, too.

6.2 Movements

Recall that the description of motion with respect to an observer requires a synchronization; in other words, a motion can be described only with respect to a reference frame (consisting of an observer and a synchronization) by giving when and where a material point is. In the nonrelativistic spacetime model, a single synchronization exists, the absolute one; consequently, an observer determines a unique reference frame.

An inertial frame $\boldsymbol{u}$ perceives the history of a material point (a world line) as a motion and describes it by assigning to an absolute instant t (a hyperplane directed by $\mathbf{S}$) the $\boldsymbol{u}$-space point (straight line directed by $\boldsymbol{u}$), which meets the material point (the world line) at t. Therefore, by formula (5.8), the ***u*-motion** corresponding to the world line function r is

$$\mathrm{T}_a \to \mathrm{S}_{\boldsymbol{u}}, \quad t \mapsto r_{\boldsymbol{u}}(t) := \pi_{\boldsymbol{u}}(t \cap \mathrm{Ran}(r)) = \pi_{\boldsymbol{u}}(r(t)) = r(t) + \mathbb{T}\boldsymbol{u},$$

illustrated by Figure 6.2.

The ***u*-relative velocity** of the material point is the time derivative of the $\boldsymbol{u}$-motion; using the formula (5.9), we get

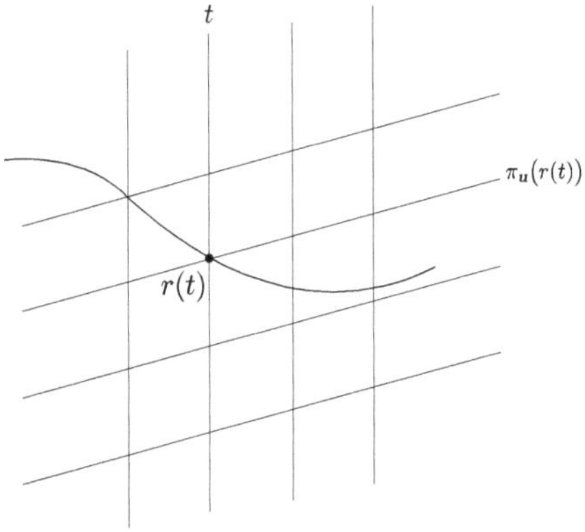

Figure 6.2: Description of $\boldsymbol{u}$-motion.

$$\dot{r}_{\boldsymbol{u}}(t) = \boldsymbol{\pi}_{\boldsymbol{u}} \cdot \dot{r}(t) = \dot{r}(t) - \boldsymbol{u} = \mathbf{v}_{\dot{r}(t)\boldsymbol{u}} \in \frac{\mathbb{S}}{\mathbb{T}}.$$

This is a generalization of our result (5.29) concerning inertial world lines. It follows immediately that

$\boldsymbol{u}$-relative acceleration equals absolute acceleration:

$$\ddot{r}_{\boldsymbol{u}}(t) = \ddot{r}(t) \in \frac{\mathbb{S}}{\mathbb{T} \otimes \mathbb{T}}.$$

6.3 Absolute and relative momentum

The unit of mass, kg in everyday practice is independent of the time unit s and the distance unit m. To treat the fundamentals of mechanics in the nonrelativistic spacetime model, we have to introduce the measure line of masses, which is an object outside the model. Quantum mechanics, however, found out that the Planck constant $\hbar$, a quantity distinguished by Nature, establishes a relation among the units of mass, time, and distance:

$$\hbar = \frac{6.62607015\ldots 10^{-34}}{2\pi}\frac{\mathrm{m}^2\,\mathrm{kg}}{\mathrm{s}} = (1.0545701\ldots)10^{-34}\,\frac{\mathrm{m}^2\,\mathrm{kg}}{\mathrm{s}}.$$

As a consequence, we can choose $\frac{\mathbb{T}}{\mathbb{L}\otimes\mathbb{L}}$ for the measure line of masses in such a way that

$$\mathrm{kg} := \frac{2\pi}{6.62607015\ldots}10^{34}\,\frac{\mathrm{s}}{\mathrm{m}^2} = (9.482536\ldots)10^{33}\,\frac{\mathrm{s}}{\mathrm{m}^2};$$

then the Planck constant becomes the real number 1.

It is evident that this choice makes the theoretical exposition easier because we need not deal with a new measure line and so the formulas become simpler.

Of course, for practical application the everyday unit kg is more advantageous.

The following list will be important for a comparison with the relativistic point of view.

Let us consider a material point having constant mass m and absolute velocity $\dot{x}$ (this notation will be explained in the next section). Then we accept that:

1. Absolute momentum = mass × absolute velocity:

$$m\dot{x},$$

which involves that
proper time derivative of absolute momentum = mass × absolute acceleration:

$$(m\dot{x})^{\cdot} = m\ddot{x}.$$

Further, we have that:

2. Time-like component of absolute momentum = mass,

$$\tau \cdot (m\dot{x}) = m.$$

3. $\boldsymbol{u}$-space-like component of absolute momentum = mass × ($\boldsymbol{u}$-relative velocity):

$$\pi_{\boldsymbol{u}} \cdot (m\dot{x}) = m\boldsymbol{v}_{\dot{x}\boldsymbol{u}}.$$

4. Time derivative of $\boldsymbol{u}$-space-like component of absolute momentum = time derivative of mass × ($\boldsymbol{u}$-relative)velocity = mass × ($\boldsymbol{u}$-relative acceleration) = mass × absolute acceleration:

$$(\pi_{\boldsymbol{u}} \cdot (m\dot{x}))^{\cdot} = m(\boldsymbol{v}_{\dot{x}\boldsymbol{u}})^{\cdot} = m\ddot{x}.$$

6.4 Newton equation

6.4.1 Absolute Newton equation

The usual Newton equation (regarding relative quantities) is a balance of momentum: the action of a (relative) force results in a change of (relative) momentum, which is the product of mass and (relative) velocity. We get the **absolute Newton equation** simply by substituting “absolute” for “relative,” i. e.,

absolute time derivative of absolute momentum = action of absolute force,

$$(x : \mathrm{T}_a \to \mathrm{M})? \quad (m\dot{x})^{\cdot} = m\ddot{x} = \boldsymbol{f}(x, \dot{x}), \tag{6.3}$$

which is a second-order ordinary differential equation *to determine* the world line functions of a material point with **given mass** m under the action of a **given absolute force** $\boldsymbol{f}$.

To have a unique solution of that differential equation, the initial spacetime position and absolute velocity of the mass point must be given. That is why, if $t \mapsto r(t)$ is a solution, it is suitable to consider the pair $(r, \dot{r})$ the **process** of the mass point: its value in an arbitrary instant determines the whole function.

The **evolution space** of a mass point is the set in which the processes take values: $\mathrm{M} \times \mathrm{V}(1)$.

In what follows, we accept the notation (widely used in physics) that:

(i) The elements of the evolution space are written in the form $(x, \dot{x})$ (note that then $\dot{x}$ is independent of x, and it can denote an arbitrary absolute velocity).
(ii) An arbitrary ("abstract") process—i. e., a time function—is also denoted by $(x, \dot{x})$.
(iii) An actual process is denoted by $(r, \dot{r})$.

Certainly, the double meaning of $(x, \dot{x})$ is objectionable in terms of precision but it makes simpler a lot of formulas and with due attention we can avoid misunderstanding.

An absolute force $\boldsymbol{f}$ is a function defined in the evolution space; its values are determined by that $m\ddot{x}$ has values in $\frac{\mathbb{T}}{\mathbb{L}\otimes\mathbb{L}} \otimes \frac{\mathbf{S}}{\mathbb{T}\otimes\mathbb{T}} \equiv \frac{\mathbf{S}}{\mathbb{L}\otimes\mathbb{L}\otimes\mathbb{T}} \equiv \frac{\mathbf{S}^*}{\mathbb{T}}$. This is why

$$\boldsymbol{f} : \mathrm{M} \times \mathrm{V}(1) \to \frac{\mathbf{S}^*}{\mathbb{T}}. \tag{6.4}$$

6.4.2 Relative Newton equation

$\boldsymbol{u}$-motion is obtained from the **$\boldsymbol{u}$-relative Newton equation**

time derivative of **u***-momentum = action of the* **u***-relative force;*

The question is, however, what is $\boldsymbol{u}$-relative momentum and what is the $\boldsymbol{u}$-relative force.

The answer is not difficult: in usual treatments, $\boldsymbol{u}$-relative momentum is the product of mass and $\boldsymbol{u}$-relative velocity, which equals the $\boldsymbol{u}$-space-like component of absolute momentum according to item 2 of Section 6.3. That is why the $\boldsymbol{u}$-relative Newton equation is the $\boldsymbol{u}$-space-like component of the absolute Newton equation, which determines the $\boldsymbol{u}$-relative force, too, only the elements of the absolute evolution space need to be expressed by the relative variables, time points (t), $\boldsymbol{u}$-space points (q), and $\boldsymbol{u}$-relative velocities ($\dot{q}$):

$$(q : \mathrm{T}_a \to \mathrm{S}_{\boldsymbol{u}})? \quad (m\dot{q})^{\cdot} = m\ddot{q} = \boldsymbol{f}_{\boldsymbol{u}}(t, q, \dot{q}).$$

This form is well understandable though is not precise because t is involved only on the right side. We allow this little inaccuracy for the sake of simpler formulas.

An absolute force has space-like values; thus, the $\boldsymbol{u}$-relative force is "essentially" the same as the absolute one, and only the absolute variables must be expressed by the relative ones: $\boldsymbol{f}_{\boldsymbol{u}} : \mathrm{T}_a \times \mathrm{S}_{\boldsymbol{u}} \times \frac{\mathbf{S}}{\mathbb{T}} \to \frac{\mathbf{S}^*}{\mathbb{T}}$,

$$\boldsymbol{f}_{\boldsymbol{u}}(t, q, \dot{q}) = \boldsymbol{f}(t \cap q, \boldsymbol{u} + \dot{q}).$$

The absolute force is obtained from the relative one by the equality

$$\boldsymbol{f}(x, \dot{x}) = \boldsymbol{f}_{\boldsymbol{u}}(\tau(x), \pi_{\boldsymbol{u}}(x), \boldsymbol{v}_{\dot{x}\boldsymbol{u}}). \tag{6.5}$$

6.5 Some special absolute forces

6.5.1 Forces having a potential

A special type of absolute forces comes from **absolute field strength** $\boldsymbol{F} : \mathrm{M} \to \mathbf{M}^* \wedge \mathbf{M}^*$:

$$\boldsymbol{f}(x, \dot{x}) := \boldsymbol{i}^* \cdot \boldsymbol{F}(x) \cdot \dot{x}.$$

According to (5.45), $((\boldsymbol{i}^* \cdot \boldsymbol{F} \cdot \boldsymbol{u}, \boldsymbol{i}^* \cdot \boldsymbol{F} \cdot \boldsymbol{i})) =: ((\boldsymbol{E}_{\boldsymbol{u}}, \boldsymbol{B}))$, the $\boldsymbol{u}$-split form of the field strength gives

$$\boldsymbol{i}^* \cdot \boldsymbol{F}(x) \cdot \dot{x} = \boldsymbol{i}^* \cdot \boldsymbol{F}(x) \cdot \boldsymbol{u} + \boldsymbol{i}^* \cdot \boldsymbol{F}(x) \cdot (\dot{x} - \boldsymbol{u}) = \boldsymbol{E}_{\boldsymbol{u}} + \boldsymbol{B} \cdot \boldsymbol{v}_{\dot{x}\boldsymbol{u}},$$

which is the well-known form of an electromagnetic force (note that here the magnetic field is an antisymmetric space-like tensor instead of which the corresponding axial vector is used in practice). Of course, the above formulas are valid not only for electromagnetic forces

It frequently occurs—as in electromagnetism—that the field strength is the antisymmetric derivative of an **absolute potential** $\boldsymbol{K} : \mathrm{M} \to \mathbf{M}^*$,

$$\boldsymbol{F} = \mathcal{D} \wedge \boldsymbol{K}.$$

The $\boldsymbol{u}$-split form of $\boldsymbol{K}$ is denoted by $(-V_{\boldsymbol{u}}, \boldsymbol{A})$ where $V_{\boldsymbol{u}}$ is the $\boldsymbol{u}$-scalar potential and $\boldsymbol{A}$ is the vector potential; then the $\boldsymbol{u}$-split form of $\mathcal{D} \wedge \boldsymbol{K}$ (see Section 7.2) is

$$((-\nabla V_{\boldsymbol{u}} - \mathcal{D}_{\boldsymbol{u}}\boldsymbol{A}, \nabla \wedge \boldsymbol{A})).$$

A potential is **absolute time-like** (see Subsection 5.1.5) if there is an **absolute scalar potential** $V : \mathrm{M} \to \mathbb{T}^*$ such that

$$\boldsymbol{K} = -V\boldsymbol{\tau};$$

then the absolute force is $-\nabla V$, independent of absolute velocity. Gravitational forces and elastic forces are of this type.

6.5.2 The simplest special forces

1) Force **depending only on absolute velocity**: there is a function $\boldsymbol{h} : \mathrm{V}(1) \to \frac{\mathbf{S}^*}{\mathbb{T}}$ and

$$\boldsymbol{f}(x, \dot{x}) = \boldsymbol{h}(\dot{x})).$$

The corresponding $\boldsymbol{u}$-relative force is

$$\boldsymbol{f}_{\boldsymbol{u}}(t, q, \dot{q}) = \boldsymbol{h}(\boldsymbol{u} + \dot{q}).$$

2) Force **independent of absolute velocity**: there is a function $\boldsymbol{b} : \mathrm{M} \to \frac{\mathbf{S}^*}{\mathbb{T}}$ and

$$\boldsymbol{f}(x, \dot{x}) = \boldsymbol{b}(x).$$

The corresponding $\boldsymbol{u}$-relative force is

$$\boldsymbol{f}_{\boldsymbol{u}}(t, q, \dot{q}) = \boldsymbol{b}(t \cap q).$$

3) Force **depending only on time**: there is a function $\boldsymbol{b} : \mathrm{T}_a \to \frac{\mathbf{S}^*}{\mathbb{T}}$ and

$$\boldsymbol{f}(x, \dot{x}) = \boldsymbol{b}(\tau(x)).$$

The corresponding $\boldsymbol{u}$-relative force is

$$\boldsymbol{f}_{\boldsymbol{u}}(t, q, \dot{q}) = \boldsymbol{b}(t).$$

4) **Constant** force: there is a $\boldsymbol{b} \in \frac{\mathbf{S}^*}{\mathbb{T}}$ and

$$\boldsymbol{f}(x, \dot{x}) = \boldsymbol{b}.$$

The gravitational force near the Earth surface is modeled by such a force. This force has an absolute scalar potential,

$$V(x) = -\boldsymbol{b}(\pi_{\boldsymbol{u}} \cdot (x - o))$$

for arbitrary $o \in \mathrm{M}$ and $\boldsymbol{u} \in \mathrm{V}(1)$; the potential itself is

$$\boldsymbol{K}(x) = \boldsymbol{b}(\pi_{\boldsymbol{u}} \cdot (x - o))\boldsymbol{\tau}. \tag{6.6}$$

Of course, the corresponding $\boldsymbol{u}$-relative force is $\boldsymbol{b}$, too.

5) **Static** force: In mechanics, one often takes static forces, i. e., forces "independent of time," in other words, "depending only on space." Of course, this definition regards reference frames and relative forces, and it is trivial that "depending only on $\boldsymbol{u}$-space" cannot hold for all $\boldsymbol{u}$ except the constant forces. It may happen, however,

that the relative force depends only on the space points of a given observer $\boldsymbol{u}_s$. This means that there is a function $\tilde{\boldsymbol{f}}$ such that

$$\boldsymbol{f}(x, \dot{x}) = \tilde{\boldsymbol{f}}(\pi_{\boldsymbol{u}_s}(x), \dot{x}). \tag{6.7}$$

Then we say that the absolute force is **$\boldsymbol{u}_s$-static.**

For an arbitrary world point o, $\pi_{\boldsymbol{u}_s}(x) = \pi_{\boldsymbol{u}_s}(o) + \boldsymbol{\pi}_{\boldsymbol{u}_s} \cdot (x - o)$ and

$$\boldsymbol{\pi}_{\boldsymbol{u}_s} \cdot (x - o) = \boldsymbol{\pi}_{\boldsymbol{u}} \cdot (x - o) + (\boldsymbol{\pi}_{\boldsymbol{u}_s} \cdot (x - o) - \boldsymbol{\pi}_{\boldsymbol{u}} \cdot (x - o)) = \boldsymbol{\pi}_{\boldsymbol{u}} \cdot (x - o) - \boldsymbol{\tau} \cdot (x - o)(\boldsymbol{u}_s - \boldsymbol{u});$$

with

$$\tau(x - o) = \tau(x) - \tau(o) =: t - t_o, \quad \boldsymbol{\pi}_{\boldsymbol{u}} \cdot (x - o) = \pi_{\boldsymbol{u}}(x) - \pi_{\boldsymbol{u}}(o) =: q - q_o,$$

we obtain that the corresponding $\boldsymbol{u}$-relative force is of the form

$$\boldsymbol{f}_{\boldsymbol{u}}(t, q, \dot{q}) = \tilde{\boldsymbol{f}}(\pi_{\boldsymbol{u}_s}(o) + (q - q_o) - (t - t_o)\boldsymbol{v}_{\boldsymbol{u}_s \boldsymbol{u}}), \boldsymbol{u} + \dot{q}).$$

It is evident that this is time independent if and only if $\boldsymbol{u} = \boldsymbol{u}_s$.

6) **Central** force: The center means a material point, i. e., a world line in spacetime. The value of the force at a world point is determined by and is parallel to the vector between the world point and the simultaneous occurrence of the world line of the center, as it is illustrated by Figure 6.3. So, a central force is given by a world line function r_c and a map $a : \mathbb{L} \to \frac{\mathbb{R}}{\mathbb{L} \otimes \mathbb{L} \otimes \mathbb{T}}$ in the following way:

$$\boldsymbol{f}(x, \dot{x}) := a(|x - r_c(\tau(x))|)(x - r_c(\tau(x))).$$

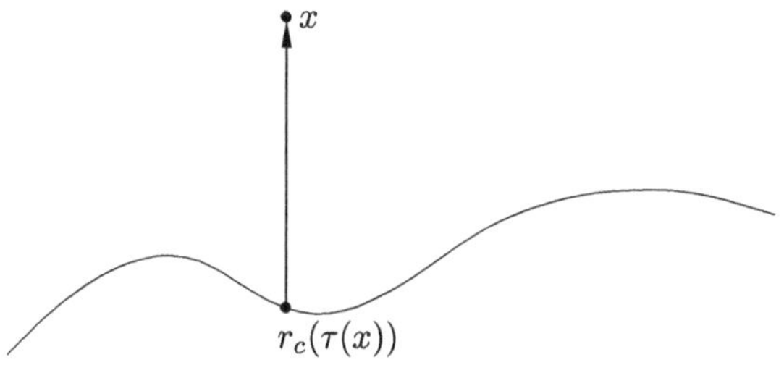

Figure 6.3: Central force at x is parallel to $x - r_c(\tau(x))$.

The corresponding $\boldsymbol{u}$-relative force is

$$\boldsymbol{f}_{\boldsymbol{u}}(t, q, \dot{q}) = a(|q - q_c(t)|)(q - q_c(t))$$

where $q_c(t) := \pi_{\boldsymbol{u}}(r_c(t))$ is the position of the center in $\boldsymbol{u}$-space.

In particular, such forces are

- The **gravitational** force of a mass point:

$$f(x,\dot{x}) = -\frac{\gamma m_c m}{|x - r_c(\tau(x))|^3}(x - r_c(\tau(x))$$

  where γ is the gravitational constant, m_c is the mass of the center, and m is the mass of the material point on which the center acts.
- An **elastic** force:

$$f(x,\dot{x}) = -k(x - r_c(\tau(x)))$$

  where $k > 0$ is a constant.

A special case is when the center is inertial with absolute velocity $\boldsymbol{u}_c$. Then with an arbitrary occurrence o of the world line of the center

$$f(x,\dot{x}) = a(|\pi_{\boldsymbol{u}_c} \cdot (x - o)|)(\pi_{\boldsymbol{u}_c} \cdot (x - o)).$$

This force is $\boldsymbol{u}_c$-static and has an absolute scalar potential

$$V(x) := -b(|\pi_{\boldsymbol{u}_c} \cdot (x - o)|)$$

where b is a primitive function of a (i. e., a is the derivative of b). The potential itself is

$$\boldsymbol{K}(x) := b(|\pi_{\boldsymbol{u}_c} \cdot (x - o)|)\boldsymbol{\tau}. \tag{6.8}$$

The $\boldsymbol{u}_c$-relative form of this force is

$$f_{\boldsymbol{u}_c}(t, q, \dot{q}) = a(|q - q_c|)(q - q_c)$$

where $q_c := \pi_{\boldsymbol{u}_c}(o)$ is the (resting) position of the center in $\boldsymbol{u}_c$-space.

6.5.3 Exercises

1. The absolute forces $\boldsymbol{f}$ and $\boldsymbol{f}'$ are physically equivalent by the definition in Section 3.3 if there is a proper Noether transformation (over the proper Galilei transformation $\boldsymbol{L}$) such that $\boldsymbol{f}'(L(x), \boldsymbol{L} \cdot \dot{x}) = \boldsymbol{L} \cdot \boldsymbol{f}(x, \dot{x})$.
   When are the elastic forces given by k, r_c, and k', $r_{c'}$ are physically equivalent? (Use (5.26).)
2. An absolute force $\boldsymbol{f}$ is invariant for the $(\boldsymbol{u}, o)$-time reversal if and only if $\boldsymbol{f}(T_{\boldsymbol{u},o}(x), -\boldsymbol{T}_{\boldsymbol{u}} \cdot \dot{x}) = \boldsymbol{T}_{\boldsymbol{u}} \cdot \boldsymbol{f}(x, \dot{x})$. The force has absolute space-like values; thus, $\boldsymbol{T}_{\boldsymbol{u}}$ can be left from the right side. Prove that if $\boldsymbol{f}$ is invariant:

(i) For the $(\boldsymbol{u}, o)$-time reversal with a given $\boldsymbol{u}$ and for all o then it is constant on every straight line directed by $\boldsymbol{u}$, i. e., it depends only on $\boldsymbol{u}$-space points and $\boldsymbol{u}$-relative velocities.
(ii) For all $(\boldsymbol{u}, o)$-time reversal, then it is constant.

6.6 Kinetic energy and power

The **$\boldsymbol{u}$-kinetic energy**, by definition, of a material point with mass m and absolute velocity $\dot{x}$ is

$$\frac{m|\mathbf{v}_{\dot{x}\boldsymbol{u}}|^2}{2} \in \frac{\mathbb{R}}{\mathbb{T}} \equiv \mathbb{T}^*.$$

This quantity evidently depends on the inertial frame $\boldsymbol{u}$. Since its value is in $\mathbb{T}^*$, we could think that it is the $\boldsymbol{u}$-time-like component of a covector but this is not true: the above expression is quadratic in $\boldsymbol{u}$ (not linear), so

*there is no covector whose **u**-time-like component is the **u**-kinetic energy.*

Multiplying the absolute Newton equation (6.3) by the relative velocity $\mathbf{v}_{\dot{x}\boldsymbol{u}} = \dot{x} - \boldsymbol{u}$ and taking into account that $\ddot{x} = (\dot{x} - \boldsymbol{u})^{\cdot}$, we get

$$\left(\frac{m|\mathbf{v}_{\dot{x}\boldsymbol{u}}|^2}{2}\right)^{\cdot} = \boldsymbol{f}(x, \dot{x}) \cdot \mathbf{v}_{\dot{x}\boldsymbol{u}}.$$

The left side is the time derivative of the $\boldsymbol{u}$-kinetic energy; the quantity on the right side is called the **$\boldsymbol{u}$-relative power** of the force $\boldsymbol{f}$.

Now it is worth making a little digression to better understand the relativistic case. An absolute force has values in $\frac{\mathbf{S}^*}{\mathbb{T}}$; that is why we can think that it is the absolute space-like component of a covector valued function; indeed, we find that

$$\overline{\boldsymbol{f}} : \mathrm{M} \times \mathrm{V}(1) \to \frac{\mathbf{M}^*}{\mathbb{T}}, \quad \overline{\boldsymbol{f}}(x, \dot{x}) := \boldsymbol{f}(x, \dot{x}) \cdot (\mathbf{1} - \dot{x} \otimes \boldsymbol{\tau})$$

is a uniquely determined function for which

$$\boldsymbol{f}(x, \dot{x}) = \overline{\boldsymbol{f}}(x, \dot{x}) \cdot \boldsymbol{i}, \quad \overline{\boldsymbol{f}}(x, \dot{x}) \cdot \dot{x} = 0$$

holds. Then

$$-\overline{\boldsymbol{f}}(x, \dot{x}) \cdot \boldsymbol{u} = \overline{\boldsymbol{f}}(x, \dot{x}) \cdot (\dot{x} - \boldsymbol{u}) = \boldsymbol{f}(x, \dot{x}) \cdot \mathbf{v}_{\dot{x}\boldsymbol{u}};$$

*there is a unique absolute covector valued function such that its negative **u**-time-like component is the **u**-power and its space-like component is the force.*

Warning: $\overline{\boldsymbol{f}}$ is never independent of $\dot{x}$, it depends even if $\boldsymbol{f}$ is independent.

6.7 Conservation laws

6.7.1 Action–reaction

Let us consider two material points briefly: particles, which without touching, exert a force one to other (action at a distance).

According to Newton's **law of action–reaction**, the interaction between the particles is **instantaneous** and the forces are **opposite** to each other.

In formula:

- If $\boldsymbol{f}_{12}(x_1, \dot{x}_1, x_2, \dot{x}_2)$ is the force acting on the "first" particle by the "second" particle when they are at the world points x_1 and x_2 and have absolute velocities $\dot{x}_1$ and $\dot{x}_2$, respectively.
- If $\boldsymbol{f}_{21}(x_2, \dot{x}_2, x_1, \dot{x}_1)$ is the other force, then

$$(\tau(x_1) = \tau(x_2)), \quad \boldsymbol{f}_{21}(x_2, \dot{x}_2, x_1, \dot{x}_1) = -\boldsymbol{f}_{12}(x_1, \dot{x}_1, x_2, \dot{x}_2)$$

must hold.

If the particles have masses m_1 and m_2, then the Newton equations for them are

$$((x_1, x_2) : \mathrm{T}_a \to \mathrm{M} \times \mathrm{M})? \quad \begin{aligned} m_1\ddot{x}_1 &= \boldsymbol{f}_{12}(x_1, \dot{x}_1, x_2, \dot{x}_2), \\ m_2\ddot{x}_2 &= \boldsymbol{f}_{21}(x_2, \dot{x}_2, x_1, \dot{x}_1). \end{aligned}$$

Then the law of action–reaction yields immediately that

$$(m_1\dot{x}_1 + m_2\dot{x}_2)^{\cdot} = 0:$$

the *total absolute momentum is conserved*, i. e., does not change in the course of the interaction.

It is a simple consequence that the $\boldsymbol{u}$-relative momentum is also conserved for all $\boldsymbol{u}$.

Let us suppose that the interaction force has an absolute scalar potential which depends only on the distance of the particles, i. e., there is continuously differentiable function $V : \mathbb{L} \to \mathbb{T}^*$ such that

$$\boldsymbol{f}_{12}(x_1, \dot{x}_1, x_2, \dot{x}_2) = -\frac{\partial V(|x_1 - x_2|)}{\partial x_1} = -V'(|x_1 - x_2|)\frac{x_1 - x_2}{|x_1 - x_2|},$$

where the prime denotes differentiation with respect to the variable of V. $\boldsymbol{f}_{21}$ is the same by changing x_2 and x_1.

Multiplying the Newton equations by the $\boldsymbol{u}$-relative velocities $(\dot{x}_1 - \boldsymbol{u})$ and $(\dot{x}_2 - \boldsymbol{u})$, respectively, we get

$$m_1\ddot{x}_1 \cdot (\dot{x}_1 - \boldsymbol{u}) = -V'(|x_1 - x_2|)\frac{(x_1 - x_2)\cdot(\dot{x}_1 - \boldsymbol{u})}{|x_1 - x_2|},$$

$$m_2\ddot{x}_2 \cdot (\dot{x}_2 - \boldsymbol{u}) = V'(|x_2 - x_1|)\frac{(x_2 - x_1)\cdot(\dot{x}_2 - \boldsymbol{u})}{|x_2 - x_1|};$$

since $|x_2 - x_1| = |x_1 - x_2|$, the sum of the two equalities gives the time derivative of the total $\boldsymbol{u}$-kinetic energy on the left side and the time derivative of the function $-V(|x_1 - x_2|)$ on the right side.

Considering V as a **potential energy**, we can state that the **$\boldsymbol{u}$-mechanical energy**

$$\frac{m_1|\mathbf{v}_{\dot{x}_1\boldsymbol{u}}|^2}{2} + \frac{m_2|\mathbf{v}_{\dot{x}_2\boldsymbol{u}}|^2}{2} + V(|x_1 - x_2|)$$

is conserved for all $\boldsymbol{u}$.

6.7.2 Collisions

We accept as a fundamental physical law that *absolute momentum is also conserved* in processes that cannot be described by a Newton equation, e. g., in collisions.

Let two particles meet and join together (inelastic collision). Let the particles have masses m_1 and m_2 and absolute velocities $\boldsymbol{u}_1$ and $\boldsymbol{u}_2$, respectively, and let the arising new particle have mass m_3 and absolute velocity $\boldsymbol{u}_3$. The conservation of total momentum gives

$$m_1\boldsymbol{u}_1 + m_2\boldsymbol{u}_2 = m_3\boldsymbol{u}_3. \tag{6.9}$$

The $\boldsymbol{u}$-space-like component of this equality is the **conservation of $\boldsymbol{u}$-relative momentum** for arbitrary $\boldsymbol{u}$:

$$m_1\mathbf{v}_{\boldsymbol{u}_1\boldsymbol{u}} + m_2\mathbf{v}_{\boldsymbol{u}_2\boldsymbol{u}} = m_3\mathbf{v}_{\boldsymbol{u}_3\boldsymbol{u}}.$$

The time-like component is just the **conservation of mass**:

$$m_1 + m_2 = m_3.$$

The total kinetic energy of the particles, relative to the inertial frame in which the arising particle is at rests, before collision is

$$\frac{m_1|\mathbf{v}_{\boldsymbol{u}_1\boldsymbol{u}_3}|^2}{2} + \frac{m_2|\mathbf{v}_{\boldsymbol{u}_2\boldsymbol{u}_3}|^2}{2}$$

and after collision is 0. If $\boldsymbol{u}_1$ and $\boldsymbol{u}_2$ are not equal to $\boldsymbol{u}_3$ (this is a proper collision), then the $\boldsymbol{u}_3$-kinetic energy vanishes in the collision.

It is experienced, however, that the arising particle has some property—a chemical bond, higher temperature, etc.—which explains this lost of energy. That is why we introduce the notion of **internal energy** and we conceive that the kinetic energy before collision is transformed into internal energy in the collision, i. e., the total energy is conserved. Internal energy is absolute, independent of reference frames.

More precisely, accepting the definition

$\boldsymbol{u}$-**relative** *energy* := $\boldsymbol{u}$-*kinetic energy* + *internal energy*,

we state the **conservation of $\boldsymbol{u}$-relative energy** for all $\boldsymbol{u}$: if $\underline{E}$ denotes internal energy, then

$$\frac{m_1|\boldsymbol{v}_{\boldsymbol{u}_1\boldsymbol{u}}|^2}{2} + \underline{E}_1 + \frac{m_2|\boldsymbol{v}_{\boldsymbol{u}_2\boldsymbol{u}}|^2}{2} + \underline{E}_2 = \frac{m_3|\boldsymbol{v}_{\boldsymbol{u}_3\boldsymbol{u}}|^2}{2} + \underline{E}_3. \tag{6.10}$$

Of course, we must show that this is right: if it is true for some—say, for $\boldsymbol{u}_3$ above—then it is true for an arbitrary $\boldsymbol{u}$. Simply, the difference between the $\boldsymbol{u}$-kinetic energy before collision and the $\boldsymbol{u}$-kinetic energy after collision equals the similar difference regarding $\boldsymbol{u}_3$.

This is true because $|\boldsymbol{u}_1 - \boldsymbol{u}|^2 = |\boldsymbol{u}_1 - \boldsymbol{u}_3|^2 + 2(\boldsymbol{u}_1 - \boldsymbol{u}_3) \cdot (\boldsymbol{u}_3 - \boldsymbol{u}) + |\boldsymbol{u}_3 - \boldsymbol{u}|^2$ and the same equality holds for the absolute velocity $\boldsymbol{u}_2$, too; furthermore, using $m_1(\boldsymbol{u}_1 - \boldsymbol{u}_3) + m_2(\boldsymbol{u}_2 - \boldsymbol{u}_3) = m_1\boldsymbol{u}_1 + m_2\boldsymbol{u}_2 - (m_1 + m_2)\boldsymbol{u}_3 = 0$, we get

$$\frac{m_1|\boldsymbol{u}_1 - \boldsymbol{u}|^2}{2} + \frac{m_2|\boldsymbol{u}_2 - \boldsymbol{u}|^2}{2} - \frac{(m_1 + m_2)|\boldsymbol{u}_3 - \boldsymbol{u}|^2}{2} = \frac{m_1|\boldsymbol{u}_1 - \boldsymbol{u}_3|^2}{2} + \frac{m_2|\boldsymbol{u}_2 - \boldsymbol{u}_3|^2}{2}.$$

6.8 Newton equation with varying mass

It may occur that a material body loses or gains mass during its existence: the most simple examples are a rocket or a raindrop passing through a cloud.

Let $m(t)$ be the mass of the material body and let $\boldsymbol{v}(t)$ be the relative velocity, with respect to the body, of the leaving or joining mass at the instant t.

If r is the world line function of the body, then the change of absolute momentum in a "short" time interval $\boldsymbol{t}$ is

$$m(t + \boldsymbol{t})\dot{r}(t + \boldsymbol{t}) - m(t)\dot{r}(t) = \big(m(t)\ddot{r}(t) + \dot{m}(t)\dot{r}(t)\big)\boldsymbol{t} + \operatorname{ordo}(\boldsymbol{t}), \tag{6.11}$$

which is due to the force acting on the body,

$$\boldsymbol{f}\big(r(t) + \operatorname{Ordo}(\boldsymbol{t}), \dot{r}(t) + \operatorname{Ordo}(\boldsymbol{t})\big)\boldsymbol{t} \tag{6.12}$$

and the momentum change due to the leaving or joining mass,

$$\big(m(t + \boldsymbol{t}) - m(t)\big)\big(\dot{r}(t) + \boldsymbol{v}(t) + \operatorname{Ordo}(\boldsymbol{t})\big). \tag{6.13}$$

Dividing by $\boldsymbol{t}$ the equality (6.11) = (6.12) + (6.13) and then taking the limit $\boldsymbol{t} \to 0$, we get that r satisfies the absolute Newton equation

$$(x : \mathrm{T}_a \mapsto \mathrm{M})? \quad m\ddot{x} = \boldsymbol{f}(x, \dot{x}) + \dot{m}\boldsymbol{v}$$

where
- The mass of the body is a given function of time, $m : \mathrm{T}_a \to \frac{\mathbb{T}}{\mathbb{L}\otimes\mathbb{L}}$.
- The relative velocity of the leaving or joining mass with respect to the body is a given function of time, $\boldsymbol{v} : \mathrm{T}_a \to \frac{\mathbf{S}}{\mathbb{T}}$.

The $\boldsymbol{u}$-relative form of this equation is

$$m\ddot{q} = \boldsymbol{f}_{\boldsymbol{u}}(t, q, \dot{q}) + \dot{m}\boldsymbol{v}.$$

Remark. We can find in some books treating mechanics that the two forms of the relative Newton equation given in Section 6.3 (the absolute Newton equation is not known there) are not equivalent if the mass varies in time; then the second one is accepted: the time derivative of the relative momentum equals the relative force. So, in the case of varying mass, the relative Newton equation would be $(m\dot{q})^{\cdot} = \dot{m}\dot{q} + m\ddot{q} = \boldsymbol{f}_{\boldsymbol{u}}(t, q, \dot{q})$, which seems meaningful.

The trouble becomes evident if we look for the corresponding absolute equation. The left side, with the aid of the explicit splitting, has the form $(m\pi_{\boldsymbol{u}}\dot{x})^{\cdot} = \pi_{\boldsymbol{u}}(\dot{m}\dot{x} + m\ddot{x})$; therefore, the absolute Newton equation would read $(m\dot{x})^{\cdot} = \dot{m}\dot{x} + m\ddot{x} = \boldsymbol{f}(x, \dot{x})$. Applying τ to this equation, we get (since $\boldsymbol{f}$ and $m\ddot{x}$ have space-like values) that $\dot{m} = 0$: the mass does not change. This means that “the time derivative of relative momentum equals the action of relative force” is not the correct form of the relative Newton equation if mass changes.

This is an excellent example how relative objects—and even coordinates— can mislead one.

7 Elements of electromagnetism in the spacetime model

Electromagnetism is somewhat more complicated than mechanics but it has a well-elaborated theory in the usual framework of inertial frames and even in coordinates. Just because of its more complicated nature, it offers further possibilities to deepen our knowledge on the nonrelativistic spacetime model. Moreover, electromagnetism played a fundamental role in composing relativity theory; our treatment helps us to better understand that role.

7.1 Splitting of spacetime functions

We consider a spacetime splitting by an inertial frame $\boldsymbol{u}$ with origin o (see Subsection 5.8.1).

7.1.1 Vector fields, covector fields

The **half-split form** by $\boldsymbol{u}$ of a vector field $\boldsymbol{J} : \mathrm{M} \to \mathbf{M}$ is obtained by splitting its range,

$$\mathrm{M} \xrightarrow[\boldsymbol{J}]{} \mathbf{M} \xrightarrow[\boldsymbol{\xi}_{\boldsymbol{u}},]{} \mathbb{T} \times \mathbf{S} \tag{7.1}$$

$$\boldsymbol{\xi}_{\boldsymbol{u}} \cdot \boldsymbol{J} = \begin{pmatrix} \boldsymbol{\tau} \cdot \boldsymbol{J} \\ \boldsymbol{\pi}_{\boldsymbol{u}} \cdot \boldsymbol{J} \end{pmatrix} =: \begin{pmatrix} \rho \\ \boldsymbol{j}_{\boldsymbol{u}} \end{pmatrix}.$$

The **completely split form** by $(\boldsymbol{u}, o)$ of $\boldsymbol{J}$ is obtained by splitting its domain, too,

$$\mathbb{T} \times \mathbf{S} \xrightarrow[\xi_{\boldsymbol{u},o}^{-1}]{} \mathrm{M} \xrightarrow[\boldsymbol{J}]{} \mathbf{M} \xrightarrow[\boldsymbol{\xi}_{\boldsymbol{u}},]{} \mathbb{T} \times \mathbf{S}$$

$$\boldsymbol{\xi}_{\boldsymbol{u}} \cdot \boldsymbol{J}(\xi_{\boldsymbol{u},o}^{-1}) =: \begin{pmatrix} \hat{\rho} \\ \hat{\boldsymbol{j}}_{\boldsymbol{u}} \end{pmatrix}.$$

A **coordinated form** of $\boldsymbol{J}$ is obtained by a coordinate system adapted to the $(\boldsymbol{u}, o)$-splitting (see (5.62)):

$$\mathbb{R}^4 \xrightarrow[C^{-1}]{} \mathbb{T} \times \mathbf{S} \xrightarrow[\xi_{\boldsymbol{u},o}^{-1}]{} \mathrm{M} \xrightarrow[\boldsymbol{J}]{} \mathbf{M} \xrightarrow[\boldsymbol{\xi}_{\boldsymbol{u}}]{} \mathbb{T} \times \mathbf{S} \xrightarrow[\boldsymbol{C},]{} \mathbb{R}^4$$

$$\boldsymbol{C} \cdot \boldsymbol{\xi}_{\boldsymbol{u}} \cdot \boldsymbol{J}(\xi_{\boldsymbol{u},o}^{-1}(C^{-1})) =: \begin{pmatrix} \tilde{\rho} \\ \tilde{\boldsymbol{j}}_{\boldsymbol{u}} \end{pmatrix} =: J^k \quad (k = 0, 1, 2, 3).$$

https://doi.org/10.1515/9783112219553-009

For perspicuity, using the symbol $\sim$ to hint at the different forms of splitting, we write

$$J \sim \begin{pmatrix} \rho \\ j_u \end{pmatrix} \sim \begin{pmatrix} \hat{\rho} \\ \hat{j}_u \end{pmatrix} \sim \begin{pmatrix} \tilde{\rho} \\ \tilde{j}_u \end{pmatrix} \sim J^k \quad (k = 0, 1, 2, 3). \tag{7.2}$$

The **half-split form** by $\boldsymbol{u}$ of a covector field $\boldsymbol{K} : \mathrm{M} \to \mathbf{M}^*$ is obtained by splitting its range,

$$\begin{array}{ccccc} \mathrm{M} & \xrightarrow{\ \boldsymbol{K}\ } & \mathbf{M} & \xrightarrow{\ (\xi_u^{-1})^*\ } & \mathbb{T}^* \times \mathbf{S}^*, \end{array} \tag{7.3}$$
$$(\xi_u^{-1})^* \cdot \boldsymbol{K} = (\boldsymbol{u} \cdot \boldsymbol{K},\ \boldsymbol{i}^* \cdot \boldsymbol{K}) =: (-V_u,\ \boldsymbol{A}).$$

The **completely split form** by $(\boldsymbol{u}, o)$ of $\boldsymbol{K}$ is obtained by splitting its domain, too,

$$\mathbb{T} \times \mathbf{S} \xrightarrow{\ \xi_{u,o}^{-1}\ } \mathrm{M} \xrightarrow{\ \boldsymbol{K}\ } \mathbf{M} \xrightarrow{\ (\xi_u^{-1})^*\ } \mathbb{T}^* \times \mathbf{S}^*,$$
$$(\xi_u^{-1})^* \cdot \boldsymbol{J}(\xi_{u,o}^{-1}) =: (-\widehat{V}_u,\ \widehat{\boldsymbol{A}}).$$

A **coordinated form** of $\boldsymbol{K}$ is obtained by a coordinate system adapted to the $(\boldsymbol{u}, o)$-splitting:

$$\mathbb{R}^4 \xrightarrow{\ C^{-1}\ } \mathbb{T} \times \mathbf{S} \xrightarrow{\ \xi_{u,o}^{-1}\ } \mathrm{M} \xrightarrow{\ \boldsymbol{K}\ } \mathbf{M} \xrightarrow{\ (\xi_u^{-1})^*\ } \mathbb{T}^* \times \mathbf{S}^* \xrightarrow{\ (C^{-1})^*\ } \mathbb{R}^4,$$
$$(C^{-1})^* \cdot (\xi_u^{-1})^* \cdot \boldsymbol{K}(\xi_{u,o}^{-1}(C^{-1})) =: (-\widetilde{V}_u,\ \widetilde{\boldsymbol{A}}) =: K_k \quad (k = 0, 1, 2, 3).$$

Concisely,

$$\boldsymbol{K} \sim (-V_u,\ \boldsymbol{A}) \sim (-\widehat{V}_u,\ \widehat{\boldsymbol{A}}) \sim (-\widetilde{V}_u,\ \widetilde{\boldsymbol{A}}) \sim K_k \quad (k = 0, 1, 2, 3). \tag{7.4}$$

7.1.2 Tensor fields, cotensor fields

Now we consider split forms of an antisymmetric tensor field $\boldsymbol{G} : \mathrm{M} \to \mathbf{M} \wedge \mathbf{M}$ and an antisymmetric cotensor field $\boldsymbol{F} : \mathrm{M} \to \mathbf{M}^* \wedge \mathbf{M}^*$ applying the formulas of Subsection 5.7.1 and the notation of the previous subsection, according to the sense.

Attention!
$\boldsymbol{G}$ and $\boldsymbol{F}$ in the cited subsection are elements of $\mathbf{M} \wedge \mathbf{M}$ and $\mathbf{M}^* \wedge \mathbf{M}^*$, respectively, while now the same symbols denote tensor/cotensor valued functions defined in spacetime.

Then we have

$$G \sim ((D, H_u)) \sim ((\widehat{D}, \widehat{H}_u)) \sim ((\widetilde{D}, \widetilde{H}_u)) \sim G^{ik} \quad (i \neq k = 0, 1, 2, 3),$$

and

$$F \sim ((E_u, B)) \sim ((\widehat{E}_u, \widehat{B})) \sim ((\widetilde{E}_u, \widetilde{B})) \sim F_{ik} \quad (i \neq k = 0, 1, 2, 3),$$

7.2 Differentiation

The mathematical notions included in this subsection are treated in Subsection 18.6 of the mathematical supplement.

7.2.1 Scalar fields

The derivative of a scalar field $f : \mathrm{M} \to \mathbb{R}$ at x is the covector field $x \mapsto \mathcal{D}f[x]$.

The **half-split form by u** of $\mathcal{D}f$ is

$$(\boldsymbol{u} \cdot \mathcal{D}f,\ \boldsymbol{i}^* \cdot \mathcal{D}f) =: (\mathcal{D}_u f,\ \nabla f);$$

$\mathcal{D}_u$ is the **u-time-like derivative** of f and ∇f is the **space-like derivative** of f; they have direct meanings as follows.

Let us restrict f onto the straight line passing through x and directed by $\boldsymbol{u}$, i. e., let us consider the function $\mathbb{T} \to \mathbb{R}$, $\boldsymbol{t} \mapsto f(x + \boldsymbol{t}\boldsymbol{u})$. The derivative at zero of this function—according to the chain rule of differentiation of composite functions—is $(\mathcal{D}f[x]) \cdot \boldsymbol{u} = \boldsymbol{u} \cdot (\mathcal{D}f[x])$.

Let us restrict f onto the hyperplane passing through x and directed by $\mathbf{S}$, i. e., let us consider the function $\mathbf{S} \to \mathbb{R}$, $\boldsymbol{q} \mapsto f(x + \boldsymbol{q}) = f(x + \boldsymbol{i} \cdot \boldsymbol{q})$. The derivative at zero of this function—according to the chain rule of differentiation of composite functions—is $(\mathcal{D}f[x]) \cdot \boldsymbol{i} = \boldsymbol{i}^* \cdot (\mathcal{D}f[x])$.

The completely split form by $(\boldsymbol{u}, o)$ of $\mathcal{D}f$ is obtained by the differentiation of the function (see equality (5.57))

$$\mathbb{T} \times \mathbf{S} \to \mathbb{R}, \quad (\boldsymbol{t}, \boldsymbol{q}) \mapsto \hat{f}(\boldsymbol{t}, \boldsymbol{q}) := f(o + \boldsymbol{t}\boldsymbol{u} + \boldsymbol{q}).$$

The “partial differentiation with respect to time” and the “partial differentiation with respect to space,” denoted by $\partial_{\mathbb{T}}$ and $\boldsymbol{\nabla}$, respectively, give the completely split form by $(\boldsymbol{u}, o)$:

$$(\partial_{\mathbb{T}} \hat{f},\ \boldsymbol{\nabla} \hat{f}).$$

A coordinated form of $\mathcal{D}f$ is obtained by the partial derivatives of the function (see Section 5.10),

$$\mathbb{R}^4 \to \mathbb{R}, \quad (\xi^0, \xi^1, \xi^2, \xi^3) \mapsto \tilde{f}(\xi^0, \xi^1, \xi^2, \xi^3) := f(o + \xi^i \boldsymbol{e}_i),$$

(Einstein summation), which are

$$\partial_k \hat{f} \quad (k = 0, 1, 2, 3).$$

Concisely,

$$\mathcal{D}f \sim (\mathcal{D}_u f,\ \nabla f) \sim (\partial_{\mathbb{T}} \hat{f},\ \boldsymbol{\nabla} \hat{f}) \sim \partial_k \tilde{f} \quad (k = 0, 1, 2, 3).$$

7.2.2 Vector fields, tensor fields

In general,

the differentiation $\mathcal{D}$ can be considered a symbolic covector for which we have

$$\mathcal{D} \sim (\mathcal{D}_u,\ \nabla) \sim (\partial_{\mathbb{T}},\ \boldsymbol{\nabla}) \sim \partial_k \quad (k = 0, 1, 2, 3).$$

1. The derivative of a vector field $\boldsymbol{J} : \mathrm{M} \to \mathbf{M}$ is the mixed tensor field $\mathcal{D}\boldsymbol{J}$. According to the mathematical supplement (see Subsection 18.6.3), it is more suitable to use its transpose, $\mathcal{D} \otimes \boldsymbol{J} : \mathrm{M} \to \mathbf{M}^* \otimes \mathbf{M}$. Its half-split form by $\boldsymbol{u}$ is (see (5.46)) the function $\mathrm{M} \to (\mathbb{T}^* \times \mathbf{S}^*) \otimes (\mathbb{T} \times \mathbf{S})$,

$$\begin{pmatrix} \boldsymbol{u} \cdot (\mathcal{D} \otimes \boldsymbol{J}) \cdot \boldsymbol{\tau}^* & \boldsymbol{u} \cdot (\mathcal{D} \otimes \boldsymbol{J}) \cdot \boldsymbol{\pi}_{\boldsymbol{u}}^* \\ \boldsymbol{i}^* \cdot (\mathcal{D} \otimes \boldsymbol{J}) \cdot \boldsymbol{\tau}^* & \boldsymbol{i}^* \cdot (\mathcal{D} \otimes \boldsymbol{J}) \cdot \boldsymbol{\pi}_{\boldsymbol{u}}^* \end{pmatrix} = \begin{pmatrix} \mathcal{D}_{\boldsymbol{u}}(\boldsymbol{\tau} \cdot \boldsymbol{J}) & \mathcal{D}_{\boldsymbol{u}}(\boldsymbol{\pi}_{\boldsymbol{u}} \cdot \boldsymbol{J}) \\ \nabla(\boldsymbol{\tau} \cdot \boldsymbol{J}) & \nabla \otimes (\boldsymbol{\pi}_{\boldsymbol{u}} \cdot \boldsymbol{J}) \end{pmatrix}.$$

Then with (7.1) we have

$$\mathcal{D} \otimes \boldsymbol{J} \sim \begin{pmatrix} \mathcal{D}_{\boldsymbol{u}} \rho & \mathcal{D}_{\boldsymbol{u}} \boldsymbol{j}_{\boldsymbol{u}} \\ \nabla \rho & \nabla \otimes \boldsymbol{j}_{\boldsymbol{u}} \end{pmatrix} \sim \begin{pmatrix} \partial_{\mathbb{T}} \hat{\rho} & \partial_{\mathbb{T}} \hat{\boldsymbol{j}}_{\boldsymbol{u}} \\ \boldsymbol{\nabla} \hat{\rho} & \boldsymbol{\nabla} \otimes \hat{\boldsymbol{j}}_{\boldsymbol{u}} \end{pmatrix} \sim \partial_i \tilde{j}^k \quad (i, k = 0, 1, 2, 3).$$

The trace of the derivative of the vector field is meaningful; in this way, we define the **divergence** of $\boldsymbol{J}$:

$$(\mathcal{D} \cdot \boldsymbol{J})[x] := \operatorname{Tr}((\mathcal{D} \otimes \boldsymbol{J})[x])$$

for which

$$\mathcal{D} \cdot \boldsymbol{J} \sim \mathcal{D}_{\boldsymbol{u}} \rho + \nabla \cdot \boldsymbol{j}_{\boldsymbol{u}} \sim \partial_{\mathbb{T}} \hat{\rho} + \boldsymbol{\nabla} \cdot \hat{\boldsymbol{j}}_{\boldsymbol{u}} \sim \partial_i \tilde{j}^i.$$

The derivative of a covector field $\boldsymbol{K} : \mathrm{M} \to \mathbf{M}^*$ is the cotensor field $\mathcal{D}\otimes\boldsymbol{K}$. Its half-split form by $\boldsymbol{u}$ is the function $\mathrm{M} \to (\mathbb{T}^* \times \mathbf{S}^*) \otimes (\mathbb{T}^* \times \mathbf{S}^*)$,

$$\begin{pmatrix} \boldsymbol{u}\cdot(\mathcal{D}\otimes\boldsymbol{K})\cdot\boldsymbol{u} & \boldsymbol{u}\cdot(\mathcal{D}\otimes\boldsymbol{K})\cdot\boldsymbol{i} \\ \boldsymbol{i}^*\cdot(\mathcal{D}\otimes\boldsymbol{K})\cdot\boldsymbol{u} & \boldsymbol{i}^*\cdot(\mathcal{D}\otimes\boldsymbol{K})\cdot\boldsymbol{i} \end{pmatrix} = \begin{pmatrix} \mathcal{D}_{\boldsymbol{u}}(\boldsymbol{K}\cdot\boldsymbol{u}) & \mathcal{D}_{\boldsymbol{u}}(\boldsymbol{K}\cdot\boldsymbol{i}) \\ \nabla(\boldsymbol{K}\cdot\boldsymbol{u}) & \nabla\otimes(\boldsymbol{K}\cdot\boldsymbol{i}) \end{pmatrix}.$$

Then with (7.3) we have

$$\mathcal{D}\otimes\boldsymbol{K} \sim \begin{pmatrix} -\mathcal{D}_{\boldsymbol{u}}V_{\boldsymbol{u}} & \mathcal{D}_{\boldsymbol{u}}\boldsymbol{A} \\ -\nabla V_{\boldsymbol{u}} & \nabla\otimes\boldsymbol{A} \end{pmatrix} \sim \begin{pmatrix} -\partial_{\mathbb{T}}\widehat{V}_{\boldsymbol{u}} & \partial_{\mathbb{T}}\widehat{\boldsymbol{A}} \\ -\boldsymbol{\nabla}\widehat{V}_{\boldsymbol{u}} & \boldsymbol{\nabla}\otimes\widehat{\boldsymbol{A}} \end{pmatrix} \sim \partial_i\tilde{K}_k \quad (i,k = 0,1,2,3).$$

The antisymmetric part of the derivative of the covector field is meaningful; in this way, we define the **exterior derivative** of $\boldsymbol{K}$:

$$\mathcal{D}\wedge\boldsymbol{K} := \mathcal{D}\otimes\boldsymbol{K} - (\mathcal{D}\otimes\boldsymbol{K})^*,$$

for which (see the notation in Subsection 5.7.1)

$$\begin{aligned} \mathcal{D}\wedge\boldsymbol{K} &\sim ((-\nabla V_{\boldsymbol{u}} - \mathcal{D}_{\boldsymbol{u}}\boldsymbol{A}, \nabla\boldsymbol{A})) \\ &\sim ((-\boldsymbol{\nabla}\widehat{V}_{\boldsymbol{u}} - \partial_{\mathbb{T}}\widehat{\boldsymbol{A}}, \boldsymbol{\nabla}\wedge\widehat{\boldsymbol{A}})) \sim \partial_i\tilde{K}_k - \partial_k\tilde{K}_i \quad (i,k = 0,1,2,3). \end{aligned}$$

Keep in mind:
- *A vector field has divergence and has no exterior derivative.*
- *A covector field has exterior derivative and has no divergence.*

2. An antisymmetric tensor field $\boldsymbol{G} : \mathrm{M} \to \mathbf{M}\wedge\mathbf{M}$, too, has divergence having values in $\mathbf{M}$; if (see Section 5.10)

$$\boldsymbol{G} \sim ((\boldsymbol{D}, \boldsymbol{H}_{\boldsymbol{u}})) \sim ((\widehat{\boldsymbol{D}}, \widehat{\boldsymbol{H}}_{\boldsymbol{u}})) \sim \tilde{G}^{ik},$$

then

$$\mathcal{D}\cdot\boldsymbol{G} \sim (\nabla\cdot\boldsymbol{D} - \mathcal{D}_{\boldsymbol{u}}\boldsymbol{D} + \nabla\cdot\boldsymbol{H}_{\boldsymbol{u}}) \sim (\boldsymbol{\nabla}\cdot\widehat{\boldsymbol{D}} - \partial_{\mathbb{T}}\widehat{\boldsymbol{D}} + \boldsymbol{\nabla}\cdot\widehat{\boldsymbol{H}}_{\boldsymbol{u}}) \sim \partial_i G^{ik}. \qquad (7.5)$$

An antisymmetric cotensor field $\boldsymbol{F} : \mathrm{M} \to \mathbf{M}^*\wedge\mathbf{M}^*$ has its exterior derivative $\mathcal{D}\wedge\boldsymbol{F}$ having values in $\mathbf{M}^*\wedge\mathbf{M}^*\wedge\mathbf{M}^*$; if

$$\boldsymbol{F} \sim ((\boldsymbol{E}_{\boldsymbol{u}}, \boldsymbol{B})) \sim ((\widehat{\boldsymbol{E}}_{\boldsymbol{u}}, \widehat{\boldsymbol{B}})) \sim \tilde{F}_{ik},$$

then

$$\begin{aligned} \mathcal{D}\wedge\boldsymbol{F} &\sim (((\nabla\wedge\boldsymbol{E}_{\boldsymbol{u}} + \mathcal{D}_{\boldsymbol{u}}\boldsymbol{B}, \nabla\wedge\boldsymbol{B}))) \\ &\sim (((\boldsymbol{\nabla}\wedge\widehat{\boldsymbol{E}}_{\boldsymbol{u}} + \partial_{\mathbb{T}}\widehat{\boldsymbol{B}}, \boldsymbol{\nabla}\wedge\widehat{\boldsymbol{B}}))) \sim \partial_j F_{ik} + \partial_k F_{ji} + \partial_i F_{kj} \end{aligned} \qquad (7.6)$$

where the three parentheses means that the two quantities inside determine the whole antisymmetric tensor.

Keep in mind:
- *An antisymmetric tensor field has divergence and has no exterior derivative.*
- *An antisymmetric cotensor field has exterior derivative and has no divergence.*

7.3 Maxwell equations

7.3.1 Relative Maxwell equations

The usual form of Maxwell equations is

$$\operatorname{div}\boldsymbol{D} = \rho, \quad -\frac{\partial \boldsymbol{D}}{\partial t} + \operatorname{curl}\underline{\boldsymbol{H}} = \boldsymbol{j}, \tag{7.7}$$

$$\frac{\partial \underline{\boldsymbol{B}}}{\partial t} + \operatorname{curl}\boldsymbol{E} = 0, \quad \operatorname{div}\underline{\boldsymbol{B}} = 0 \tag{7.8}$$

where the quantities are functions of "time" and "space," so it is obvious that these are relative equations with respect an inertial frame.

Now we look for whether there are absolute Maxwell equations.

First step:
According to Subsection 18.6.4, we replace

$$\operatorname{div}\boldsymbol{D} \text{ by } \nabla\cdot\boldsymbol{D}, \quad \operatorname{curl}\boldsymbol{E} \text{ by } \nabla\times\boldsymbol{E},$$
$$\operatorname{div}\underline{\boldsymbol{B}} \text{ by } \nabla\cdot\underline{\boldsymbol{B}}, \quad \operatorname{curl}\underline{\boldsymbol{H}} \text{ by } \nabla\times\underline{\boldsymbol{H}}.$$

Second step:
The values of the magnetic quantities $\underline{\boldsymbol{H}}$ and $\underline{\boldsymbol{B}}$ are "axial vectors," which are, in fact, antisymmetric three-tensors (see Subsection 18.3.3). Without doubt, the antisymmetric tensors instead of axial vectors make some formulas more complicated but they are correct from a theoretical point of view. $\boldsymbol{H}$ and $\boldsymbol{B}$ will denote the corresponding antisymmetric tensors. Then we replace

$$\nabla\times\underline{\boldsymbol{H}} \text{ by } \nabla\cdot\boldsymbol{H}, \quad \nabla\cdot\underline{\boldsymbol{B}} \text{ by } \nabla\wedge\boldsymbol{B}.$$

Accordingly, $\frac{\partial \boldsymbol{B}}{\partial t}$ is an antisymmetric three-tensor, so

$$\nabla\times\boldsymbol{E}\ (\operatorname{curl}\boldsymbol{E}) \text{ must be replaced by } \nabla\wedge\boldsymbol{E}.$$

Third step:
ρ is the charge density and $\boldsymbol{j} = \rho\boldsymbol{v}$ is the charge current density where $\boldsymbol{v}$ is the relative velocity field of the charges. Putting them together, we have $\rho(1, \boldsymbol{v})$, which shows clearly that it is the split form of an absolute current density $\boldsymbol{J} = \rho\boldsymbol{U}$ where $\boldsymbol{U}$ is the absolute velocity field of the charge flow.

Fourth step:
Now we name the reference frame to which (7.7) and (7.8) apply; let it be $(\boldsymbol{u}, o)$. Then every quantity in those equations is a completely $(\boldsymbol{u}, o)$-split component of some absolute quantity. To make simpler the formula, we will back to half-split components by $\boldsymbol{u}$.

$$\rho = \tau \cdot \boldsymbol{J} \quad \text{is the time-like component of } \boldsymbol{J},$$

and $\boldsymbol{j}$ is, in fact,

$$\boldsymbol{j}_u = \pi_u \cdot \boldsymbol{J} \quad \text{is the } \boldsymbol{u}\text{-space-like component of } \boldsymbol{J}.$$

As a consequence, the left sides of (7.7) and (7.8) must be a time-like component and a $\boldsymbol{u}$-space-like component of an absolute vector field, respectively. This means that $\boldsymbol{D}$ and $\boldsymbol{H}$ should somehow come from some absolute quantity. We find from (5.41) that this absolute quantity is an antisymmetric tensor field $\boldsymbol{G}$:

$$\boldsymbol{D} = -\tau \cdot \boldsymbol{G} \ \text{ is the time-like component of } \boldsymbol{G},$$

and $\boldsymbol{H}$ is, in fact,

$$\boldsymbol{H}_u = \boldsymbol{G} - \boldsymbol{u} \wedge (\tau \cdot \boldsymbol{G}) \ \text{ is the } \boldsymbol{u}\text{-space-like component of } \boldsymbol{G}.$$

Similarly, we expect that $\boldsymbol{E}$ and $\boldsymbol{B}$ together come from an absolute quantity, and we find from (5.44) that this absolute quantity is an antisymmetric covector field $\boldsymbol{F}$:

$$\boldsymbol{B} = \boldsymbol{i}^* \cdot \boldsymbol{F} \cdot \boldsymbol{i} \ \text{ is the space-like component of } \boldsymbol{F},$$

and $\boldsymbol{E}$ is, in fact,

$$\boldsymbol{E}_u = \boldsymbol{i}^* \cdot \boldsymbol{F} \cdot \boldsymbol{u} \ \text{ is the } \boldsymbol{u}\text{-time-like component of } \boldsymbol{F},$$

$\frac{\partial}{\partial t}$ in (7.7) and (7.8) means the $\boldsymbol{u}$-time-like derivative; we get from (7.5) and (7.6) that the precise form of the $\boldsymbol{u}$-relative Maxwell equations is

$$\nabla \cdot \boldsymbol{D} = \rho, \quad -\mathcal{D}_u \boldsymbol{D} + \nabla \cdot \boldsymbol{H}_u = \boldsymbol{j}_u, \tag{7.9}$$

$$\mathcal{D}_u \boldsymbol{B} + \nabla \wedge \boldsymbol{E}_u = 0, \qquad \nabla \wedge \boldsymbol{B} = 0. \tag{7.10}$$

7.3.2 Absolute Maxwell equations

Then we can state that equations (7.9) and (7.10) are the $\boldsymbol{u}$-split form of the **absolute Maxwell equations**

$$\mathcal{D} \cdot \boldsymbol{G} = \boldsymbol{J}, \quad \mathcal{D} \wedge \boldsymbol{F} = 0. \tag{7.11}$$

$\boldsymbol{F}$ is the **electromagnetic field**, $\boldsymbol{G}$ is the **electromagnetic displacement**, and $\boldsymbol{J}$ is the **absolute current density**.

According to the second equation in (7.11), the electromagnetic field, under some "regularity" conditions, is the exterior derivative of a covector field, called **absolute potential:** $\boldsymbol{F} = \mathcal{D} \wedge \boldsymbol{K}$.

Our next task is to find the measure lines for the electromagnetic quantities.

The existence of the elementary electric charge makes it possible to measure electric charges by real numbers, i. e., we can choose the real line as the measure line of electric charges; though this does not fit the everyday practice, it makes simpler the formulas in the theory. Accordingly, the charge density is a function having values $\frac{\text{charge}}{\text{volume}}$, i. e., the measure line of ρ is $\frac{\mathbb{R}}{\mathbb{L}^{(3)}}$. Accordingly, the absolute current density, $\boldsymbol{J} = \rho \boldsymbol{U}$ is a vector field

$$\boldsymbol{J} : \mathrm{M} \to \frac{\mathbf{M}}{\mathbb{T} \otimes \mathbb{L}^{(3)}} \tag{7.12}$$

and

$$\rho = \tau \cdot \boldsymbol{J} : \mathrm{M} \to \frac{\mathbb{R}}{\mathbb{L}^{(3)}}, \quad \boldsymbol{j}_{\boldsymbol{u}} = \pi_{\boldsymbol{u}} \cdot \boldsymbol{J} : \mathrm{M} \to \frac{\mathbf{S}}{\mathbb{T} \otimes \mathbb{L}^{(3)}},$$

are the **absolute charge density** and the ***u*-current density**, respectively.

Then (7.12) and the first equation in (7.11) fix that the electromagnetic displacement is a function

$$\boldsymbol{G} : \mathrm{M} \to \frac{\mathbf{M} \wedge \mathbf{M}}{\mathbb{T} \otimes \mathbb{L}^{(3)}};$$

consequently, we have that

$$\boldsymbol{D} : \mathrm{M} \to \frac{\mathbf{S}}{\mathbb{L}^{(3)}} \quad \text{and} \quad \boldsymbol{H}_{\boldsymbol{u}} : \mathrm{M} \to \frac{\mathbf{S} \wedge \mathbf{S}}{\mathbb{T} \otimes \mathbb{L}^{(3)}}.$$

$\boldsymbol{E}_{\boldsymbol{u}}$ is the electric force acting on a unit charge and $\boldsymbol{B} \cdot \boldsymbol{v}$ ($\boldsymbol{v} \times \underline{\boldsymbol{B}}$) is the Lorentz force acting on a unit charge moving with velocity $\boldsymbol{v}$ with respect to the inertial frame $\boldsymbol{u}$. Correspondingly (see (6.4)),

$$\boldsymbol{E}_{\boldsymbol{u}} : \mathrm{M} \to \frac{\mathbf{S}^*}{\mathbb{T}} = \frac{\mathbf{S}}{\mathbb{T} \otimes \mathbb{L}^{(2)}} \quad \text{and} \quad \boldsymbol{B} : \mathrm{M} \to \mathbf{S}^* \wedge \mathbf{S}^* = \frac{\mathbf{S} \wedge \mathbf{S}}{\mathbb{L}^{(4)}},$$

which fix that the electromagnetic field is a function,

$$\boldsymbol{F} : \mathrm{M} \to \mathbf{M}^* \wedge \mathbf{M}^*.$$

Note the important fact that

$\boldsymbol{G}$ is a **tensor** *valued function,*

whereas

$\boldsymbol{F}$ is a **cotensor** *valued function,*

i. e., the electromagnetic displacement and the electromagnetic field are of different types.

7.4 Constitutive relations

7.4.1 Influence of a medium

(7.11) are differential equations for the electromagnetic quantities $\boldsymbol{G}$ and $\boldsymbol{F}$ determined by a **given** absolute current density $\boldsymbol{J}$.

The equation concerning $\boldsymbol{G}$ has four components in coordinates which, however, are not independent because the divergence of $\boldsymbol{J}$ is zero; so there are three independent components.

The equation concerning $\boldsymbol{F}$ has three components in coordinates.

Both $\boldsymbol{G}$ and $\boldsymbol{F}$ have six independent components.

As a consequence, (7.11) gives $3+3 = 6$ independent component equations for $6+6 = 12$ field components.

It is known, however, that $\boldsymbol{G}$ and $\boldsymbol{F}$ are related somehow by a material medium in which electromagnetic phenomena exist. Such relations are usually described for an infinite, homogeneous inertial medium as follows in our terms. If $\boldsymbol{u}_m$ is the absolute velocity of the medium, then

$$\boldsymbol{D} = \frac{1}{\epsilon\epsilon_0}\boldsymbol{E}_{\boldsymbol{u}_m}, \quad \boldsymbol{H}_{\boldsymbol{u}_m} = \mu\mu_0\boldsymbol{B}, \tag{7.13}$$

where ϵ_0 and μ_0 are positive elements of $\frac{\mathbb{L}}{\mathbb{T}}$, which play the only role that the measure lines be in order in these equalities.

As a generalization, we accept that a **constitutive relation**,

$$\boldsymbol{G} = \Gamma(\boldsymbol{F}), \tag{7.14}$$

can reflect how a material medium in spacetime influences electromagnetism.

Then we get the **absolute Maxwell equations with a constitutive relation** in which only the electromagnetic field are included:

$$\mathcal{D} \cdot \Gamma(\boldsymbol{F}) = \boldsymbol{J}, \quad \mathcal{D} \wedge \boldsymbol{F} = 0.$$

7.4.2 What is the trouble?

Let us emphasize:

there are nonrelativistic absolute Maxwell equations

and the relative Maxwell equations (7.8), the split forms of the absolute ones (7.11), are the same for all inertial frames.

This is important because it is usually stated that the Maxwell equations "do not transform in a correct way" in the nonrelativistic case, i. e., they have different forms for different inertial frames, and this discrepancy led to the theory of relativity.

Our result is exact; what about then with the incorrect transformation?

Note that in both equalities of (7.13) one of the sides is independent of $\boldsymbol{u}_m$ and the other side depends on $\boldsymbol{u}_m$. It is trivial then that for any $\boldsymbol{u} \neq \boldsymbol{u}_m$ the $\boldsymbol{u}$-relative constitutive relation differs from the $\boldsymbol{u}_m$-relative one, $\boldsymbol{E}_{\boldsymbol{u}} \neq \frac{1}{\epsilon\epsilon_0}\boldsymbol{D}$ and $\boldsymbol{H}_{\boldsymbol{u}} \neq \mu\mu_0\boldsymbol{B}$. This is evident from a physical point of view: the observer $\boldsymbol{u}_m$ is distinguished by the really existing medium.

Let us consider the vacuum: there is no medium. It is usually stated that the permittivity and permeability of vacuum is 1. Therefore, one writes

$$\boldsymbol{E} = \frac{1}{\epsilon_0}\boldsymbol{D}, \quad \boldsymbol{H} = \mu_0\boldsymbol{B}.$$

Relative quantities are included in these equalities but the symbol referring to an inertial frame is missing, so this formula is not right; the correct constitutive relation would be

$$\boldsymbol{E}_{\boldsymbol{u}_v} = \frac{1}{\epsilon_0}\boldsymbol{D}, \quad \boldsymbol{H}_{\boldsymbol{u}_v} = \mu_0\boldsymbol{B},$$

where $\boldsymbol{u}_v$ is the "absolute velocity of the vacuum."

The trouble is that vacuum (no medium) has no absolute velocity. To make sense to such a constitutive relation, a fictitious "neutral" medium, the **ether** was invented to replace vacuum.

Resuming:

1. The constitutive relation (7.14) establishes a relation between the electromagnetic displacement and the electromagnetic field where Γ is defined by an existing medium.
2. If the medium is homogeneous, then Γ is a linear map.
3. In vacuum there is no medium, so there is no Γ, i. e., there is no relation between the tensor valued $\boldsymbol{G}$ and the cotensor valued $\boldsymbol{F}$.
4. To make sense for the Maxwell equations in vacuum, the ether was invented to replace vacuum.

Item 3 is the consequence of that there is no distinguished linear bijection—identification—between $\mathbf{M}$ and $\mathbf{M}^*$ as it is emphasized in Subsection 5.1.5.

Then we can state:

The ether in nonrelativistic electromagnetism is necessary because of a simple algebraic fact: covectors cannot be identified with vectors.

8 Noninertial observers

Noninertial observers can be well treated in the nonrelativistic spacetime model, in particular, rigid observers, uniformly accelerated observers, uniformly rotating observers, and corresponding notions such as inertial forces, centrifugal forces, Coriolis forces. All these can be found in the book *Spacetime Without Reference Frames*.

https://doi.org/10.1515/9783112219553-010

Part III: **Absolute light propagation**

In this part, we treat a fair flat spacetime model, the **special relativistic spacetime model** arising from everyday experiences and an important experimental fact on light propagation. The way to it is somewhat more complicated than to the nonrelativistic model but is highly instructive and it is worth going through; everything is well understandable even disregarding the proofs in small size.

Since we remain in the framework of flat spacetime models, we omit the adjective special.

https://doi.org/10.1515/9783112219553-011

9 Construction of the relativistic spacetime model

The construction is somewhat more complicated than the nonrelativistic one but it is highly instructive and it is worth going through. Everything is well understandable even disregarding the proofs in small size.

9.1 Basic experiences

9.1.1 Light signals

Up to now, we have only considered histories of material points leading us to the notion of world lines in the model. According to our experience, a "point-like light packet," let us call it a **light signal**, is similar to a material point in some respect: it moves in our space. In other respect, a light signal is different from a material point: it "never" can be at rest in the space of any observer.

The history of a light signal, too, is modeled by a curve in spacetime. The tangent vectors of such a curve cannot be time-like because then the light signal would be momentarily at rest in the space of an inertial observer.

As a consequence, the direction vectors of light signals in the nonrelativistic spacetime model would be in **S**, so the light signals would be simultaneously in every point of its path with respect to any observer; as a consequence, a light signal reflected on a mirror would arrive back at its source at the very moment of its start. This contradicts our experience. We can state:

Absolute simultaneity and the right description of light phenomena exclude each other.

9.1.2 Movement of light signals

Fundamental facts about light signals in vacuum are the following:

($\mathcal{L}$1). **Every straight line** *in an inertial space can be the* **path of a light signal**.

($\mathcal{L}$2). **The path of an unhindered light signal** *in an inertial space is a* **straight line**.

($\mathcal{L}$3). *A light signal is* **faster** *than any material point* **along the same path in the same direction**.

($\mathcal{L}$4). *The motion of any light signal can be* **arbitrarily approximated** *by the motion of a material point.*

The first and second statement do not require a comment. The third statement makes sense without a synchronization, as it is known (see Subsection 1.3.1).

https://doi.org/10.1515/9783112219553-012

The fourth statement makes the following sense (without a synchronization). Let us consider the motion of a light signal on a path between a start and a finish line. For an arbitrary (short) proper time interval t, there is a material body, which starts together with the light signal and arrives at the finish line t time after the light signal.

Though the third and fourth properties are simple, together they have a vital importance. First of all, they imply:

Light signals moving along the same path in the same direction are equally fast.

Indeed, if one of them were faster than an other then a material point whose motion approximates sufficiently the first light signal would be faster than the second light signal.

Then it follows:

The history of a light signal is independent of its source.

This can be illustrated simply as follows. There is a lamp on a train and a lamp beside the track on which the train rushes. When the lamps meet, they flash sending light along the track in the direction of the movement of the train: the two light signals will advance together.

9.1.3 Homogeneous and isotropic round-way light speed

The homogeneous and isotropic light propagation is the usual starting point of relativity theory "... the velocity of propagation of light in vacuo must be the same constant value $c = 3 \cdot 10^8$ m/s in every system of inertia" [12], which means that light signals in the space of an arbitrary inertial observer, starting anywhere and propagating in any direction, have the same speed. This assertion, however, refers to the speed of movements along different paths, which only makes sense if a synchronization is given.

Since absolute synchronization and light signals exclude each other, one would have to specify the synchronization, which the homogeneous and isotropic light propagation holds for (because if it holds for a synchronization then it does not hold for another one).

It is clear then that the homogeneous and isotropic light propagation without specifying a synchronization, cannot be accepted as a fundamental fact.

By the way, the homogeneous and isotropic light propagation is astonishing from the point of view of everyday thinking: if you move relative to me and something moves relative to me in the same direction, then that thing "must" move slower relative to you than to me.

Nevertheless, the homogenous and isotropic light propagation, in a convenient sense, is an experimental fact, which does not refer to a synchronization. This should be well known for a long time [13] but unfortunately it is not generally known. Namely, experiments regarding light speed concern **two-way** light propagation: a light signal starts at a space point (source) of an inertial observer, reflects at an other space point (mirror) and returns to the source. Knowing the distance between the source and the mirror, and measuring at the source the time elapsed between the start and the arrival, we can calculate the **two-way speed of light.**[1] It is evident that this procedure uses only the proper time in one space point and does not require a synchronization.

Varying the place of the source and the mirror, we can measure the two-way light speed in all directions and over all distances. Similarly, when the signal returns to the source after more reflections or even running round in a path, the round-way speed of light can be measured. The experimental facts about **homogeneous and isotropic round-way light propagation** can be stated as follows:

($\mathcal{L}$5). The round-way light speed *in an arbitrary inertial space is* **the same over all paths**.

This round-way light speed is $c := (2.99792458\ldots)10^8$ m/s.

The round-way speed does not imply anything on the one-way speed which requires a synchronization.

9.1.4 Measuring distances by time intervals

The homogeneous and isotropic round-way light propagation gives us a possibility to measure the distance between two points in the space of an inertial observer by the half of the elapsed proper time in one of the space points between the start and the return of a light signal reflecting from the other space point. This method is common in astronomy and nowadays it is used even in the everyday practice applying laser beams.

The formulas of the spacetime model will be simpler with this method of measuring distances. Consequently, in the relativistic spacetime model the measure line $\mathbb{L}$ of distances will coincide with the measure line $\mathbb{T}$ of time intervals in such a way that

$$\mathrm{m} := \frac{1}{(2.99792458\ldots)10^8}\,\mathrm{s} = 3.3335641\ldots 10^{-9}\,\mathrm{s};$$

according to this choice the round-way speed of light will be the real number 1.

9.2 Light-like vectors

We cannot state that some proper time is passing for a light signal (there is no experience in this respect). The relation earlier–later of material points, however, allows us to define such a relation for light signals, too: the reflected signal returns after the start. The relation earlier–later means an orientation, so we accept:

1 Both Fizeau and Foucault measured the light speed in century XIX essentially in this manner.

The life (history) of a light signal is a **connected, oriented curve** in spacetime, called a **light line**.

Similar to world lines, properties ($\mathcal{L}$**1**) and ($\mathcal{L}$**2**), together with the affine structure of spacetime, suggest us to accept:

The light line of a free light signal is an oriented straight line.

As said in connection with world lines, according to the affine structure of spacetime, arbitrary translation of a light line is a light line as well.

An oriented straight line is determined uniquely by an arbitrary point of its and a direction vector. According to the affine structure of spacetime, if an oriented straight line is a light line then its translations, too, are light lines, i. e., every straight line parallel to that and oriented by the same direction vector is a light line, too.

Accordingly, in order to determine the set of free light lines, we need to specify the set of the possible direction vectors.

The direction vectors of free light lines are called **future light-like.**

The set of future light-like vectors is denoted by $\mathrm{L}^{\rightarrow}$. Similar to future-like vectors, the positive multiples of future light-like vectors must be future light-like (in other words, the set of future light-like vectors is a cone whose apex is the zero vector) and the negative multiples of future light-like vectors must not be future light-like.

Properties of ($\mathcal{L}$**3**) and ($\mathcal{L}$**4**), similar to Subsection 2.5.2, can be formulated as follows:

($\mathcal{L}$**3**):
- For all $\boldsymbol{u} \in V(1)$
- For all $\boldsymbol{u}' \in V(1)$, $\boldsymbol{u}' \neq \boldsymbol{u}$, and for all positive $\boldsymbol{t}' \in \mathbb{T}$
- There are an $\boldsymbol{a} \in \mathrm{L}^{\rightarrow}$ and a positive $\boldsymbol{t} \in \mathbb{T}$

such that $\boldsymbol{a} + \boldsymbol{t}\boldsymbol{u} = \boldsymbol{t}'\boldsymbol{u}'$.

($\mathcal{L}$**4**):
- For all $\boldsymbol{u} \in V(1)$ and for all positive $\boldsymbol{t} \in \mathbb{T}$
- For all $\boldsymbol{a} \in \mathrm{L}^{\rightarrow}$
- There are a $\boldsymbol{u}' \in V(1)$ and a positive $\boldsymbol{t}' \in \mathbb{T}$

such that $\boldsymbol{a} + \boldsymbol{t}\boldsymbol{u} = \boldsymbol{t}'\boldsymbol{u}'$.

Then we can state:

The set of future light-like vectors is the boundary of the set of future-like vectors, except zero:

$$\mathrm{L}^{\rightarrow} = \partial \mathrm{T}^{\rightarrow} \setminus \{0\}.$$

$t'u'$ in ($\mathcal{L}4$) is a future-like vector depending on t, which will be denoted by $x(t)$; then $a+tu = x(t)$. The left side has a limit as t tends to zero; thus, the right side also has a limit and we find that $0 \neq a \in L^{\rightarrow}$ is in the boundary of $T^{\rightarrow}$, i. e., $L^{\rightarrow} \subset \partial T^{\rightarrow} \setminus \{0\}$.

Let $c \in \partial T^{\rightarrow} \setminus \{0\}$. Then there is a sequence $n \mapsto z_n \in T^{\rightarrow}$ such that $c = \lim_n z_n$. $T^{\rightarrow}$ being convex, $z_n + T^{\rightarrow}$ is a subset of $T^{\rightarrow}$; thus, $c + T^{\rightarrow}$ is a subset of the closure of $T^{\rightarrow}$ and $c + T^{\rightarrow} \subset T^{\rightarrow}$ holds as well because $c + T^{\rightarrow}$ is open. In particular, $a + T^{\rightarrow} \subset T^{\rightarrow}$ for $a \in L^{\rightarrow}$.

Taking an absolute velocity u and a positive s, applying statement ($\mathcal{L}3$) for $c + su$ in the role of $t'u'$, we have that there are a future light-like vector a—i. e., an element of $\partial T^{\rightarrow} \setminus \{0\}$—and a positive t such that $a + tu = c + su$. If $s > t$, then $a = c + (s - t)u$: the right side is in $T^{\rightarrow}$, the left side is in $\partial T^{\rightarrow}$, which is impossible. If $t > s$, then $a + (t - s)u = c$, which is impossible, too, because $a + T^{\rightarrow} \subset T^{\rightarrow}$. Consequently, $s = t$ and c is future light-like.

Then $L^{\leftarrow} := -L^{\rightarrow}$ is the collection of **past light-like vectors**, and $L := L^{\leftarrow} \cup L^{\rightarrow}$ is the collection of **light-like vectors.**

It is clear then that $L^{\rightarrow}$ is also a cone with apex at zero, i. e., if $a \in L^{\rightarrow}$ and α is a positive number, then $\alpha a \in L^{\rightarrow}$.

Further, $L^{\rightarrow} \cup T^{\rightarrow}$ is the closure of the convex set $T^{\rightarrow}$, except zero, so $L^{\rightarrow} \cup T^{\rightarrow}$, too, is a convex cone with apex at zero.

9.3 Round-way light propagation in the model

Let us take an inertial observer u. A light signal launched from a u-space point—source—and reflected in another u-space point—mirror—returns to the source. Let a and a' be the future light-like vectors of the light signal from the source to the mirror and from the mirror the source, respectively. Let t be the proper time period of the source between the start and the arrival of the light signals; then $tu = a + a'$ (see Figure 9.1).

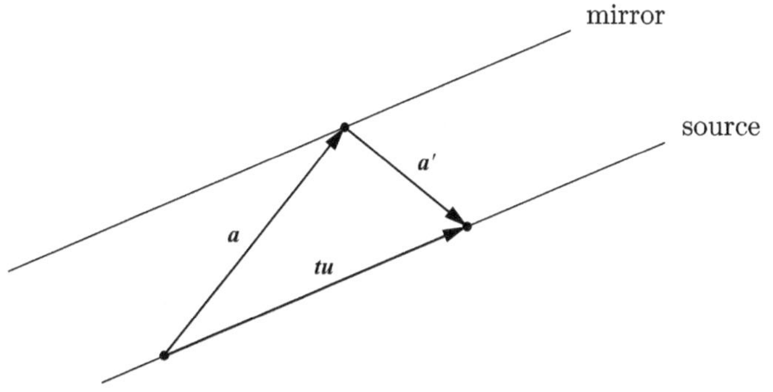

Figure 9.1: Two-way light propagation.

The distance between the source and the mirror is $\sqrt{b_u(a, a)} = \sqrt{b_u(a', a')}$ (see Section 2.4.4). According to the homogeneous and isotropic two-way light propagation—

property ($\mathcal{L}5$)—the time elapsed at the source between the departure and the return of the light signal equals

$$t = \frac{\sqrt{b_u(a,a)} + \sqrt{b_u(a',a')}}{c}$$

where c is the round-way speed of light. According to our agreement, we use measure lines such that $c = 1$; thus, the round-way speed will not be included in the following formulas.

Then we have that the homogeneous and isotropic two-way light propagation is characterized in our model as follows:

If $u \in V(1)$ and a, a' are elements of $L^{\to}$, such that for some $t \in \mathbb{T}$,

$$tu = a + a'. \quad \text{Then} \quad t = \sqrt{b_u(a,a)} + \sqrt{b_u(a',a')}. \tag{9.1}$$

In a similar way, considering a light signal reflected from more mirrors, we can formulate the homogeneous and isotropic round-way propagation of light in the model as follows:

If $u \in V(1)$, $n \geq 2$ is a natural number and $a_1, \ldots a_n$ are elements of $L^{\to}$, such that for some $t \in \mathbb{T}$,

$$tu = \sum_{k=1}^{n} a_k, \quad \text{then} \quad t = \sum_{k=1}^{n} \sqrt{b_u(a_k,a_k)}. \tag{9.2}$$

Then we have

$$\sum_{k=1}^{n} \left(a_k - \sqrt{b_u(a_k,a_k)}u \right) = 0. \tag{9.3}$$

As a consequence, the semi-Euclidean forms cannot be given independently of the future light-like vectors.

9.4 Space-like vectors

Formula (9.3) holds only for $n \geq 2$; n cannot be 1 because a light-like vector cannot be equal a time-like vector.

Nevertheless, this formula suggests us investigating the set of vectors given by a similar formula for $n = 1$, more closely, the set

$$\mathbf{S}_u := \left\{ a - \sqrt{b_u(a,a)}u \mid a \in L^{\to} \right\} \cup \{0\}. \tag{9.4}$$

We have the following important fact:

$\mathbf{S}_u$ is a three-dimensional linear subspace, which does not contain either time-like or light-like vectors; in particular, it is transverse to $\mathbb{T}\boldsymbol{u}$.

For the proof, we reformulate (**$\mathcal{L}$3**) and (**$\mathcal{L}$4**) in Section 9.2 in such a way that future-like vectors are considered instead of absolute velocities.

(A) *For all nonparallel*
- $\boldsymbol{x}, \boldsymbol{y} \in \mathrm{T}^{\rightarrow}$ $(\boldsymbol{x} := t'\boldsymbol{u}', \boldsymbol{y} := t'\boldsymbol{u})$,
- *there is a positive real number* α $(\alpha := \frac{t}{t'})$, *such that*

$$\boldsymbol{x} - \alpha\boldsymbol{y} \in \mathrm{L}^{\rightarrow}.$$

(B) *For all*
- $\boldsymbol{a} \in \mathrm{L}^{\rightarrow}$,
- $\boldsymbol{x} \in \mathrm{T}^{\rightarrow}$ $(\boldsymbol{x} := t'\boldsymbol{u})$,
- *positive real numbers* α $(\alpha := \frac{t}{t'})$,

$$\boldsymbol{a} + \alpha\boldsymbol{x} \in \mathrm{T}^{\rightarrow}.$$

Further, we have:

(C) *For all*
- $\boldsymbol{x} \in \mathrm{T}^{\rightarrow}$,
- $\boldsymbol{z} \in \mathbf{M}$ *nonparallel to* $\boldsymbol{x}$,
- *there are positive real numbers* α, β, *such that*

$$\boldsymbol{x} + \alpha\boldsymbol{z} \in \mathrm{L}^{\rightarrow} \quad \textit{and} \quad \beta\boldsymbol{x} + \boldsymbol{z} \in \mathrm{L}^{\rightarrow}.$$

Indeed, since $\mathrm{T}^{\rightarrow}$ is open, there is a positive λ, such that $\boldsymbol{x} + \lambda\boldsymbol{z} \in \mathrm{T}^{\rightarrow}$ but this cannot hold for every λ because $\mathrm{T}^{\rightarrow}$ does not contain a whole straight line. As a consequence, the supremum of such λ-s must be finite; let it be α. $\boldsymbol{x}+\alpha\boldsymbol{z}$ is in the boundary of $\mathrm{T}^{\rightarrow}$ and cannot be zero. Then $\boldsymbol{x}+\alpha\boldsymbol{z} = \alpha(\frac{1}{\alpha}\boldsymbol{x}+\boldsymbol{z})$; the vector in the parenthesis is also future light-like, and putting $\beta := \frac{1}{\alpha}$, we get the desired result.

Then we prove the statement as a collection of claims.

First, we claim that $\mathbf{S}_u$ is homogeneous, i. e., if $\boldsymbol{q} \in \mathbf{S}_u$ and $\alpha \in \mathbb{R}$, then $\alpha\boldsymbol{q} \in \mathbf{S}_u$. If $\alpha = 0$ or $\boldsymbol{q} = 0$, then the statement is trivial. In the following, we assume $\alpha \neq 0$ and $\boldsymbol{q} \neq 0$. If $\alpha > 0$, then the claim is implied by positive homogeneity of $\mathrm{L}^{\rightarrow}$. If $\alpha < 0$, then $\alpha\boldsymbol{q} = -|\alpha|\boldsymbol{q}$; thus, it suffices to show that $-\boldsymbol{q} \in \mathbf{S}_u$ whenever $\boldsymbol{q} \in \mathbf{S}_u$. Since $\boldsymbol{q} \neq 0$, there is an $\boldsymbol{a} \in \mathrm{L}^{\rightarrow}$ such that $\boldsymbol{q} = \boldsymbol{a} - \sqrt{\boldsymbol{b}_u(\boldsymbol{a},\boldsymbol{a})}\boldsymbol{u}$. Then by statement (C) there is a $\boldsymbol{t} \in \mathbb{T}^+$, such that $\boldsymbol{a}' := \boldsymbol{t}\boldsymbol{u} - \boldsymbol{a} \in \mathrm{L}^{\rightarrow}$ is future light-like, or equivalently $\boldsymbol{t}\boldsymbol{u} = \boldsymbol{a} + \boldsymbol{a}'$. Now (9.1) implies $-\boldsymbol{q} = -\boldsymbol{a} + \sqrt{\boldsymbol{b}_u(\boldsymbol{a},\boldsymbol{a})}\boldsymbol{u} = \boldsymbol{a}' - \sqrt{\boldsymbol{b}_u(\boldsymbol{a}',\boldsymbol{a}')}\boldsymbol{u} \in \mathbf{S}_u$.

Next, we claim that $\mathbf{S}_u$ is closed under addition, i. e., if $\boldsymbol{q}_1, \boldsymbol{q}_2 \in \mathbf{S}_u$ then $\boldsymbol{q}_1 + \boldsymbol{q}_2 \in \mathbf{S}_u$ as well. If $\boldsymbol{q}_2 = -\boldsymbol{q}_1$, then this is trivial so we will assume from now on that $\boldsymbol{q}_2 \neq -\boldsymbol{q}_1$. There are $\boldsymbol{a}_1, \boldsymbol{a}_2 \in \mathrm{L}^{\rightarrow}$, such that $\boldsymbol{a}_1 - \sqrt{\boldsymbol{b}_u(\boldsymbol{a}_1,\boldsymbol{a}_1)}\boldsymbol{u} = \boldsymbol{q}_1$ and $\boldsymbol{a}_2 - \sqrt{\boldsymbol{b}_u(\boldsymbol{a}_2,\boldsymbol{a}_2)}\boldsymbol{u} = \boldsymbol{q}_2$ and

$$\boldsymbol{q}_1 + \boldsymbol{q}_2 = \boldsymbol{a}_1 + \boldsymbol{a}_2 - \left(\sqrt{\boldsymbol{b}_u(\boldsymbol{a}_1,\boldsymbol{a}_1)} + \sqrt{\boldsymbol{b}_u(\boldsymbol{a}_2,\boldsymbol{a}_2)}\right)\boldsymbol{u} \neq 0.$$

Then $\boldsymbol{a}_1 + \boldsymbol{a}_2$ is not parallel to $\boldsymbol{u}$, so by statement (C) there is a $\boldsymbol{t} \in \mathbb{T}^+$ such that $\boldsymbol{a}_3 := \boldsymbol{t}\boldsymbol{u} - (\boldsymbol{a}_1 + \boldsymbol{a}_2) \in \mathrm{L}^{\rightarrow}$, i. e., $\boldsymbol{t}\boldsymbol{u} = \boldsymbol{a}_1 + \boldsymbol{a}_2 + \boldsymbol{a}_3$ and $\boldsymbol{t} = \sqrt{\boldsymbol{b}_u(\boldsymbol{a}_1,\boldsymbol{a}_2)} + \sqrt{\boldsymbol{b}_u(\boldsymbol{a}_2,\boldsymbol{a}_2)} + \sqrt{\boldsymbol{b}_u(\boldsymbol{a}_3,\boldsymbol{a}_3)}$ by (9.2). Consequently,

$$\boldsymbol{q}_1 + \boldsymbol{q}_2 = -\left(\boldsymbol{a}_3 - \sqrt{\boldsymbol{b}_u(\boldsymbol{a}_3,\boldsymbol{a}_3)}\right) \in -\mathbf{S}_u = \mathbf{S}_u.$$

So far, we have established that $\mathbf{S}_u$ is a linear subspace of $\mathbf{M}$. Now we claim $\dim \mathbf{S}_u \geq 3$. Since $\mathrm{T}^{\rightarrow}$ is an open set, there are $x_1, x_2, x_3 \in \mathrm{T}^{\rightarrow}$ which, together with a positive multiple of $\boldsymbol{u}$, form a basis in $\mathbf{M}$. Statement (A) implies the existence of positive elements $\boldsymbol{t}_1$, $\boldsymbol{t}_2$, $\boldsymbol{t}_3$ of $\mathbb{T}$, such that $\boldsymbol{a}_1 := x_1 - \boldsymbol{t}_1\boldsymbol{u}$, $\boldsymbol{a}_2 := x_2 - \boldsymbol{t}_2\boldsymbol{u}$, $\boldsymbol{a}_3 := x_3 - \boldsymbol{t}_3\boldsymbol{u}$ are in $\mathrm{L}^{\rightarrow}$. Then $\boldsymbol{b}_u(\boldsymbol{a}_k, \boldsymbol{a}_k) = \boldsymbol{b}_u(x_k, x_k)$ and the elements $\boldsymbol{a}_k - \sqrt{\boldsymbol{b}_u(\boldsymbol{a}_k, \boldsymbol{a}_k)}\boldsymbol{u} = x_k - (t_k + \sqrt{\boldsymbol{b}_u(x_k, x_k)})\boldsymbol{u}$ $(k = 1, 2, 3)$ in $\mathbf{S}_u$ are linearly independent.

Lastly, to show that $\mathbf{S}_u$ does not contain time-like and light-like vectors, it suffices to consider future-like and future light-like vectors. If $\boldsymbol{x} \in \mathrm{T}^{\rightarrow} \cup \mathrm{L}^{\rightarrow}$ were equal to a vector $\boldsymbol{a} - \sqrt{\boldsymbol{b}_u(\boldsymbol{a}, \boldsymbol{a})}\boldsymbol{u}$ in $\mathbf{S}_u$, then $\boldsymbol{a} \in \mathrm{L}^{\rightarrow}$ would be equal to $\boldsymbol{x} + \sqrt{\boldsymbol{b}_u(\boldsymbol{a}, \boldsymbol{a})}\boldsymbol{u}$, which is an element of $\mathrm{T}^{\rightarrow}$ according to statement (B).

The kernel of $\boldsymbol{b}_u$ is spanned by $\boldsymbol{u}$, therefore, we have

$$\boldsymbol{b}_u\Big(\boldsymbol{a} - \sqrt{\boldsymbol{b}_u(\boldsymbol{a}, \boldsymbol{a})}\boldsymbol{u}, \boldsymbol{a} - \sqrt{\boldsymbol{b}_u(\boldsymbol{a}, \boldsymbol{a})}\boldsymbol{u}\Big) = \boldsymbol{b}_u(\boldsymbol{a}, \boldsymbol{a}) > 0.$$

Consequently,

The restriction of $\boldsymbol{b}_u$ onto $\mathbf{S}_u$, being positive definite, is a Euclidean form.

Formula (9.4) actually says that

$$\mathrm{L}^{\rightarrow} \to \mathbf{S}_u \setminus \{0\}, \quad \boldsymbol{a} \mapsto \boldsymbol{a} - \sqrt{\boldsymbol{b}_u(\boldsymbol{a}, \boldsymbol{a})}\boldsymbol{u}, \tag{9.5}$$

is a surjection. It is an injection as well, so it is a bijection; its inverse is

$$\mathbf{S}_u \setminus \{0\} \to \mathrm{L}^{\rightarrow}, \quad \boldsymbol{q} \mapsto \boldsymbol{q} + \sqrt{\boldsymbol{b}_u(\boldsymbol{q}, \boldsymbol{q})}\boldsymbol{u}. \tag{9.6}$$

Suppose that $\boldsymbol{a} - \sqrt{\boldsymbol{b}_u(\boldsymbol{a}, \boldsymbol{a})}\boldsymbol{u} = \boldsymbol{a}' - \sqrt{\boldsymbol{b}_u(\boldsymbol{a}', \boldsymbol{a}')}\boldsymbol{u}$. Then $\sqrt{\boldsymbol{b}_u(\boldsymbol{a}, \boldsymbol{a})} = \sqrt{\boldsymbol{b}_u(\boldsymbol{a}', \boldsymbol{a}')}$ because if, say, $\sqrt{\boldsymbol{b}_u(\boldsymbol{a}', \boldsymbol{a}')} > \sqrt{\boldsymbol{b}_u(\boldsymbol{a}, \boldsymbol{a})}$, then $\boldsymbol{a}' = \boldsymbol{a} + (\sqrt{\boldsymbol{b}_u(\boldsymbol{a}', \boldsymbol{a}')} - \sqrt{\boldsymbol{b}_u(\boldsymbol{a}, \boldsymbol{a})})\boldsymbol{u}$, which is impossible because of statement (C) in Section 9.3. As a consequence, $\boldsymbol{a}' = \boldsymbol{a}$, (9.5) is a bijection. Let $\boldsymbol{q} := \boldsymbol{a} - \sqrt{\boldsymbol{b}_u(\boldsymbol{a}, \boldsymbol{a})}\boldsymbol{u}$. Since the kernel of $\boldsymbol{b}_u$ is spanned by $\boldsymbol{u}$, we have

$$\boldsymbol{b}_u(\boldsymbol{q}, \boldsymbol{q}) = \boldsymbol{b}_u(\boldsymbol{a}, \boldsymbol{a}), \tag{9.7}$$

so $\boldsymbol{q} + \sqrt{\boldsymbol{b}_u(\boldsymbol{q}, \boldsymbol{q})}\boldsymbol{u} = \boldsymbol{a}$: (9.6) is the inverse of (9.5).

Since $\mathbf{S}_u$ and $\mathbb{T}\boldsymbol{u}$ are complementary subspaces, there is a unique linear map

$$\boldsymbol{\tau}_u : \mathbf{M} \to \mathbb{T},$$

such that

$$\boldsymbol{\tau}_u \cdot \boldsymbol{u} = 1$$

and

$$\mathbf{S}_u = \{\boldsymbol{q} \in \mathbf{M} \mid \tau_u \cdot \boldsymbol{q} = 0\}.$$

If $\boldsymbol{u}$ and $\boldsymbol{u}'$ are different absolute velocities then

- The linear subspaces $\mathbf{S}_u$ and $\mathbf{S}_{u'}$ of $\mathbf{M}$ are different.
- Their intersection $\mathbf{S}_u \cap \mathbf{S}_{u'}$ is two-dimensional.
- The orthogonal complement of $\mathbf{S}_u \cap \mathbf{S}_{u'}$

$$\text{in } \mathbf{S}_u \text{ is } \mathbb{T}\big(\boldsymbol{u}' - (\tau_u \cdot \boldsymbol{u}')\boldsymbol{u}\big), \quad \text{in } \mathbf{S}_{u'} \text{ is } \mathbb{T}\big(\boldsymbol{u} - (\tau_{u'} \cdot \boldsymbol{u})\boldsymbol{u}'\big).$$

- The $\boldsymbol{u}$-length of $\boldsymbol{u}' - (\tau_u \cdot \boldsymbol{u}')\boldsymbol{u}$ and the $\boldsymbol{u}'$-length of $\boldsymbol{u} - (\tau_{u'} \cdot \boldsymbol{u})\boldsymbol{u}'$ are equal.

$\boldsymbol{a} \in \mathrm{L}^{\rightarrow}$ is not in $\mathbf{S}_u$, so $\tau_u(\boldsymbol{a}) \neq 0$ and (9.4) involves

$$\tau_u \cdot \boldsymbol{a} = \sqrt{\boldsymbol{b}_u(\boldsymbol{a}, \boldsymbol{a})} > 0. \tag{9.8}$$

If $\boldsymbol{q}$ is in $\mathbf{S}_u \cap \mathbf{S}_{u'}$, then $-\boldsymbol{q}$, too, is in this subspace; according to (9.6), $\pm\boldsymbol{q} + \sqrt{\boldsymbol{b}_{u'}(\boldsymbol{q}, \boldsymbol{q})}\boldsymbol{u}' \in \mathrm{L}^{\rightarrow}$; thus, applying (9.8) and then taking the square of the equality we get

$$\boldsymbol{b}_{u'}(\boldsymbol{q}, \boldsymbol{q})\big(\tau_u \cdot \boldsymbol{u}')\big)^2 = \boldsymbol{b}_u(\boldsymbol{q}, \boldsymbol{q}) \pm 2\sqrt{\boldsymbol{b}_{u'}(\boldsymbol{q}, \boldsymbol{q})}\boldsymbol{b}_u\big(\boldsymbol{q}, \boldsymbol{u}'\big) + \boldsymbol{b}_{u'}(\boldsymbol{q}, \boldsymbol{q})\boldsymbol{b}_u\big(\boldsymbol{u}', \boldsymbol{u}'\big), \tag{9.9}$$

which yields $\boldsymbol{b}_u(\boldsymbol{q}, \boldsymbol{u}') = 0$ (because $2\sqrt{\boldsymbol{b}_{u'}(\boldsymbol{q}, \boldsymbol{q})} \neq 0$), so $\boldsymbol{b}_u(\boldsymbol{q}, \boldsymbol{u}' - (\tau_u \cdot \boldsymbol{u}')\boldsymbol{u}) = 0$, too, i. e., $\boldsymbol{u}' - (\tau_u \cdot \boldsymbol{u}')\boldsymbol{u}$ is orthogonal to $\mathbf{S}_u \cap \mathbf{S}_{u'}$. Interchanging the roles of the absolute velocities, we get a similar result for $\boldsymbol{u}'$. $\mathbf{S}_u$ and $\mathbf{S}_{u'}$ are different three-dimensional linear subspaces in a four- dimensional vector space, so their intersection is necessarily two-dimensional.

Finally, according to (3.11),

$$\boldsymbol{b}_u\big(\boldsymbol{u}' - (\tau_u \cdot \boldsymbol{u}')\boldsymbol{u}, \boldsymbol{u}' - (\tau_u \cdot \boldsymbol{u}')\boldsymbol{u}\big) = \boldsymbol{b}_u\big(\boldsymbol{u}', \boldsymbol{u}'\big) = \boldsymbol{b}_{u'}(\boldsymbol{u}, \boldsymbol{u}) = \boldsymbol{b}_{u'}\big(\boldsymbol{u} - (\tau_{u'} \cdot \boldsymbol{u})\boldsymbol{u}', \boldsymbol{u} - (\tau_{u'} \cdot \boldsymbol{u})\boldsymbol{u}'\big).$$

Later we shall use

$$\boldsymbol{b}_{u'}(\boldsymbol{q}, \boldsymbol{q})(\tau_u \cdot \boldsymbol{u}')^2 = \boldsymbol{b}_u(\boldsymbol{q}, \boldsymbol{q}) + \boldsymbol{b}_{u'}(\boldsymbol{q}, \boldsymbol{q})\boldsymbol{b}_u(\boldsymbol{u}', \boldsymbol{u}') \quad (\boldsymbol{q} \in \mathbf{S}_u \cap \mathbf{S}_{u'}), \tag{9.10}$$

which is obtained from (9.9) because the middle term in its right side is zero.

9.5 Direction of movements of light signals

Every tangent vector of a light line is future light-like. It is convenient to introduce the set

$$\mathrm{V}^{\rightarrow} := \left\{\boldsymbol{w} \in \frac{\mathbf{M}}{\mathbb{T}} \;\middle|\; \mathbb{T}^{+}\boldsymbol{w} \subset \mathrm{L}^{\rightarrow}\right\};$$

its elements are called **absolute light directions** and are considered to be tangent vectors of light lines, as absolute velocities are considered to be tangents of world lines. Contrary to absolute velocities, however, if $\boldsymbol{w}$ is an absolute light direction, then $\lambda\boldsymbol{w}$ is an absolute light direction, too, for all positive numbers λ.

We have the following important property of light propagation:

A light signal that
- moves in $\boldsymbol{u}$-space in the same direction as $\boldsymbol{u}'$ moves,
- moves in $\boldsymbol{u}'$-space opposite to the direction as $\boldsymbol{u}$ moves.

According to Subsection 2.5.1, the path of a $\boldsymbol{u}'$-space point in $\boldsymbol{u}$-space is a straight line directed by $\boldsymbol{u}' + \mathbb{R}\boldsymbol{u} = \boldsymbol{u}' - (\boldsymbol{\tau}_{\boldsymbol{u}} \cdot \boldsymbol{u}')\boldsymbol{u} + \mathbb{R}\boldsymbol{u}$. Similar formula holds for the path of a $\boldsymbol{u}$-space point in $\boldsymbol{u}'$-space and even for the paths of a light line directed by $\boldsymbol{w}$. Since the absolute light directions are determined only up to a positive multiple, we have to prove: if

$$\boldsymbol{w} - (\boldsymbol{\tau}_{\boldsymbol{u}} \cdot \boldsymbol{w})\boldsymbol{u} = \boldsymbol{u}' - \left(\boldsymbol{\tau}_{\boldsymbol{u}} \cdot \boldsymbol{u}'\right)\boldsymbol{u}, \tag{9.11}$$

then there is a positive λ such that

$$\lambda\boldsymbol{w} - (\boldsymbol{\tau}_{\boldsymbol{u}'} \cdot \lambda\boldsymbol{w})\boldsymbol{u}' = -\left(\boldsymbol{u} - (\boldsymbol{\tau}_{\boldsymbol{u}'} \cdot \boldsymbol{u})\boldsymbol{u}'\right). \tag{9.12}$$

According to (9.11), $w = \boldsymbol{u}' + (\boldsymbol{\tau}_{\boldsymbol{u}} \cdot \boldsymbol{w} - \boldsymbol{\tau}_{\boldsymbol{u}} \cdot \boldsymbol{u}')\boldsymbol{u}$; thus,

$$\boldsymbol{\tau}_{\boldsymbol{u}} \cdot \boldsymbol{w} - \boldsymbol{\tau}_{\boldsymbol{u}} \cdot \boldsymbol{u}' < 0, \tag{9.13}$$

otherwise the right side would be a future-like vector, which cannot equal a future light-like vector. The $\boldsymbol{b}_{\boldsymbol{u}'}$ product of (9.11) and (9.12) by $\boldsymbol{w}$ results in

$$\boldsymbol{b}_{\boldsymbol{u}'}(\boldsymbol{w}, \boldsymbol{w}) = \left(\boldsymbol{\tau}_{\boldsymbol{u}} \cdot \boldsymbol{w} - \boldsymbol{\tau}_{\boldsymbol{u}} \cdot \boldsymbol{u}'\right)\boldsymbol{b}_{\boldsymbol{u}'}(\boldsymbol{w}, \boldsymbol{u}),$$
$$\lambda\boldsymbol{b}_{\boldsymbol{u}'}(\boldsymbol{w}, \boldsymbol{w}) := -\boldsymbol{b}_{\boldsymbol{u}'}(\boldsymbol{w}, \boldsymbol{u}).$$

Neither quantity above is zero, so we get $\lambda(\boldsymbol{\tau}_{\boldsymbol{u}} \cdot \boldsymbol{w} - \boldsymbol{\tau}_{\boldsymbol{u}} \cdot \boldsymbol{u}') = -1$ and (9.13) shows that $\lambda > 0$.

9.6 Boosts

Absolute light propagation suggests that the boosts be realized by the following physical procedure.

Let us consider two (sufficiently large) spaceships S and S' as two observers. Let:
(i) S' move in the direction $\boldsymbol{d}$ with respect to S.
(ii) S move in the direction $\boldsymbol{d}'$ with respect to S'.

According to the previous subsection, a light signal sent by S in the direction $\boldsymbol{d}$ arrives to S' in the direction $-\boldsymbol{d}'$. This simple fact gives the idea of making a correspondence (boost) between the space vectors of the two spaceships.

Let:
(i) P be a plane, orthogonal to $\boldsymbol{d}$, in the space of S.
(ii) P' be a plane, orthogonal to $\boldsymbol{d}'$, in the space of S'.

Any vector in the space of S' can be given as a sum of two components, one of them lying in P' and the other being parallel to $\boldsymbol{d}'$; thus, it suffices to give how these components are boosted.

Taking a vector $\boldsymbol{z}'$ in P', S' sends, in the direction of $\boldsymbol{d}'$:
(1) A low frequency light signal from the base point of $\boldsymbol{z}'$.
(2) A high frequency light signal from the end point of $\boldsymbol{z}'$.

These light signals hit the plane P and the vector $\boldsymbol{z}$ determined by the points of impact will correspond to $\boldsymbol{z}'$.

Then S' sends two high frequency light signals at once from the end point of $\boldsymbol{d}'$:
(1) One of them in the direction of $\boldsymbol{d}'$.
(2) The other in the direction of $-\boldsymbol{d}'$, which reflects at the base point of $\boldsymbol{d}'$.

The two light signals hit the plane P at a single point with a time delay from which S can calculate the length $|\boldsymbol{d}'|_{S'}$ of the vector $\boldsymbol{d}'$ (it does not matter now how this calculation runs); the vector $-\frac{|\boldsymbol{d}'|_{S'}}{|\boldsymbol{d}|_S}\boldsymbol{d}$ will correspond to $\boldsymbol{d}'$.

Similarly, the roles of the base point and the end point interchanged, the vector corresponding to $-\boldsymbol{d}$ is obtained by two law frequency light signals.

According to the described procedure and the previous subsection, if $\boldsymbol{u}$ and $\boldsymbol{u}'$ represent S and S' in the model, then:
(i) $\boldsymbol{u}' - (\boldsymbol{\tau}_{\boldsymbol{u}} \cdot \boldsymbol{u}')\boldsymbol{u}$ gives $\boldsymbol{d}$
(ii) $\boldsymbol{u} - (\boldsymbol{\tau}_{\boldsymbol{u}'} \cdot \boldsymbol{u})\boldsymbol{u}'$ gives $\boldsymbol{d}'$

and $\mathbf{S}_{\boldsymbol{u}} \cap \mathbf{S}_{\boldsymbol{u}'}$ corresponds to both P and P'.

Then the boost from $\boldsymbol{u}'$ to $\boldsymbol{u}$, supported also by the form of nonrelativistic boost, is defined by

$$\boldsymbol{B}_{\boldsymbol{u}\boldsymbol{u}'} \cdot \boldsymbol{u} := \boldsymbol{u}', \quad \boldsymbol{B}_{\boldsymbol{u}\boldsymbol{u}'} \cdot \boldsymbol{q} := \boldsymbol{q} \quad (\boldsymbol{q} \in \mathbf{S}_{\boldsymbol{u}} \cap \mathbf{S}_{\boldsymbol{u}'}), \tag{9.14}$$

$$\boldsymbol{B}_{\boldsymbol{u}\boldsymbol{u}'} \cdot (\boldsymbol{u} - (\boldsymbol{\tau}_{\boldsymbol{u}'} \cdot \boldsymbol{u})\boldsymbol{u}') = -(\boldsymbol{u}' - (\boldsymbol{\tau}_{\boldsymbol{u}} \cdot \boldsymbol{u}')\boldsymbol{u}). \tag{9.15}$$

It is trivial that this choice satisfies requirements (3.4), (3.7), and (3.8). (3.10) is satisfied with $\alpha_{\boldsymbol{u},\boldsymbol{u}'} = \boldsymbol{\tau}_{\boldsymbol{u}'} \cdot \boldsymbol{u}$, so

$$\boldsymbol{\tau}_{\boldsymbol{u}} \cdot \boldsymbol{u}' = \boldsymbol{\tau}_{\boldsymbol{u}'} \cdot \boldsymbol{u}. \tag{9.16}$$

As concerns (3.2) and (3.3), the inertial time progress and the semi-Euclidean forms will defined just on those formulas.

9.7 The absolute Lorentz form

We infer from the previous results:

There is an arrow oriented Lorentz form $\boldsymbol{g} : \mathbf{M} \times \mathbf{M} \to \mathbb{T} \otimes \mathbb{T}$ such that

$$\boldsymbol{g}(\boldsymbol{x},\boldsymbol{y}) = -(\boldsymbol{\tau}_{\boldsymbol{u}} \cdot \boldsymbol{x})(\boldsymbol{\tau}_{\boldsymbol{u}} \cdot \boldsymbol{y}) + \boldsymbol{b}_{\boldsymbol{u}}(\boldsymbol{x},\boldsymbol{y})$$

for all $\boldsymbol{u} \in \mathrm{V}(1)$.

Equalities (3.3) and (9.14) imply

$$\boldsymbol{b}_{u'}(\boldsymbol{q},\boldsymbol{q}) = \boldsymbol{b}_u(\boldsymbol{q},\boldsymbol{q}) \quad (\boldsymbol{q} \in \mathrm{S}_u \cap \mathrm{S}_{u'}) \tag{9.17}$$

which, together with equality (9.10) result in

$$-\left(\tau_u \cdot \boldsymbol{u}'\right)^2 + \boldsymbol{b}_u\left(\boldsymbol{u}',\boldsymbol{u}'\right) = -1. \tag{9.18}$$

An arbitrary vector $\boldsymbol{x} \in \mathbf{M}$ can be written in the form

$$\boldsymbol{x} = \boldsymbol{q} + t\boldsymbol{u} + t'\boldsymbol{u}'$$

where $\boldsymbol{q} \in \mathrm{S}_u \cap \mathrm{S}_{u'}$ and $\boldsymbol{t},\boldsymbol{t}' \in \mathbb{T}$.

Let $|\boldsymbol{q}|^2$ denote the quantity in (9.17). From (9.10), we infer

$$\boldsymbol{b}_u(\boldsymbol{x},\boldsymbol{x}) = |\boldsymbol{q}|^2 + \left(\boldsymbol{t}'\right)^2 \boldsymbol{b}_u\left(\boldsymbol{u}',\boldsymbol{u}'\right).$$

Further,

$$\tau_u \cdot \boldsymbol{x} = \boldsymbol{t} + \boldsymbol{t}'\tau_u \cdot \boldsymbol{u}'.$$

Finally, with (9.18) we arrive at

$$-(\tau_u \cdot \boldsymbol{x})^2 + \boldsymbol{b}_u(\boldsymbol{x},\boldsymbol{x}) = |\boldsymbol{q}|^2 - \boldsymbol{t}^2 - \left(\boldsymbol{t}'\right)^2 - 2\boldsymbol{t}\boldsymbol{t}'\tau_u \cdot \boldsymbol{u}'.$$

Interchanging the roles of $\boldsymbol{u}$ and $\boldsymbol{u}'$, we obtain a similar equality. As a consequence of (9.16),

$$-(\tau_u \cdot \boldsymbol{x})^2 + \boldsymbol{b}_u(\boldsymbol{x},\boldsymbol{x}) = -(\tau_{u'} \cdot \boldsymbol{x})^2 + \boldsymbol{b}_{u'}(\boldsymbol{x},\boldsymbol{x})$$

for all absolute velocities $\boldsymbol{u}$ and $\boldsymbol{u}'$ and world vectors $\boldsymbol{x}$.

This equality for $\boldsymbol{y}$ and $(\boldsymbol{x}+\boldsymbol{y})$ results in

$$-(\tau_u \cdot \boldsymbol{x})(\tau_u \cdot \boldsymbol{y}) + \boldsymbol{b}_u(\boldsymbol{x},\boldsymbol{y}) = -(\tau_{u'} \cdot \boldsymbol{x})(\tau_{u'} \cdot \boldsymbol{x}) + \boldsymbol{b}_{u'}(\boldsymbol{x},\boldsymbol{y}),$$

which holds for arbitrary $\boldsymbol{u}$ and $\boldsymbol{u}'$.

The Lorentz form is usually called the metric tensor.

For an absolute velocity $\boldsymbol{u}$, $\tau_u \cdot \boldsymbol{u} = 1$, and $\boldsymbol{b}_u(\boldsymbol{u},\boldsymbol{u}) = 0$; thus,

$$\mathrm{V}(1) = \left\{\boldsymbol{u} \in \frac{\mathbf{M}}{\mathbb{T}} \;\middle|\; \boldsymbol{g}(\boldsymbol{u},\boldsymbol{u}) = -1,\ \boldsymbol{u} \text{ has positive arrow}\right\},$$

which implies that

$\mathrm{T}^{\rightarrow} = \{\boldsymbol{x} \in \mathbf{M} \mid \boldsymbol{g}(\boldsymbol{x},\boldsymbol{x}) < 0,\ \boldsymbol{x}$ has positive arrow$\}$.

Then, because $\frac{\boldsymbol{x}}{P(\boldsymbol{x})} \in \mathrm{V}(1)$, i. e., $\boldsymbol{g}(\frac{\boldsymbol{x}}{P(\boldsymbol{x})}, \frac{\boldsymbol{x}}{P(\boldsymbol{x})}) = -1$ for $\boldsymbol{x} \in \mathrm{T}^{\rightarrow}$,

---

$P(\boldsymbol{x}) = \sqrt{-\boldsymbol{g}(\boldsymbol{x},\boldsymbol{x})}$.

It is trivial that $\boldsymbol{q}$ is in $\mathbf{S}_u$ if and only if $\boldsymbol{g}(\boldsymbol{u},\boldsymbol{q}) = 0$, which implies for an arbitrary world vector $\boldsymbol{x}$ that $\boldsymbol{\tau}_u \cdot \boldsymbol{x} = -\boldsymbol{g}(\boldsymbol{u},\boldsymbol{x})$; as a consequence,

---

$$\boldsymbol{b}_u(\boldsymbol{x},\boldsymbol{y}) = \boldsymbol{g}(\boldsymbol{x} - \boldsymbol{u}(\boldsymbol{\tau}_u \cdot \boldsymbol{x}), \boldsymbol{y} - \boldsymbol{u}(\boldsymbol{\tau}_u \cdot \boldsymbol{y})).$$

---

10 Structure of the relativistic spacetime model

The fair flat spacetime model constructed in the previous sections is based on ($\mathcal{L}1$)–($\mathcal{L}5$) besides the general assumptions.

In the following sections, we define and treat the relativistic spacetime model without referring to the way how we obtained it, so that even those who have skipped the previous sections can perfectly understand this model. That is why some previous formulas will appear again.

10.1 Fundamental properties of the model

10.1.1 New notation

Instead of the general notation $(\mathrm{M}, \mathbb{T}, \mathbb{L}, \mathrm{T}^{\rightarrow}, \boldsymbol{P}, \boldsymbol{b})$, a relativistic spacetime model will be referred to by the symbol

---

$$(\mathrm{M}, \mathbb{T}, \boldsymbol{g})$$

where

- M is spacetime, a four-dimensional oriented affine space (over the vector space **M**).
- $\mathbb{T}$ is the measure line of time periods and space distances.
- $\boldsymbol{g} : \mathbf{M} \times \mathbf{M} \to \mathbb{T} \otimes \mathbb{T}$ is an arrow oriented Lorentz form, which will be written as a dot product,

$$\boldsymbol{x} \cdot \boldsymbol{y} := \boldsymbol{g}(\boldsymbol{x}, \boldsymbol{y}).$$

Properties of the Lorentz form can be found in Section 18.4 of the mathematical supplement; most important relations are the reversed Cauchy inequality and the reversed triangle inequality for pseudolength.

Then

- The set of future-like vectors

$$\mathrm{T}^{\rightarrow} := \{\boldsymbol{x} \in \mathbf{M} \mid \boldsymbol{x} \cdot \boldsymbol{x} < 0,\ \boldsymbol{x} \text{ has positive arrow}\}$$

  is an open, convex cone with apex at zero.
- The inertial time progress:

$$\boldsymbol{P}(\boldsymbol{x}) := \sqrt{-\boldsymbol{x} \cdot \boldsymbol{x}} =: |\boldsymbol{x}| \quad (\boldsymbol{x} \in \mathrm{T}^{\rightarrow}) \tag{10.1}$$

  is positive homogeneous and continuous (even smooth).
- The set of absolute velocities

$$\mathrm{V}(1) = \left\{\boldsymbol{u} \in \frac{\mathbf{M}}{\mathbb{T}} \middle| \boldsymbol{u} \cdot \boldsymbol{u} = -1,\ \boldsymbol{u} \text{ has positive arrow}\right\} \tag{10.2}$$

https://doi.org/10.1515/9783112219553-013

is a three-dimensional submanifold; the reversed Cauchy inequality results in

$$-\boldsymbol{u}\cdot\boldsymbol{u}' \geq 1 \quad (\boldsymbol{u},\boldsymbol{u}' \in V(1)) \tag{10.3}$$

where equality occurs if and only if $\boldsymbol{u} = \boldsymbol{u}'$.

- The semi-Euclidean forms of inertial observers are given by

$$\boldsymbol{b}_{\boldsymbol{u}}(\boldsymbol{x},\boldsymbol{y}) := \boldsymbol{x}\cdot\boldsymbol{y} + (\boldsymbol{u}\cdot\boldsymbol{x})(\boldsymbol{u}\cdot\boldsymbol{y}) \quad (\boldsymbol{u} \in \mathrm{V}(1),\ \boldsymbol{x},\boldsymbol{y} \in \mathbf{M}); \tag{10.4}$$

we can rewrite this formula by introducing the notation

$$\tau_{\boldsymbol{u}}\cdot\boldsymbol{x} := -\boldsymbol{u}\cdot\boldsymbol{x} \tag{10.5}$$

and returning to $\boldsymbol{g}$ for a moment:

$$\boldsymbol{b}_{\boldsymbol{u}}(\boldsymbol{x},\boldsymbol{y}) = \boldsymbol{g}(\boldsymbol{x} - \boldsymbol{u}(\tau_{\boldsymbol{u}}\cdot\boldsymbol{x}), \boldsymbol{y} - \boldsymbol{u}(\tau_{\boldsymbol{u}}\cdot\boldsymbol{y})) \quad (\boldsymbol{u} \in \mathrm{V}(1),\ \boldsymbol{x},\boldsymbol{y} \in \mathbf{M}),$$

which is similar to (5.5).

- The set of **past-like vectors** and **time-like vectors** are

$$\mathrm{T}^{\leftarrow} := -\mathrm{T}^{\rightarrow}, \quad \mathrm{T}^{\leftarrow} \cup \mathrm{T}^{\rightarrow} = \{\boldsymbol{x} \in \mathbf{M} \mid \boldsymbol{x}\cdot\boldsymbol{x} < 0\}.$$

Note that the time progress and the Euclidean structures in the nonrelativistic model are given by the independent mathematical objects $\boldsymbol{\tau}$ and $\boldsymbol{h}$, whereas in the relativistic model they are derived from a single object, the Lorentz form $\boldsymbol{g}$.

Further,

-

$$\mathrm{L}^{\rightarrow} = \{\boldsymbol{x} \in \mathbf{M} \mid \boldsymbol{x}\cdot\boldsymbol{x} = 0,\ \boldsymbol{x} \neq \mathbf{0},\ \boldsymbol{x} \text{ has positive arrow}\} = \partial\mathrm{T}^{\rightarrow} \setminus \{0\} \tag{10.6}$$

is the set of **future light-like** vectors.

-

$$\mathrm{L}^{\leftarrow} := -\mathrm{L}^{\rightarrow}, \quad \mathrm{L} := \mathrm{L}^{\leftarrow} \cup \mathrm{L}^{\rightarrow} = \{\boldsymbol{x} \in \mathbf{M} \mid \boldsymbol{x}\cdot\boldsymbol{x} = 0,\ \boldsymbol{x} \neq \mathbf{0}\} \tag{10.7}$$

are the set of **past light-like** vectors and **light-like** vectors, respectively,

$$\boldsymbol{x}\cdot\boldsymbol{y} < 0 \quad (\boldsymbol{x} \in \mathrm{T}^{\rightarrow}, \boldsymbol{y} \in \mathrm{L}^{\rightarrow}). \tag{10.8}$$

- For an absolute velocity $\boldsymbol{u}$,

$$\mathbf{S}_{\boldsymbol{u}} := \{\boldsymbol{q} \in \mathbf{M} \mid \boldsymbol{u}\cdot\boldsymbol{q} = 0\} \tag{10.9}$$

is the set of **u-space-like** vectors which is a three-dimensional linear subspace; finally,

$$\mathrm{S} := \bigcup_{u \in \mathrm{V}(1)} \mathbf{S}_u = \{x \in \mathbf{M} \mid x \cdot x > 0\} \cup \{\mathbf{0}\}$$

is the set of **space-like** vectors.

10.1.2 Standard space vectors of inertial observers

Let us recall (see Subsection 2.4.3) that the space $\mathrm{S}_u = \mathrm{M}/\mathbb{T}u$ of the inertial observer u, the set of straight lines in M directed by u, is a three-dimensional affine space in a mathematically perfect way but the underlying vector space $\mathbf{M}/\mathbb{T}u$ cannot be illustrated in an eye-catching way.

Now we have a possibility to improve the situation with the aid of u-space-like vectors.

First of all, we note the orientation of $\mathbf{M}$ determines an orientation of $\mathbf{S}_u$ in a natural way: let an ordered basis (q_1, q_2, q_3) of $\mathbf{S}_u$ be positively oriented if (x, q_1, q_2, q_3) is a positively oriented basis of $\mathbf{M}$ for an arbitrary future-like vector x. It is not hard to demonstrate that this orientation is well-defined.

Elements of $\mathbf{M}/\mathbb{T}u$ are straight lines in $\mathbf{M}$, directed by u. Every such a line has a unique point in $\mathbf{S}_u$. The point $x + tu$ of $x + \mathbb{T}u$ is in $\mathbf{S}_u$ if $0 = -u \cdot (x + tu) = -u \cdot x + t$, i. e., $t = u \cdot x = -\tau_u \cdot x$. That is why it is convenient to represent the u-space vector $x + \mathbb{T}u$ by the element $x - (\tau_u \cdot x)u = x + (u \cdot x)u$ of $\mathbf{S}_u$.

In other words, introducing the projection onto $\mathbf{S}_u$ along $\mathbb{T}u$,

$$\pi_u := \mathbf{1} - u \otimes \tau_u = \mathbf{1} + u \otimes u : \mathbf{M} \to \mathbf{S}_u, \quad x \mapsto x + u(u \cdot x), \tag{10.10}$$

we make the identification

$$\mathbf{M}/\mathbb{T}u \equiv \mathbf{S}_u, \quad x + \mathbb{T}u \equiv \pi_u \cdot x.$$

This identification is correct, i. e., it is an orientation preserving linear bijection: the proof in Subsection 5.1.2 can be copied word by word.

Accordingly, the elements of $\mathbf{S}_u$ are called the **standard space vectors** of the inertial observer u.

With this identification, the subtraction in u-space S_u becomes

$$(y + \mathbb{T}u) - (x + \mathbb{T}u) := \pi_u \cdot (y - x) \in \mathbf{S}_u. \tag{10.11}$$

We have nice formulas by introducing the notation

$$\pi_u : \mathrm{M} \to \mathrm{S}_u, \quad \pi_u(x) := x + \mathbb{T}u; \tag{10.12}$$

then π_u is an affine map over π_u:

$$\pi_u(y) - \pi_u(x) := \pi_u \cdot (y - x). \tag{10.13}$$

Let u and u' be different absolute velocities. Then $\pi_u \cdot u'$ represents $u' + \mathbb{R}u$, the direction of motion of u' with respect to u (see Subsection 2.5.1). For later convenience, we introduce

$$v_{u'u} := \frac{\pi_u \cdot u'}{-u \cdot u'} = \frac{u'}{-u \cdot u'} - u \in \frac{\mathbf{S}_u}{\mathbb{T}} \tag{10.14}$$

representing the direction of motion of u' with respect to u,

$$v_{uu'} := \frac{\pi_{u'} \cdot u}{-u' \cdot u} = \frac{u}{-u' \cdot u} - u' \in \frac{\mathbf{S}_{u'}}{\mathbb{T}}$$

representing the direction of motion of u with respect to u'.

$v_{u'u}$ is not in $\frac{\mathbf{S}_{u'}}{\mathbb{T}}$ because $u' \cdot v_{u'u} \neq 0$ and $v_{uu'}$ is not in $\frac{\mathbf{S}_u}{\mathbb{T}}$ because $u \cdot v_{uu'} \neq 0$.

As a consequence, $\mathbf{S}_u \neq \mathbf{S}_{u'}$; their intersection $\mathbf{S}_u \cap \mathbf{S}_{u'}$ is a two-dimensional linear subspace.

If $q \in \mathbf{S}_u \cap \mathbf{S}_{u'}$, then $u \cdot q = u' \cdot q = 0$; therefore, both $v_{u'u}$ and $v_{uu'}$ are orthogonal to $\mathbf{S}_u \cap \mathbf{S}_{u'}$:

$$v_{u'u} \perp (\mathbf{S}_u \cap \mathbf{S}_{u'}) \perp v_{uu'}. \tag{10.15}$$

Then we can assert:

The spaces S_u and $S_{u'}$ of different inertial observers u and u' are **different** three-dimensional affine spaces over the **different** vector spaces $\mathbf{S}_u$ and $\mathbf{S}_{u'}$.

Finally, again for later convenience, we derive the following equalities:

$$|v_{u'u}|^2 = 1 - \frac{1}{(-u \cdot u')^2} = |v_{uu'}|^2$$

and

$$-u \cdot u' = \frac{1}{\sqrt{1 - |v_{u'u}|^2}} = \frac{1}{\sqrt{1 - |v_{uu'}|^2}} \tag{10.16}$$

is called the **relativistic factor** between u and u'.

10.1.3 Illustrations

The arithmetic spacetime model (see Section 10.2) suggests the following illustrations:

1. The world vectors are shown in Figure 10.1.

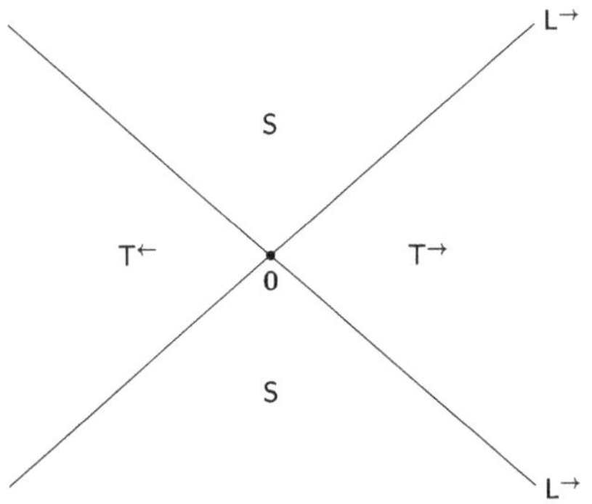

Figure 10.1: World vectors.

This figure is somewhat misleading because the set of time-like vectors seems similar to the set of space-like vectors; by rotating the figure around the "horizontal" axis, we get a better picture in three dimensions, which shows that the set of time-like vectors consists of two disjoint parts but the space-like vectors form a connected set.

2. V(1) is depicted as it is seen in Figure 10.2.

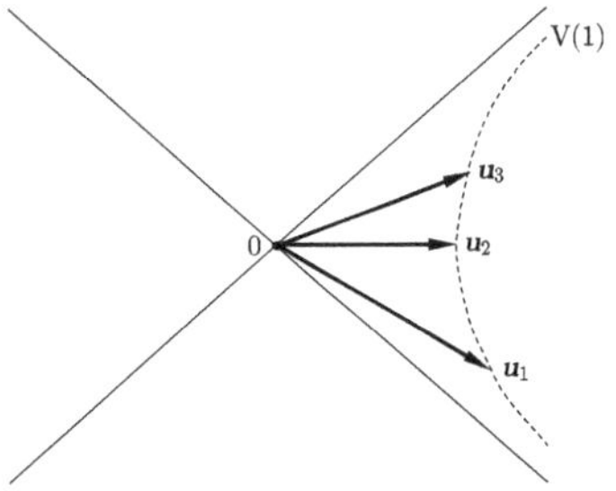

Figure 10.2: Absolute velocities.

Now we call attention again to that

- There is no zero absolute velocity
- An absolute velocity has no magnitude; in particular, it makes no sense that an absolute velocity is higher or lower than another one
- The angle between two absolute velocities is not meaningful

the illustration must not mislead us: $\boldsymbol{u}_1$ in Figure 5.2 is not higher than $\boldsymbol{u}_2$, the angle between $\boldsymbol{u}_1$ and $\boldsymbol{u}_2$ does not exist, and $\boldsymbol{u}_2$ is not a bisector.

Important note:

The equal time passed on straight lines with different absolute velocities are represented, in general, by sections of different lengths in our figures. The larger the angle between the horizontal line and the lines in question, the larger sections correspond to equal time periods; equal time periods correspond to equal sections in case of two absolute velocities that have the same angles—over and under—with the horizontal one. That is why two absolute velocities when treating their relations will be illustrated in this way. We emphasize that *"horizontal" and "angle" are not meaningful in the model, they are properties of the illustrations.*

3. $\mathbf{S}_{u}$ ($\mathbf{S}_{u'}$) is illustrated by a line, which does the same angle with the light cone as $\boldsymbol{u}$ ($\boldsymbol{u}'$) does (see Figure 10.3); it is emphasized that, of course, those angles have no physical meaning. Note that $\mathbf{S}_{u}$ and $\mathbf{S}_{u'}$ are different for different $\boldsymbol{u}$ and $\boldsymbol{u}'$.

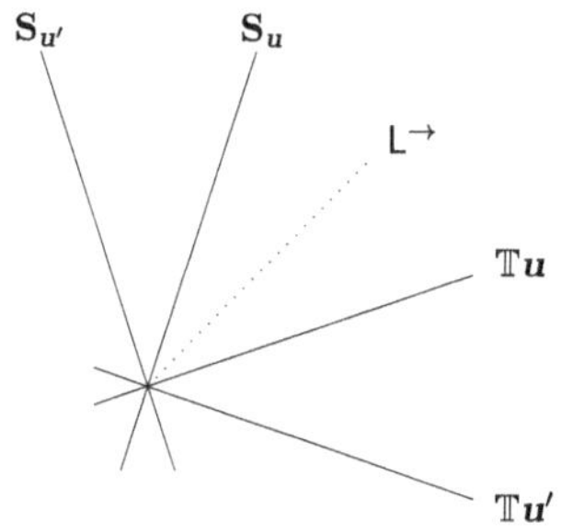

Figure 10.3: Space-like vectors.

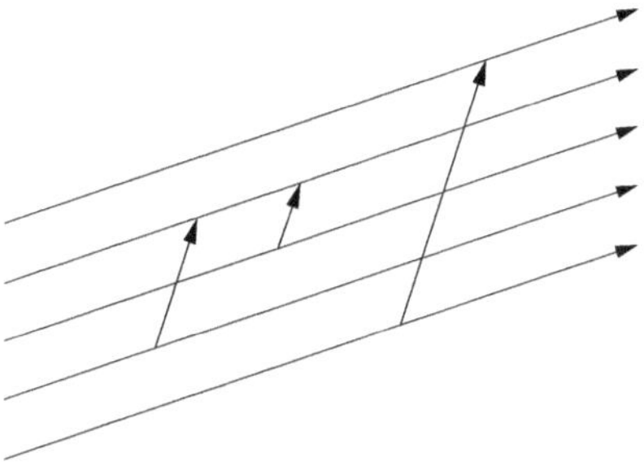

Figure 10.4: Standard space vectors of an inertial observer.

Figure 10.4 shows standard space vectors between space points of an inertial observer.

10.1.4 Boosts

The relativistic spacetime model is fair. The boost from $\boldsymbol{u}$ to $\boldsymbol{u}'$ is

$$\boldsymbol{B}_{\boldsymbol{u}\boldsymbol{u}'} := \mathbf{1} + \frac{(\boldsymbol{u}+\boldsymbol{u}')\otimes(\boldsymbol{u}+\boldsymbol{u}')}{1-\boldsymbol{u}\cdot\boldsymbol{u}'} - 2\boldsymbol{u}\otimes\boldsymbol{u}'; \quad (\boldsymbol{u}',\boldsymbol{u}\in \mathrm{V}(1)), \tag{10.17}$$

which depends continuously on $\boldsymbol{u}$ and $\boldsymbol{u}'$ and direct computations show that (the tensor product of absolute velocities is a linear map acting on vectors by Lorentz multiplication, e. g., $(\boldsymbol{u}\otimes\boldsymbol{u}')\cdot\boldsymbol{x} = \boldsymbol{u}(\boldsymbol{u}'\cdot\boldsymbol{x})$),

$$\boldsymbol{B}_{\boldsymbol{u}\boldsymbol{u}'} \cdot \boldsymbol{u}' = \boldsymbol{u}, \quad \boldsymbol{B}_{\boldsymbol{u}\boldsymbol{u}'} \cdot \boldsymbol{q} = \boldsymbol{q} \quad (\boldsymbol{q} \in \mathbf{S}_{\boldsymbol{u}} \cap \mathbf{S}_{\boldsymbol{u}'}), \tag{10.18}$$

$$\boldsymbol{B}_{\boldsymbol{u}\boldsymbol{u}'} \cdot \left(\frac{\boldsymbol{u}}{-\boldsymbol{u}' \cdot \boldsymbol{u}} - \boldsymbol{u}' \right) = -\left(\frac{\boldsymbol{u}'}{-\boldsymbol{u} \cdot \boldsymbol{u}'} - \boldsymbol{u} \right) \tag{10.19}$$

$$\boldsymbol{B}_{\boldsymbol{u}'\boldsymbol{u}} = \boldsymbol{B}_{\boldsymbol{u}\boldsymbol{u}'}^{-1}.$$

1. (10.18) and (10.19) involve that $\boldsymbol{B}_{\boldsymbol{u}\boldsymbol{u}'}$ preserves orientation and arrow orientation.
2. $\boldsymbol{B}_{\boldsymbol{u}\boldsymbol{u}'}$ preserves the Lorentz form because it maps Lorentz orthogonal vectors into Lorentz orthogonal vectors having the same length, consequently, it preserves the inertial time progress and the Euclidean structures.
3. It is trivial that $\boldsymbol{B}_{\boldsymbol{u}\boldsymbol{u}'}$ satisfies requirements (3.4), (3.7), and (3.8). (3.10) is satisfied with $\alpha_{\boldsymbol{u},\boldsymbol{u}'} = \boldsymbol{\tau}_{\boldsymbol{u}'} \cdot \boldsymbol{u} = -\boldsymbol{u}' \cdot \boldsymbol{u}$.

Contrary to the nonrelativistic case, however, *the boosts are not transitive*, in general:

$$\boldsymbol{B}_{\boldsymbol{u}''\boldsymbol{u}'} \cdot \boldsymbol{B}_{\boldsymbol{u}'\boldsymbol{u}} \neq \boldsymbol{B}_{\boldsymbol{u}''\boldsymbol{u}} \tag{10.20}$$

for the absolute velocities $\boldsymbol{u}$, $\boldsymbol{u}'$, and $\boldsymbol{u}''$, and *equality occurs if and only if the three absolute velocities are coplanar.*

The left side of (10.20) maps $\boldsymbol{u}$ to $\boldsymbol{u}''$: $\boldsymbol{B}_{\boldsymbol{u}'\boldsymbol{u}} \cdot \boldsymbol{u} = \boldsymbol{u}'$ and $\boldsymbol{B}_{\boldsymbol{u}\boldsymbol{u}''\boldsymbol{u}'} \cdot \boldsymbol{u}' = \boldsymbol{u}''$; thus, it is the boost from $\boldsymbol{u}$ to $\boldsymbol{u}''$ if and only if it is the identity on $\mathbf{S}_{\boldsymbol{u}} \cap \mathbf{S}_{\boldsymbol{u}''}$.

First of all, note that if $\boldsymbol{q} \in \mathbf{S}_{\boldsymbol{u}'} \cap \mathbf{S}_{\boldsymbol{u}} \cap \mathbf{S}_{\boldsymbol{u}''}$, then $\boldsymbol{B}_{\boldsymbol{u}''\boldsymbol{u}'} \cdot \boldsymbol{B}_{\boldsymbol{u}'\boldsymbol{u}} \cdot \boldsymbol{q} = \boldsymbol{B}_{\boldsymbol{u}''\boldsymbol{u}} \cdot \boldsymbol{q} = \boldsymbol{q}$.

If the absolute velocities are coplanar, $\boldsymbol{u}'$ is a linear combination of $\boldsymbol{u}$ and $\boldsymbol{u}''$, then $\mathbf{S}_{\boldsymbol{u}} \cap \mathbf{S}_{\boldsymbol{u}''} \subset \mathbf{S}_{\boldsymbol{u}'}$, i. e., $\mathbf{S}_{\boldsymbol{u}'} \cap \mathbf{S}_{\boldsymbol{u}} \cap \mathbf{S}_{\boldsymbol{u}''} = \mathbf{S}_{\boldsymbol{u}} \cap \mathbf{S}_{\boldsymbol{u}''}$; thus, equality occurs.

If the absolute velocities are not coplanar, then there is a $\boldsymbol{q}$ such that $\boldsymbol{u} \cdot \boldsymbol{q} = \boldsymbol{u}'' \cdot \boldsymbol{q} = 0$ but $\boldsymbol{u}' \cdot \boldsymbol{q} \neq 0$. For such a $\boldsymbol{q}$, we have $\boldsymbol{B}_{\boldsymbol{u}''\boldsymbol{u}} \cdot \boldsymbol{q} = \boldsymbol{q}$ and

$\boldsymbol{B}_{\boldsymbol{u}'\boldsymbol{u}} \cdot \boldsymbol{q} = \boldsymbol{q} + \frac{\boldsymbol{u}'+\boldsymbol{u}}{1-\boldsymbol{u}\cdot\boldsymbol{u}'}(\boldsymbol{u}' \cdot \boldsymbol{q})$ and

$$\begin{aligned} &\boldsymbol{B}_{\boldsymbol{u}''\boldsymbol{u}'} \cdot \left(\boldsymbol{q} + \frac{\boldsymbol{u}' + \boldsymbol{u}}{1 - \boldsymbol{u} \cdot \boldsymbol{u}'} (\boldsymbol{u}' \cdot \boldsymbol{q}) \right) \\ &= \boldsymbol{q} + \left(\frac{\boldsymbol{u}'' + \boldsymbol{u}'}{1 - \boldsymbol{u}'' \cdot \boldsymbol{u}'} - 2\boldsymbol{u}'' + \frac{\boldsymbol{u}' + \boldsymbol{u}}{1 - \boldsymbol{u} \cdot \boldsymbol{u}'} - \frac{(\boldsymbol{u}'' + \boldsymbol{u}')(1 - \boldsymbol{u} \cdot \boldsymbol{u}' - \boldsymbol{u} \cdot \boldsymbol{u}'' - \boldsymbol{u}'' \cdot \boldsymbol{u}')}{(1 - \boldsymbol{u}'' \cdot \boldsymbol{u}')(1 - \boldsymbol{u} \cdot \boldsymbol{u}')} + 2\boldsymbol{u}'' \right) (\boldsymbol{u}' \cdot \boldsymbol{q}), \end{aligned}$$

which does not equal $\boldsymbol{q}$ because the vector in the parenthesis, a linear combination of the absolute velocities, is not zero.

We can reformulate our result:

$$\boldsymbol{R}_{\boldsymbol{u}(\boldsymbol{u}'\boldsymbol{u}'')} := \boldsymbol{B}_{\boldsymbol{u}\boldsymbol{u}''} \cdot \boldsymbol{B}_{\boldsymbol{u}''\boldsymbol{u}'} \cdot \boldsymbol{B}_{\boldsymbol{u}'\boldsymbol{u}}$$

is the identity if and only if $\boldsymbol{u}$, $\boldsymbol{u}'$, and $\boldsymbol{u}''$ are coplanar.

$\boldsymbol{R}_{\boldsymbol{u}(\boldsymbol{u}'\boldsymbol{u}'')}$ preserves the Lorentz form, the orientation, and the arrow orientation, which maps $\boldsymbol{u}$ into $\boldsymbol{u}$; therefore, its restriction to $\mathbf{S}_{\boldsymbol{u}}$ is an orientation preserving Euclidean transformation; if the absolute velocities are not coplanar, then it is a rotation, called the **Thomas-rotation** in $\boldsymbol{u}$, determined by $\boldsymbol{u}'$ and $\boldsymbol{u}''$. The axis of this rotation is $\mathbf{S}_{\boldsymbol{u}} \cap \mathbf{S}_{\boldsymbol{u}'} \cap \mathbf{S}_{\boldsymbol{u}''}$.

Let us see the importance of the nontransitivity of boosts. A boost establishes a natural correspondence between different observer spaces, which makes sense to the parallelism of straight lines in different spaces (in particular, to the usual tacit assumption that the coordinate axes of different moving spatial coordinate systems are parallel).

Let A, B, and C be different space ships;

- If a straight line l_B in B's space and a straight line l_A in A's space are parallel.
- If a straight line l_C in C's space and the straight line l_B are parallel
  then
- The straight line l_C in C's space and the straight line l_A in A's space are not parallel, in general.

A paradox of relativity theory is based on the fact that one takes it for granted that parallelism between straight lines in different observer spaces is transitive (see Section 15.3).

10.1.5 Proper times

The history of a material point is a world line in spacetime. The proper time passed on the world line C between two of its points x and y is (see Subsection 2.3.3)

$$\boldsymbol{t}_C(x,y) = \int_{p^{-1}(x)}^{p^{-1}(y)} |\dot{p}(a)| da$$

where p is an arbitrary progressive parametrization and $|\ |$ denotes the pseudolength, i. e., $|\dot{p}(a)| = \sqrt{-\dot{p}(a) \cdot \dot{p}(a)}$.

Let y be future-like with respect to x, i. e., $y - x \in T^{\rightarrow}$. Let us take an approximation of the integral with a single point z between x and y. The result is the sum of the inertial time period from x to z and the inertial time period from z to y, $\boldsymbol{P}(z - x) + \boldsymbol{P}(y - z) = |z - x| + |y - z|$; according to the reversed triangle inequality, this is less than the inertial time period $|y-x|$ from x to y (except, of course, if the three world points are on a straight line). Taking further middle points for approximation, the result becomes less and less.

Finally:

If x and y are world points, y is future-like with respect to x, then the proper time passed between them on a noninertial world line C is shorter than the inertial proper time,

$$\boldsymbol{t}_C(x,y) < |y - x|.$$

10.1.6 Duals

The dual $\mathbf{M}^*$ of $\mathbf{M}$, the set of $\mathbf{M} \to \mathbb{R}$ linear maps is a four-dimensional vector space, which is strongly related to $\mathbf{M}$ by the Lorentz form $\boldsymbol{g}$; more closely, we make the **identification**

$$\frac{\mathbf{M}}{\mathbb{T} \otimes \mathbb{T}} \equiv \mathbf{M}^*, \quad \frac{\boldsymbol{x}}{s^2} \equiv \frac{\boldsymbol{g}(\boldsymbol{x}, \cdot)}{s^2}.$$

In another way, elements of $\mathbf{M}$ can be identified with linear maps $\mathbf{M} \to \mathbb{T} \otimes \mathbb{T}$: $\boldsymbol{x} \equiv \boldsymbol{g}(\boldsymbol{x}, \cdot)$. This corresponds to our notation that $\boldsymbol{g}$ is replaced by a dot product.

This identification is a significant difference between the nonrelativistic model and the relativistic one.

10.1.7 Exercise

Take the relativistic spacetime model in which

(1) Spacetime is $\mathbb{R}^4$ as an affine space over itself as a vector space with the standard orientation; let the world points and world vectors be denoted in the form $x = (\xi^1, \xi^2, \xi^3, \xi^4)$ and $\boldsymbol{x} = (x^1, x^2, x^3, x^4)$, respectively.
(2) The measure line of time periods is $\mathbb{R}$ with the standard orientation.
(3) The Lorentz form is

$$(x^1, x^2, x^3, x^4) \cdot (y^1, y^2, y^3, y^4) := -x^1y^3 - x^3y^1 + x^2y^2 + x^4y^4.$$

(i) Demonstrate that this is indeed a Lorentz form.
(ii) Define an arrow orientation of this Lorentz form.
(iii) What are the absolute velocities?
(iv) What are the space-like vectors corresponding to a given absolute velocity?
(v) Give the Lorentz-identification of vectors and covectors.

10.2 The arithmetic spacetime model

The following relativistic spacetime model, constructed with the aid of real numbers, is called the **arithmetic** one:

- $\mathrm{M} = \mathbb{R}^4$ with the standard orientation; then $\mathbf{M} = \mathbb{R}^4$ as well. To distinguish between the world points and world vectors, they are denoted in the form $x = (\xi^0, \xi^1, \xi^2, \xi^3)$ and $\boldsymbol{x} = (x^0, x^1, x^2, x^3)$, respectively.
- $\mathbb{T} = \mathbb{R}$ with the standard orientation.
- $\boldsymbol{g}$ is the usual Lorentz form on $\mathbb{R}^4$, $\boldsymbol{g}(\boldsymbol{x}, \boldsymbol{y}) = -x^0y^0 + x^1y^1 + x^2y^2 + x^3y^3$ which, of course, is called Lorentz product, too, and is denoted by $\boldsymbol{x} \cdot \boldsymbol{y}$.

The Lorentz identification $\mathbb{R}^4 \equiv (\mathbb{R}^4)^*$ is the map

$$(x^0, x^1, x^2, x^3) \mapsto (-x^0, x^1, x^2, x^3) =: (x_0, x_1, x_2, x_3);$$

accordingly, the Lorentz product is written in the form

$$\boldsymbol{x} \cdot \boldsymbol{y} = x_i y^i$$

where the Einstein summation is applied. Further,

$$\begin{aligned}
\mathrm{T}^{\rightarrow} &:= \{(x^0, x^1, x^2, x^3) \mid x_i x^i < 0,\ x^0 > 0\}, \\
\mathrm{L}^{\rightarrow} &= \{(x^0, x^1, x^2, x^3) \mid x_i x^i = 0,\ x^0 > 0\}, \\
\mathrm{S} &= \{(x^0, x^1, x^2, x^3) \mid x_i x^i > 0\} \cup \{0\}, \\
\mathrm{V}(1) &= \{(u^0, u^1, u^2, u^3) \mid u_i u^i = -1,\ u^0 > 0\}.
\end{aligned}$$

We will take absolute velocities of the form

$$\boldsymbol{u} := \gamma(1, v, 0, 0), \quad 0 \le v < 1, \quad \gamma := \frac{1}{\sqrt{1 - v^2}}$$

and $\boldsymbol{u}' := \gamma'(1, 0, v', 0)$, etc.

A clear distinction between vectors and covectors can be made in formulas if the vectors are represented by columns and covectors are represented by rows and the action of a covector on a vector is calculated by the usual role of matrix multiplication:

$$\begin{matrix} (k_0\ k_1\ k_2\ k_3) & \\ & \begin{pmatrix} x^0 \\ x^1 \\ x^2 \\ x^3 \end{pmatrix} \end{matrix}$$

$(-x^0\ x^1\ x^2\ x^3)$ is the covector corresponding to $\begin{pmatrix} x^0 \\ x^1 \\ x^2 \\ x^3 \end{pmatrix}$.

Accordingly, the vector-vector linear maps are represented by 4×4 matrices.

In the linear map $\boldsymbol{u}' \otimes \boldsymbol{u}$, the first absolute velocity is a vector (column) and the second one is a covector (row). Accordingly, $\boldsymbol{u}' \otimes \boldsymbol{u}$ is also obtained by the rule of matrix multiplication, e. g.,

$$\begin{pmatrix} \gamma \\ \gamma v \\ 0 \\ 0 \end{pmatrix} \begin{matrix} (-1 \quad 0 \quad 0 \quad 0) \\ \\ \\ \\ \end{matrix} = \begin{pmatrix} -\gamma & 0 & 0 & 0 \\ -\gamma v & 0 & 0 & 0 \\ 0 & 0 & 0 & 0 \\ 0 & 0 & 0 & 0 \end{pmatrix}$$

and, of course, the other tensor products $\boldsymbol{u} \otimes \boldsymbol{u}$, $\boldsymbol{u}' \otimes \boldsymbol{u}$, etc. have to be treated similarly.

Since the last two coordinates are zero, we consider only the first two ones provisionally, for the sake of simplicity.

Then, taking only the first 2 × 2 components, we have

$$(\boldsymbol{u}' + \boldsymbol{u}) \otimes (\boldsymbol{u}' + \boldsymbol{u}) = \begin{pmatrix} \gamma + 1 \\ \gamma v \end{pmatrix} \begin{pmatrix} -(\gamma+1) & \gamma v \end{pmatrix} = \begin{pmatrix} -(\gamma+1)^2 & (\gamma+1)\gamma v \\ -(\gamma+1)\gamma v & \gamma^2 v^2 \end{pmatrix},$$

and the boost (10.17) is

$$\begin{pmatrix} 1 & 0 \\ 0 & 1 \end{pmatrix} + \frac{1}{1+\gamma} \begin{pmatrix} -(\gamma+1)^2 & (\gamma+1)\gamma v \\ -(\gamma+1)\gamma v & \gamma^2 v^2 \end{pmatrix} - 2 \begin{pmatrix} -\gamma & 0 \\ -\gamma v & 0 \end{pmatrix} = \gamma \begin{pmatrix} 1 & v \\ v & 1 \end{pmatrix}$$

where we used that

$$1 + \frac{\gamma^2 v^2}{\gamma + 1} = 1 + \frac{(\gamma^2 v^2 - \gamma^2) + \gamma^2}{\gamma + 1} = 1 + \frac{-1 + \gamma^2}{\gamma + 1} = 1 + (\gamma - 1).$$

Finally,

$$\text{the boost from } \begin{pmatrix} 1 \\ 0 \\ 0 \\ 0 \end{pmatrix} \text{ to } \begin{pmatrix} \gamma \\ \gamma v \\ 0 \\ 0 \end{pmatrix} \text{ is } \begin{pmatrix} \gamma & \gamma v & 0 & 0 \\ \gamma v & \gamma & 0 & 0 \\ 0 & 0 & 1 & 0 \\ 0 & 0 & 0 & 1 \end{pmatrix}; \tag{10.21}$$

its inverse,

$$\text{the boost from } \begin{pmatrix} \gamma \\ \gamma v \\ 0 \\ 0 \end{pmatrix} \text{ to } \begin{pmatrix} 1 \\ 0 \\ 0 \\ 0 \end{pmatrix} \text{ is } \begin{pmatrix} \gamma & -\gamma v & 0 & 0 \\ -\gamma v & \gamma & 0 & 0 \\ 0 & 0 & 1 & 0 \\ 0 & 0 & 0 & 1 \end{pmatrix}. \tag{10.22}$$

Compare these formulas with (5.17) and (5.18) for $v^1 = v$ and $v^2 = v^3 = 0$.
More generally, for $\boldsymbol{u} := \gamma(1, v, 0, 0)$ and $\boldsymbol{u}' := \gamma'(1, 0, v', 0)$, we obtain

$$(\boldsymbol{u}' + \boldsymbol{u}) \otimes (\boldsymbol{u}' + \boldsymbol{u}) = \begin{pmatrix} -(\gamma+\gamma')^2 & (\gamma+1)\gamma v & (\gamma+\gamma')\gamma' v' & 0 \\ -(\gamma+\gamma' 1)\gamma v & (\gamma v)^2 & \gamma\gamma' v v' & 0 \\ -(\gamma+\gamma')\gamma' v' & \gamma\gamma' v v' & (\gamma' v')^2 & 0 \\ 0 & 0 & 0 & 0 \end{pmatrix} \tag{10.23}$$

and we can imagine how complicated the boost from $\boldsymbol{u}'$ to $\boldsymbol{u}$ is.

10.3 Isomorphisms

Isomorphism of models is an important notion: two models, which are formally different, have the same physical content if and only if they are isomorphic.

10.3.1 The form of isomorphisms

Since in the relativistic case the measure line of space distances equals the measure line of time periods, $\boldsymbol{Z}$ equals $\boldsymbol{F}$ in the general definition of isomorphism (see Section 3.2), the pair $(L, \boldsymbol{F})$ is an isomorphism between the relativistic spacetime models $(\mathrm{M}, \mathbb{T}, \boldsymbol{g})$ and $(\widehat{\mathrm{M}}, \widehat{\mathbb{T}}, \hat{\boldsymbol{g}})$ if:

(i) $L : \mathrm{M} \to \widehat{\mathrm{M}}$ is an orientation and arrow orientation preserving affine bijection (over the linear bijection $\boldsymbol{L} : \mathbf{M} \to \widehat{\mathbf{M}}$)
(ii) $\boldsymbol{F} : \mathbb{T} \to \widehat{\mathbb{T}}$ is an orientation preserving linear bijection

which transform "conveniently" $\mathrm{T}^{\to}$ into $\widehat{\mathrm{T}^{\to}}$, $\boldsymbol{P}$ into $\hat{\boldsymbol{P}}$, $\boldsymbol{b}$ into $\hat{\boldsymbol{b}}$.

Since $\mathrm{T}^{\to}$, $\boldsymbol{P}$ and $\boldsymbol{b}$ are determined by $\boldsymbol{g}$, the three general requirements can be united:

(I)–(II)–(III)

$$\frac{\hat{\boldsymbol{g}}(\boldsymbol{L}\cdot\boldsymbol{x}, \boldsymbol{L}\cdot\boldsymbol{y})}{\boldsymbol{F}(\boldsymbol{t})^2} = \frac{\boldsymbol{g}(\boldsymbol{x},\boldsymbol{y})}{\boldsymbol{t}^2}$$

for all $\boldsymbol{x}, \boldsymbol{y} \in \mathbf{M}$, and $0 \neq \boldsymbol{t} \in \mathbb{T}$.

10.3.2 Relativistic spacetime models are isomorphic

We can easily demonstrate:

Arbitrary special relativistic spacetime model is isomorphic to the arithmetic one, which involves that arbitrary two special relativistic spacetime models are isomorphic.

Indeed, let $(\mathrm{M}, \mathbb{T}, \boldsymbol{g})$ be a relativistic spacetime model and let us take:

(i) A time unit $\mathrm{s} \in \mathbb{T}^+$.
(ii) An "origin" o in M.
(iii) A positively oriented $\boldsymbol{g}$-orthogonal basis $(\boldsymbol{e}_0, \boldsymbol{e}_1, \boldsymbol{e}_2, \boldsymbol{e}_3)$ in $\mathbf{M}$, normalized to s, such that $\boldsymbol{e}_0$ is future-like,

and let

$$L : \mathrm{M} \to \mathbb{R}^4, \quad x \mapsto \text{coordinates of } x - o \text{ in the basis } (\boldsymbol{e}_0, \boldsymbol{e}_1, \boldsymbol{e}_2, \boldsymbol{e}_3),$$

$$\boldsymbol{F} : \mathbb{T} \to \mathbb{R}, \quad \boldsymbol{t} \mapsto \frac{\boldsymbol{t}}{\mathrm{s}}.$$

This taking standard coordinates of spacetime will be treated thoroughly in Section 10.11; please study that section before proceeding further.

Then it is a simple fact that in this way an isomorphism is established:

if
– $x = \sum_{i=0}^{3} x^i e_i$ and $y = \sum_{k=0}^{3} y^k e_k$, then

$$\sum_{i=0}^{3} x^i y_i = \frac{g(L \cdot x, L \cdot y)}{s^2}.$$

Our result says that the physical content of any relativistic spacetime model is the same; we can use an arbitrary one. A special model, however, may contain extra properties, which

– Have nothing to do with the structure of the model.
– Hide the essential features of the model.

Extra properties of a special model may be the source of misunderstandings. From a practical point of view, to solve actual problems, the arithmetic one in a convenient coordinate system is excellent but from a theoretical point of view it is not advisable because:

(1) Spacetime M and the set of spacetime vectors **M** appear in the same form, they are $\mathbb{R}^4$, so they can be confused.
(2) All measure lines are the same $\mathbb{R}$, so different types of vectors and tensors are not distinguished.
(3) It get lost that the standard space vectors (see Subsection 10.1.2) of different observers are different (contrary to the nonrelativistic case).
(4) Equality of vectors (parallelism of straight lines) in spaces of different observers appears as a straightforward notion (resulting in the velocity addition paradox, see Section 15.3) (resulting in the light propagation paradox, see Section 15.4).

Further, it is advised to examine the remarks in 5.7.1, 5.9, 10.1.2, 10.6.3, 10.7.3, 10.10, 10.11, 11.8.

If, using the arithmetic model, we do not want to make errors. We have to check permanently whether the formulas in question have a real physical meaning; this is rather tiresome and something can escape our attention.

Exercise
Give an isomorphism from the spacetime model of Exercise 10.1.7 to the arithmetic model.

10.4 Spacetime symmetries and reversals

10.4.1 Lorentz transformations

In the relativistic spacetime model, the linear bijections $L : \mathbf{M} \to \mathbf{M}$ satisfying

$$(L \cdot x) \cdot (L \cdot y) = x \cdot y \quad \text{for all } x, y \in \mathbf{M}$$

are called **Lorentz-transformations.** The linear map $\boldsymbol{L}:\mathbf{M}\to\mathbf{M}$ is a Lorentz transformation if and only if $\boldsymbol{L}^* = \boldsymbol{L}^{-1}$.

The compositions of Lorentz transformations as well as the inverse of a Lorentz transformation are Lorentz transformations; thus, they form a group, the **Lorentz group** $\mathcal{L}$.

The orientation and arrow orientation preserving Lorentz transformations are called **proper Lorentz transformations.**

According to the general definition in Section 3.3 and the formulas in Subsection 10.3.1:

The proper Lorentz transformations are the vectorial symmetries of the relativistic spacetime model.

The boosts are proper Lorentz transformations.

In contrast to the nonrelativistic case, there is no "special subgroup" in the Lorentz group, which is shown by the fact that the composition of boosts is not a boost, in general (see (10.20)).

We emphasize that

the "three-dimensional orthogonal group" is not a subgroup of the Lorentz group.

For all absolute velocities $\boldsymbol{u}$,

$$\mathcal{O}_{\boldsymbol{u}} := \{\boldsymbol{L}\in\mathcal{L} \mid \boldsymbol{L}\cdot\boldsymbol{u}=\boldsymbol{u}\}$$

is a subgroup of the Lorentz group, isomorphic to the "three-dimensional orthogonal group." These groups are Wigner's little groups.

It is worth saying a few words about "time and space reversals" in the Lorentz group.

For every absolute velocity $\boldsymbol{u}$, there is a **$\boldsymbol{u}$-time reversal $\boldsymbol{T}_{\boldsymbol{u}}$** defined by

$$\boldsymbol{T}_{\boldsymbol{u}}\cdot\boldsymbol{u} = -\boldsymbol{u},\quad \boldsymbol{T}_{\boldsymbol{u}}\cdot\boldsymbol{q}=\boldsymbol{q}\quad (\boldsymbol{q}\in\mathbf{S}_{\boldsymbol{u}})$$

and there is a **$\boldsymbol{u}$-space reversal $\boldsymbol{P}_{\boldsymbol{u}}$** defined by

$$\boldsymbol{P}_{\boldsymbol{u}}\cdot\boldsymbol{u} = \boldsymbol{u},\quad \boldsymbol{P}_{\boldsymbol{u}}\cdot\boldsymbol{q}=-\boldsymbol{q}\quad (\boldsymbol{q}\in\mathbf{S}_{\boldsymbol{u}});$$

it is trivial that $\boldsymbol{P}_{\boldsymbol{u}} = -\boldsymbol{T}_{\boldsymbol{u}}$ and $\boldsymbol{T}_{\boldsymbol{u}}^{-1} = \boldsymbol{T}_{\boldsymbol{u}}$.

For a world vector $\boldsymbol{x}$, $\boldsymbol{x}-\boldsymbol{u}(\boldsymbol{\tau}_{\boldsymbol{u}}\cdot\boldsymbol{x}) = \boldsymbol{x}+\boldsymbol{u}(\boldsymbol{u}\cdot\boldsymbol{x})$ is in $\mathbf{S}_{\boldsymbol{u}}$; thus, $\boldsymbol{T}_{\boldsymbol{u}}\cdot\boldsymbol{x} = \boldsymbol{x}-2\boldsymbol{u}(\boldsymbol{\tau}_{\boldsymbol{u}}\cdot\boldsymbol{x}) = \boldsymbol{x}+2\boldsymbol{u}(\boldsymbol{u}\cdot\boldsymbol{x})$; briefly,

$$\boldsymbol{T}_{\boldsymbol{u}} = \mathbf{1}-2\boldsymbol{u}\otimes\boldsymbol{\tau}_{\boldsymbol{u}} = \mathbf{1}+2\boldsymbol{u}\otimes\boldsymbol{u}, \tag{10.24}$$

which exhibits that the $\boldsymbol{u}$-time-like reversal (and the $\boldsymbol{u}$-space-like reversal) is a Lorentz transformation: $\boldsymbol{T}_{\boldsymbol{u}}^* = \boldsymbol{T}_{\boldsymbol{u}} = \boldsymbol{T}_{\boldsymbol{u}}^{-1}$. Two time reflections of the same vector are illustrated by Figure 10.5.

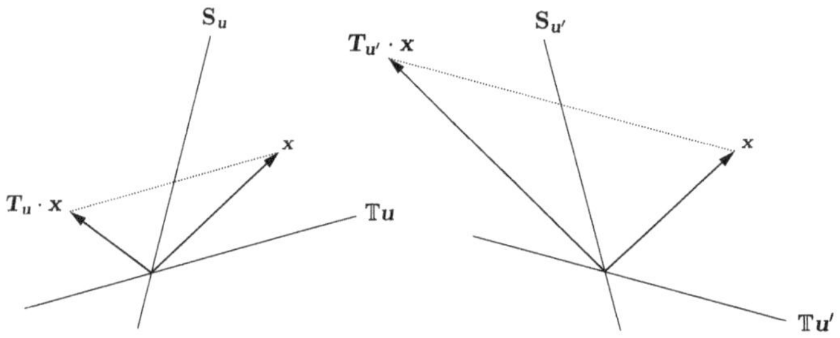

Figure 10.5: Different time reversals.

Time reversals due to different absolute velocities are different: if $\boldsymbol{u} \neq \boldsymbol{u}'$, then $\boldsymbol{T}_{\boldsymbol{u}} \neq \boldsymbol{T}_{\boldsymbol{u}'}$.

If $\boldsymbol{u}$ and $\boldsymbol{u}'$ are different absolute velocities, then

$$\boldsymbol{T}_{\boldsymbol{u}'} \cdot \boldsymbol{T}_{\boldsymbol{u}} = \mathbf{1} + 2\boldsymbol{u}' \otimes \boldsymbol{u}' + 2\boldsymbol{u} \otimes \boldsymbol{u} + 4(\boldsymbol{u}' \cdot \boldsymbol{u})\boldsymbol{u}' \otimes \boldsymbol{u}$$

is a proper Lorentz transformation. Then $\boldsymbol{T}_{\boldsymbol{u}''} \cdot \boldsymbol{T}_{\boldsymbol{u}'} \cdot \boldsymbol{T}_{\boldsymbol{u}}$—contrary to the nonrelativistic case—is not a time reversal and the product of four different time reversals is again a proper Lorentz transformation: all the time reversals generate a subgroup of the Lorentz group consisting of the proper Lorentz transformations and all the time reversals.

Arithmetic formulas

The Lorentz transformations in the arithmetic spacetime model are linear bijections $\mathbb{R}^4 \to \mathbb{R}^4$, given by 4×4 matrices. Their form is more complicated then the form of Galilei transformations given in (5.21) because there is no three-dimensional subspace invariant for all Lorentz transformations.

Nevertheless, matrices of the form (5.23) are Lorentz transformations and that is why it seems that "the three-dimensional orthogonal group" is a subgroup of the Lorentz group but this is only one of Wigner's little group, the one corresponding to the absolute velocity $(1, 0, 0, 0)$.

One usually considers "the" time reversal $x^0 \mapsto -x^0$;

$$\text{this is the} \begin{pmatrix} 1 \\ 0 \\ 0 \\ 0 \end{pmatrix}\text{-time reversal} \begin{pmatrix} -1 & 0 & 0 & 0 \\ 0 & 1 & 0 & 0 \\ 0 & 0 & 1 & 0 \\ 0 & 0 & 0 & 1 \end{pmatrix}.$$

$$\text{The } \gamma \begin{pmatrix} 1 \\ v \\ 0 \\ 0 \end{pmatrix}\text{-time reversal is} \begin{pmatrix} 1 - 2\gamma^2 & 2\gamma^2 v & 0 & 0 \\ -2\gamma^2 v & 1 + 2\gamma^2 v^2 & 0 & 0 \\ 0 & 0 & 1 & 0 \\ 0 & 0 & 0 & 1 \end{pmatrix}.$$

We see, how arithmetic formulas can be complicated.

10.4.2 Poincaré transformations

The **Poincaré transformations** are affine transformations $L : \mathrm{M} \to \mathrm{M}$ over the Lorentz transformations; they form a group, the **Poincaré group** $\mathcal{P}$.

Poincaré transformations over orientation preserving, arrow orientation preserving, and proper Lorentz transformations are called **orientation preserving, arrow orientation preserving and proper**, respectively.

The proper Poincaré transformations are the symmetries of the relativistic spacetime model.

Poincaré transformation are affine transformations of spacetime. Lorentz transformations are linear transformations of space vectors; it is trivial that

we emphasize that

the Lorentz group is not a subgroup of the Poincaré group.

If spacetime is vectorized by an "origin" o, then the Poincaré transformations are given on world vectors in the following way:

$$\boldsymbol{x} \mapsto L(o + \boldsymbol{x}) - o = L(o) - o + \boldsymbol{L} \cdot \boldsymbol{x}, \tag{10.25}$$

as a consequence, for every o,

$$\mathcal{L}_o := \{L \in \mathcal{P} \mid L(o) = o\}$$

is a subgroup of the Poincaré group, isomorphic to the Lorentz group.

In spacetime, there is no $\boldsymbol{u}$-time reversal because a reversal requires an "origin." For every world point o and absolute velocity $\boldsymbol{u}$, there is an **o-centered $\boldsymbol{u}$-time reversal**

$$T_{\boldsymbol{u},o}(x) := o + \boldsymbol{T}_{\boldsymbol{u}} \cdot (x - o) = x - 2\tau_{\boldsymbol{u}} \cdot (x - o)\boldsymbol{u} = x + 2\boldsymbol{u} \cdot (x - o)\boldsymbol{u}, \tag{10.26}$$

which is a Poincaré transformation over $\boldsymbol{T}_{\boldsymbol{u}}$.

We get similarly **o-centered $\boldsymbol{u}$-space reversal.**

A thorough examination of the Lorentz group and the Poincaré group can be found in the book *Spacetime without Reference Frames.*

Remarks.

1. If world points are not distinguished from world vectors—as in the arithmetic model—then it seems with "the" origin o that a Poincaré transformation consists of a Lorentz transformation and a translation (see (10.25)). Consequently, in usual treatments of special relativity, Poincaré transformations are called inhomogeneous Lorentz transformations and the Lorentz group is considered a subgroup (with zero translation) of the Poincaré group.
2. "**The**" three-dimensional orthogonal group as a subgroup of the Lorentz group, "**the**" time reversal, "**the**" Lorentz group as a subgroup of the Poincaré group are

errors, which show well how the arithmetic spacetime model may cause misunderstandings.

10.4.3 Exercises

1. Show that the $\boldsymbol{u}$-time reversal changes orientation and arrow orientation.
2. What is the $\boldsymbol{u}$-time reversal of the absolute velocity $\boldsymbol{u}'$?
3. Show that the $\boldsymbol{u}$-space reversal changes orientation and preserves arrow orientation.
4. What are the o-centered $\boldsymbol{u}$-space reversals?
5. Give the $\gamma'(1,0,v',0)$-time reversal in the arithmetic model. What is the matrix of the linear bijection consisting of the $\gamma(1,v,0,0)$-time reversal followed by the $\gamma'(1,0,v',0)$-time reversal?
6. Give Wigner's little group corresponding to $\gamma(1,v,0,0)$ in the arithmetic spacetime model.

10.5 Light signals

10.5.1 Light lines

As said in Section 9.1, the history of a light signal in spacetime is an oriented curve, a **light line**. Every tangent vector of a light line is future light-like. It is convenient to characterize the future light-like vectors by **absolute light directions**, elements of

$$\mathrm{V}^{\rightarrow} := \left\{ \boldsymbol{w} \in \frac{\mathbf{M}}{\mathbb{T}} \;\middle|\; \mathbb{T}^{+}\boldsymbol{w} \subset \mathrm{L}^{\rightarrow} \right\}. \tag{10.27}$$

The absolute light directions play a similar role as absolute velocities but—since their pseudolength is zero—if $\boldsymbol{w}$ is a line direction, then $a\boldsymbol{w}$ is also a line direction for all positive numbers a.

The Lorentz product of future-like and future light-like vectors is negative (see (10.8)):

$$\boldsymbol{u} \cdot \boldsymbol{w} < 0$$

for all $\boldsymbol{u} \in \mathrm{V}(1)$ and $\boldsymbol{w} \in \mathrm{V}^{\rightarrow}$.

10.5.2 Absolute light propagation

The fundamental properties of light signals (which give the base of the construction of the relativistic spacetime model) can be proved in the spacetime model:

In the space of any inertial observer

(1) *Every straight line can be the path of a light signal.*
(2) *The path of a free light signal is a straight line.*
(3) *A light signal is faster then any material point along the same path in the same direction.*
(4) *the movement of any light signal can be arbitrarily approximated by a movement of a material point.*

Let us take an inertial observer $\boldsymbol{u}$.

To prove assertion 2, we can copy Subsection 2.5.1, taking light direction $\boldsymbol{w}$ instead of $\boldsymbol{u}'$: the path of the light signal in $\boldsymbol{u}$-space is the straight line directed by $\boldsymbol{w}+\mathbb{R}\boldsymbol{u}$. According to the standard representation of $\boldsymbol{u}$-space vectors, the direction of this line is $\pi_{\boldsymbol{u}} \cdot \boldsymbol{w}$.

To prove assertion 1, we have to show that for all $0 \neq \boldsymbol{v} \in \frac{\mathbf{S}_{\boldsymbol{u}}}{\mathbb{T}}$ there is a light direction $\boldsymbol{w}$ in such a way that $\pi_{\boldsymbol{u}} \cdot \boldsymbol{w} = \boldsymbol{v}$; this is simple: $\boldsymbol{w} := |\boldsymbol{v}|\boldsymbol{u} + \boldsymbol{v}$.

A little part of an arbitrary curve can be approximated by straight line, so avoiding complications, for the sake of simplicity, the other two assertions will be proved only for straight paths.

To prove assertion 3, first we state that if a light signal with light direction $\boldsymbol{w}$ and a material point with absolute velocity $\boldsymbol{u}'$ move along the same path in the same direction in $\boldsymbol{u}$-space. Then there is a number $\beta' > 0$, such that $\pi_{\boldsymbol{u}} \cdot \boldsymbol{w} = \beta' \pi_{\boldsymbol{u}} \cdot \boldsymbol{u}'$. Rearranging the equality $\boldsymbol{w} + (\boldsymbol{u} \cdot \boldsymbol{w})\boldsymbol{u} = \beta'(\boldsymbol{u}' + (\boldsymbol{u} \cdot \boldsymbol{u}')\boldsymbol{u})$ we get

$$\boldsymbol{w} = \beta'\boldsymbol{u}' + (\beta'\boldsymbol{u} \cdot \boldsymbol{u}' - \boldsymbol{u} \cdot \boldsymbol{w})\boldsymbol{u}.$$

The coefficient of $\boldsymbol{u}$ must be negative; if no, the right side would be future-like. According to (2.9) (with $\boldsymbol{w}$ instead of $\boldsymbol{t}^{+}\boldsymbol{u}^{+}$ and $\beta'\boldsymbol{u}'$ instead of $\boldsymbol{t}^{-}\boldsymbol{u}^{-}$) this means that the light signal is faster.

To prove assertion 4, first we formulate it: for given $\boldsymbol{u}$ and $\boldsymbol{w}$ and for all $\beta > 0$, there are a $\boldsymbol{u}'$ and a number $\beta' > 0$ such that $\boldsymbol{w} + \beta\boldsymbol{u} = \beta'\boldsymbol{u}'$. This is evident because the sum of a future light-like vector and a future-like vector is future-like (see the mathematical supplement); thus, $\boldsymbol{w} + \beta\boldsymbol{u}$ is future-like and then $\boldsymbol{u}' := \frac{\boldsymbol{w}+\beta\boldsymbol{u}}{|\boldsymbol{w}+\beta\boldsymbol{u}|}$.

The assertions above have the consequence:

Two light signals meeting at a world point
- And moving in the same direction along the same path in the space of an inertial observer.
- Have the same history in spacetime.

This is a very important property of light signals, which can be shown independently of the previous proof.

Let us consider two light lines directed by $\boldsymbol{w} \in \mathrm{V}^{\to}$ and $\boldsymbol{w}' \in \mathrm{V}^{\to}$, respectively, and let the corresponding light signals move in the same direction along the same path in the space of the inertial observer $\boldsymbol{u}$. Since the light directions are determined up to positive numbers, we can take them in such a way that $\pi_{\boldsymbol{u}}\boldsymbol{w} = \pi_{\boldsymbol{u}}\boldsymbol{w}'$, which means that there is a number α such that

$$\boldsymbol{w}' = \boldsymbol{w} + \alpha\boldsymbol{u}.$$

The Lorentz square of both sides results in $0 = 2\alpha(\boldsymbol{u}\cdot\boldsymbol{w}) - \alpha^2$.

The Lorentz product of both sides by $\boldsymbol{u}$ results in $\boldsymbol{u}\cdot\boldsymbol{w}' = \boldsymbol{u}\cdot\boldsymbol{w} - \alpha$, which has the square $(\boldsymbol{u}\cdot\boldsymbol{w}')^2 = (\boldsymbol{u}\cdot\boldsymbol{w})^2 - 2\alpha(\boldsymbol{u}\cdot\boldsymbol{w}) + \alpha^2$.

The two squares together give $\boldsymbol{u}\cdot\boldsymbol{w}' = \boldsymbol{u}\cdot\boldsymbol{w}$; as a consequence, $\alpha = 0$, i. e., $\boldsymbol{w} = \boldsymbol{w}'$.

The light signals are parallel in spacetime and have a point in common, so they coincide.

10.5.3 The round-way speed of light

A further, most important experimental fact (which has a role in the construction of the relativistic spacetime model) can also be proved in the spacetime model:

(5) *The round-way light propagation for any inertial observer is homogeneous and isotropic.*

This means the following: if a light signal

- Is emitted in an arbitrary space point of an arbitrary inertial observer.
- Moves on an arbitrary path in the space of the observer.
- Returns to the space point of emission,

  then
- The length of the path divided by the proper time period measured in the space point between start and arrival is independent of the observer, the space point and the path.

For the sake of simplicity, we show this only for a light signal that moves in a straight line from a source to a mirror and backwards.

Let $\boldsymbol{u}$ be an inertial observer, let $\boldsymbol{a}$ and $\boldsymbol{a}'$ be the future light-like vectors from the source to the mirror and from the mirror to the source, respectively (see Figure 9.1).

If $2t$ is the time period passed in the source between the start and the arrival, then $\boldsymbol{a} + \boldsymbol{a}' = 2t\boldsymbol{u}$; thus, $2t = -\boldsymbol{u}\cdot(\boldsymbol{a} + \boldsymbol{a}')$.

The $\boldsymbol{u}$-space vector between the mirror and the source is $\pi_{\boldsymbol{u}}\cdot\boldsymbol{a} = \pi_{\boldsymbol{u}}\cdot(-\boldsymbol{a}')$ (see (10.11)); thus, the distance is $\boldsymbol{d} := |\pi_{\boldsymbol{u}}\cdot\boldsymbol{a}| = |\pi_{\boldsymbol{u}}\cdot\boldsymbol{a}'|$.

$|\pi_{\boldsymbol{u}}\cdot\boldsymbol{a}|^2 = |\boldsymbol{a} + (\boldsymbol{u}\cdot\boldsymbol{a})\boldsymbol{u}|^2 = (\boldsymbol{u}\cdot\boldsymbol{a})^2$ and a similar equality holds for $\boldsymbol{a}'$, too, so $2\boldsymbol{d} = -(\boldsymbol{u}\cdot\boldsymbol{a}) - (\boldsymbol{u}\cdot\boldsymbol{a}') = 2\boldsymbol{t}$, i. e., the two-way light speed is $\frac{2d}{2t} = 1$.

10.5.4 Movement of light signals

Similar to the directions of motion $\boldsymbol{v}_{\boldsymbol{u}'\boldsymbol{u}}$ and $\boldsymbol{v}_{\boldsymbol{u}\boldsymbol{u}'}$ of the inertial observers $\boldsymbol{u}'$ and $\boldsymbol{u}$ with respect to each other (see Subsection 10.1.2), the directions of motion of a light signal with absolute light direction $\boldsymbol{w}$, with respect to the observers, are

$$\boldsymbol{v}_{\boldsymbol{w}\boldsymbol{u}} := \frac{\pi_{\boldsymbol{u}}\cdot\boldsymbol{w}}{-\boldsymbol{u}\cdot\boldsymbol{w}} = \frac{\boldsymbol{w}}{-\boldsymbol{u}\cdot\boldsymbol{w}} - \boldsymbol{u}, \quad \text{and} \quad \boldsymbol{v}_{\boldsymbol{w}\boldsymbol{u}'} := \frac{\pi_{\boldsymbol{u}'}\cdot\boldsymbol{w}}{-\boldsymbol{u}'\cdot\boldsymbol{w}} = \frac{\boldsymbol{w}}{-\boldsymbol{u}'\cdot\boldsymbol{w}} - \boldsymbol{u}'$$

for which $|\boldsymbol{v}_{\boldsymbol{wu}}| = |\boldsymbol{v}_{\boldsymbol{wu}'}| = 1$ holds. Then we have:

- If a light signal and $\boldsymbol{u}'$ move in the same direction in $\boldsymbol{u}$-space
- Then the light signal and $\boldsymbol{u}$ move in opposite directions in $\boldsymbol{u}'$-space:

$$\textit{if}\quad \boldsymbol{v}_{\boldsymbol{wu}} = \frac{\boldsymbol{v}_{\boldsymbol{u}'\boldsymbol{u}}}{|\boldsymbol{v}_{\boldsymbol{u}'\boldsymbol{u}}|}\quad \textit{then}\ |\ \boldsymbol{v}_{\boldsymbol{wu}'} = -\frac{\boldsymbol{v}_{\boldsymbol{uu}'}}{|\boldsymbol{v}_{\boldsymbol{uu}'}|}.$$

Indeed, with the notation $v := |\boldsymbol{v}_{\boldsymbol{u}'\boldsymbol{u}}| = |\boldsymbol{v}_{\boldsymbol{uu}'}|$, we have by (10.16)

$$\frac{\boldsymbol{w}}{-\boldsymbol{u}\cdot\boldsymbol{w}} - \boldsymbol{u} = \frac{\sqrt{1-v^2}\boldsymbol{u}' - \boldsymbol{u}}{v}. \tag{10.28}$$

Taking the Lorentz product by $\boldsymbol{w}$ of this equality and using $\boldsymbol{w}\cdot\boldsymbol{w} = 0$, we get $-v\boldsymbol{u}\cdot\boldsymbol{w} = \sqrt{1-v^2}\boldsymbol{u}'\cdot\boldsymbol{w} - \boldsymbol{u}\cdot\boldsymbol{w}$ from which

$$\boldsymbol{u}'\cdot\boldsymbol{w}\sqrt{1+v} = \boldsymbol{u}\cdot\boldsymbol{w}\sqrt{1-v}.$$

So, $\frac{\boldsymbol{w}}{-\boldsymbol{u}'\cdot\boldsymbol{w}} = \frac{\boldsymbol{w}\sqrt{1+v}}{-\boldsymbol{u}\cdot\boldsymbol{w}\sqrt{1-v}}$, which results in the equality

$$\frac{\boldsymbol{w}}{-\boldsymbol{u}'\cdot\boldsymbol{w}} - \boldsymbol{u}' = -\frac{\sqrt{1-v^2}\boldsymbol{u} - \boldsymbol{u}'}{v}.$$

10.5.5 “Earlier” and “later” arrival of light signals

An inertial observer $\boldsymbol{u}'$ sends from a $\boldsymbol{u}'$-space point (source) two light signals a time delay $\boldsymbol{t}'$ to the approaching inertial observer $\boldsymbol{u}$, i. e., opposite to the direction of motion of $\boldsymbol{u}$ with respect to $\boldsymbol{u}'$. The light signals arrive at a $\boldsymbol{u}$-space point (relay) with a time delay $\boldsymbol{t}_-$ from the direction of motion of $\boldsymbol{u}'$ with respect to $\boldsymbol{u}$. Let the first light signal be emitted at the world point x and let it be absorbed by the relay at the world point y; then $\boldsymbol{a}_1 := y - x$ is future light-like (see Figure 10.6).

The second light signal is emitted at $x + \boldsymbol{t}'\boldsymbol{u}'$ and is absorbed at $y + \boldsymbol{t}_-\boldsymbol{u}$. Then $(y + \boldsymbol{t}_-\boldsymbol{u}) - (x + \boldsymbol{t}\boldsymbol{u}') = \boldsymbol{a}_1 + \boldsymbol{t}_-\boldsymbol{u} - \boldsymbol{t}'\boldsymbol{u}' =: \boldsymbol{a}_2$ is also future light-like and

$$\frac{\boldsymbol{a}_1 - \boldsymbol{a}_2}{\boldsymbol{t}_-} = \frac{\boldsymbol{t}'}{\boldsymbol{t}_-}\boldsymbol{u}' - \boldsymbol{u}.$$

The light signals $\boldsymbol{a}_1$ and $\boldsymbol{a}_2$ have the same absolute light direction $\boldsymbol{w}_-$ and $\boldsymbol{u}$-space direction; therefore, the left side of the second equality above is $\alpha\boldsymbol{v}_{\boldsymbol{w}_-\boldsymbol{u}}$ for some positive α. Then

$$\alpha\left(\frac{\boldsymbol{w}_-}{-\boldsymbol{u}\cdot\boldsymbol{w}_-} - u\right) = \frac{\boldsymbol{t}'}{\boldsymbol{t}_-}\boldsymbol{u}' - \boldsymbol{u},$$

which is only possible if $\alpha = 1$, and we conclude from (10.28) that

$$t_- = \sqrt{\frac{1-v}{1+v}}\, t' < t'. \tag{10.29}$$

In the similar way, if the light signals from $\boldsymbol{u}'$ are sent in the direction of motion $\boldsymbol{u}$ with respect to $\boldsymbol{u}'$ with a time delay $\boldsymbol{t}'$, then the time delay $\boldsymbol{t}_+$ of the arrivals is

$$t_+ = \sqrt{\frac{1+v}{1-v}}\, t' > t'. \tag{10.30}$$

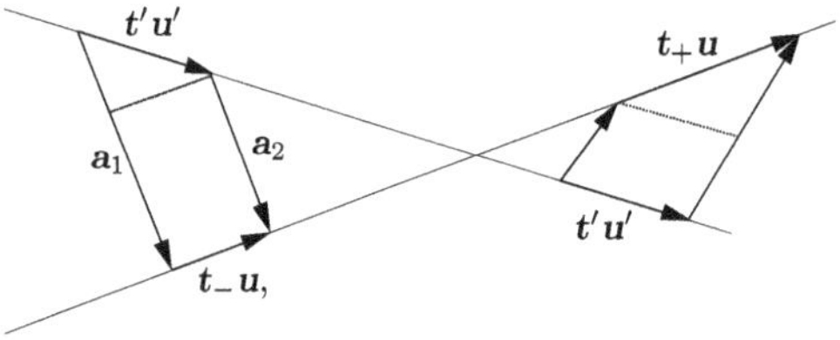

Figure 10.6: Light signals sent from $\boldsymbol{u}'$ to $\boldsymbol{u}$.

Summing up: If the light signals sent from the observer $\boldsymbol{u}'$ arrive to the observer $\boldsymbol{u}$
- In the direction of $\mathbf{v}_{u'u}$, then the time interval between the arrivals is longer than the time interval between the starts.
- In the direction opposite to $\mathbf{v}_{u'u}$, then the time interval between the arrivals is shorter than the time interval between the starts.

This is a special case of the **Doppler effect** in the directions of motion: if the frequency of the emission of the light signals is $\frac{1}{t'}$, then the frequencies of the arrival of the light signals are $\frac{1}{t_\pm}$.

10.6 Standard inertial frames, relative velocities

10.6.1 Standard synchronization of an inertial observer

The usual "axiom" of homogeneous and isotropic light propagation concerns one-way light speed without specifying a synchronization. The homogeneous and isotropic round-way light propagation is meaningful without synchronization; now we show that this allows to establish a synchronization in which the one-way light propagation is also homogeneous and isotropic.

Let us consider an inertial observer $\boldsymbol{u}$. A light signal starts at the occurrence z of a $\boldsymbol{u}$-space point ("source"), arrives at the occurrence x of another $\boldsymbol{u}$-space point ("mirror"), there it is reflected and returns to the source at the occurrence z'. Then the occurrence y

of the source is defined to be simultaneous with x if there is a proper time period $\boldsymbol{t}$, such that $y - z = z' - y = \boldsymbol{tu}$ (see Figure 10.7). $x - z$ and $z' - x$ are light-like vectors, so with the notation $\boldsymbol{q} := x - y$, $\boldsymbol{tu} + \boldsymbol{q}$, and $-\boldsymbol{q} + \boldsymbol{tu}$ have zero Lorentz square, so $\boldsymbol{q} \cdot \boldsymbol{q} \pm 2\boldsymbol{tu} \cdot \boldsymbol{q} - \boldsymbol{t}^2 = 0$, which implies $\boldsymbol{u} \cdot \boldsymbol{q} = 0$, i. e., $\boldsymbol{q} \in \mathbf{S}_{\boldsymbol{u}}$.

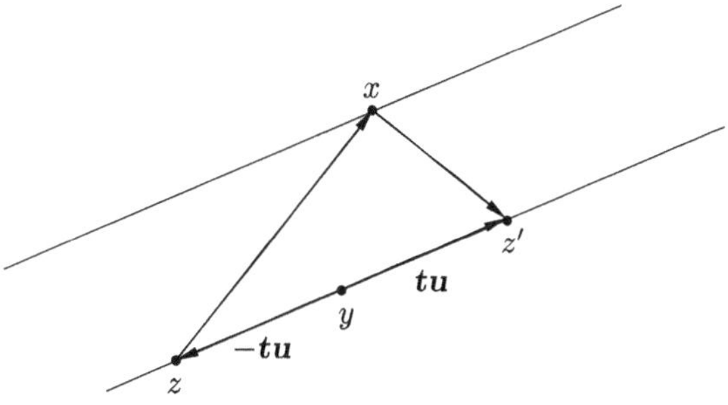

Figure 10.7: Standard synchronization: y and x are $\boldsymbol{u}$-simultaneous.

Our result shows that this synchronization is independent of the source and the mirror. According to such a synchronization established by the inertial observer $\boldsymbol{u}$, the world points x and y are simultaneous if and only if $x - y \in \mathbf{S}_{\boldsymbol{u}}$. In other words, the collection of world points simultaneous with x in such a way is $x + \mathbf{S}_{\boldsymbol{u}}$.

This uniform synchronization defined by the three-dimensional linear subspace $\mathbf{S}_{\boldsymbol{u}}$ is called the **standard synchronization** of the observer $\boldsymbol{u}$.

The standard synchronization is established by the observer in such a way that the one-way light speed be equal to the round-way speed. Since $\mathbf{S}_{\boldsymbol{u}} \neq \mathbf{S}_{\boldsymbol{u}'}$ if $\boldsymbol{u} \neq \boldsymbol{u}'$, different standard synchronizations correspond to different inertial observers.

According to Subsection 3.5.5, a uniform synchronization and an inertial observer together form an inertial frame.

The inertial observer $\boldsymbol{u}$ and its standard synchronization together, $(\boldsymbol{u}, \mathbf{S}_{\boldsymbol{u}})$ is called the **standard inertial frame $\boldsymbol{u}$**.

Figure 10.8 shows a standard inertial frame.

Remarks.

1. The method of establishing such synchronizations shows well what has been emphasized earlier: synchronization is not a physical fact, it is an artificial (human) construction. We are not obliged to use the standard synchronization. We could synchronize in such a way that the one-way light speed from Greenwich to Budapest be the half of the one-way light speed from Budapest to Greenwich. The standard synchronization is distinguished by its simple and nice feature.

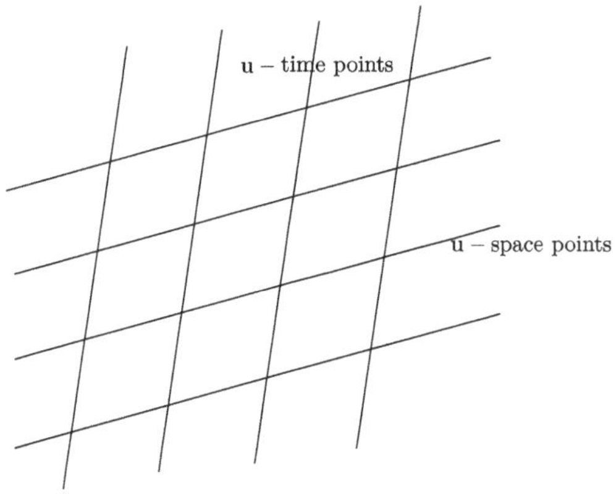

Figure 10.8: Standard inertial frame.

2. It is worth putting the following questions. Here, we considered the Earth an inertial observer though we know that it is not. Can we apply on the Earth in practice the treated light speed synchronization, and if yes, does it not lead to some problem? It can be shown that the error arising from the rotation of the Earth is negligible for everyday practice. Further, there is another, ancient method of synchronization, which is established by the position of stars (or the Sun): midnight (midday) is in Greenwich and in Budapest if the stars have (the Sun has) a prescribed position. It can be shown that these synchronizations are not the same but the difference is negligible for everyday practice [7]. It is not negligible, however, for managing spaceships and satellites.

10.6.2 Time points of a standard synchronization

The $\boldsymbol{u}$-standard time points are the hyperplanes directed by $\mathbf{S}_u$, their collection is the **$\boldsymbol{u}$-standard time**, $\mathrm{T}_u := \mathrm{M}/\mathbf{S}_u$. We introduce the **$\boldsymbol{u}$-time evaluation**

$$\tau_u : \mathrm{M} \to \mathrm{T}_u, \quad x \mapsto x + \mathbf{S}_u, \tag{10.31}$$

in accordance with the nonrelativistic one (5.13). $\boldsymbol{u}$-time evaluation assigns to every world point the corresponding standard $\boldsymbol{u}$-time point.

T_a is the factor space $\mathrm{M}/\mathbf{S}$, but now it can be endowed with a nicer affine structure than in the general case given in Subsection 3.5.4.

The difference of two $\boldsymbol{u}$-instants t and s is defined as the proper time duration in an arbitrary $\boldsymbol{u}$-space point between them. Let $y \in t$ and $x \in s$ are arbitrary; then there is an $x' \in s$ such that $y - x'$ is parallel to $\boldsymbol{u}$ and $x - x' \in \mathbf{S}_u$, so the proper time duration on the $\boldsymbol{u}$-space point $y + \mathbb{T}\boldsymbol{u}$ is $-\boldsymbol{u} \cdot (y - x') = -\boldsymbol{u} \cdot (y - x)$, therefore,

$$t - s = -\boldsymbol{u} \cdot (y - x), \quad (y \in t, x \in s). \tag{10.32}$$

In other words,

$$(y + S_u) - (x + S_u) = \tau_u(y) - \tau_u(x) := -\boldsymbol{u} \cdot (y - x) = \boldsymbol{\tau}_{\boldsymbol{u}} \cdot (y - x). \tag{10.33}$$

The subtraction above turns T_u into a one-dimensional affine space over $\mathbb{T}$ and τ_u becomes an affine map over the linear map $\boldsymbol{\tau}_{\boldsymbol{u}}$.

10.6.3 Standard relative velocities

Since in the nonrelativistic spacetime model there is a single synchronization, we need not bother about it, and a relative velocity is meaningful without specifying a synchronization. Let us take care not to do it in the relativistic case. Relative velocity makes sense only with a synchronization. Now we consider standard synchronizations of inertial observers.

In an inertial frame, an inertial world line results in a uniform motion on a straight line, as it is stated in Subsection 1.3.2. Now we examine such a motion in the relativistic spacetime model.

Let us consider a standard inertial frame $\boldsymbol{u}$ and the world line of an inertial material point, i. e., a straight line in spacetime, directed by an absolute velocity $\boldsymbol{u}'$ (which can be an arbitrary space point of the inertial observer $\boldsymbol{u}'$). If $\boldsymbol{u}' \neq \boldsymbol{u}$, then the material point moves with respect to the observer; the relative velocity of the material point with respect to the standard inertial frame is defined as follows.

Let x and y be occurrences of the world line, i. e., $y - x = \boldsymbol{t}'\boldsymbol{u}'$ for some $\boldsymbol{t}' \in \mathbb{T}$.

The material point meets the $\boldsymbol{u}$-space points $\pi_u(x)$ and $\pi_u(y)$ at the $\boldsymbol{u}$-instants $\tau_u(x)$ and $\tau_u(y)$, respectively. The **standard relative velocity** of $\boldsymbol{u}'$ with respect to $\boldsymbol{u}$, or the **standard $\boldsymbol{u}$-relative velocity** of $\boldsymbol{u}'$ is

$$\begin{aligned}\boldsymbol{v}_{\boldsymbol{u}'\boldsymbol{u}} &:= \frac{\pi_u(y) - \pi_u(x)}{\tau_u(y) - \tau_u(x)} = \frac{\boldsymbol{\pi}_{\boldsymbol{u}} \cdot (y - x)}{\boldsymbol{\tau}_{\boldsymbol{u}} \cdot (y - x)} = \frac{\boldsymbol{t}'\boldsymbol{u}' + \boldsymbol{t}'\boldsymbol{u}(\boldsymbol{u} \cdot \boldsymbol{u}')}{\boldsymbol{t}'(-\boldsymbol{u} \cdot \boldsymbol{u}')} \\ &= \frac{\boldsymbol{u}'}{-\boldsymbol{u} \cdot \boldsymbol{u}'} - \boldsymbol{u} \in \frac{\mathbf{S}_u}{\mathbb{T}}.\end{aligned} \tag{10.34}$$

The result is independent of the choice of the world points: an inertial material point moves uniformly on a straight line with respect to a standard inertial frame. The $\boldsymbol{u}$-space vector defined in (10.14) and used earlier is just the standard relative velocity.

The three-dimensional Euclidean vector space $\frac{\mathbf{S}_u}{\mathbb{T}}$ is the set of standard $\boldsymbol{u}$-relative velocities; contrary to absolute velocities,

- There is a zero standard $\boldsymbol{u}$-relative velocity.
- The magnitude of a standard $\boldsymbol{u}$-relative velocity is meaningful.
- The angle between two standard $\boldsymbol{u}$-relative velocities is meaningful.

Interchanging the roles of the absolute velocities, we have

$$\boldsymbol{v}_{\boldsymbol{u}\boldsymbol{u}'} = \frac{\boldsymbol{u}}{-\boldsymbol{u}' \cdot \boldsymbol{u}} - \boldsymbol{u}' \in \frac{\mathbf{S}_{\boldsymbol{u}'}}{\mathbb{T}}.$$

As it is shown for directions of movements in Subsection 10.1.2, $\mathbf{S}_{\boldsymbol{u}'} \neq \mathbf{S}_{\boldsymbol{u}}$ if $\boldsymbol{u}' \neq \boldsymbol{u}$, so

The standard $\boldsymbol{u}'$-relative velocity of $\boldsymbol{u}$ is not the opposite to the standard $\boldsymbol{u}$-relative velocity of $\boldsymbol{u}'$.

$$\boldsymbol{v}_{\boldsymbol{u}\boldsymbol{u}'} \neq -\boldsymbol{v}_{\boldsymbol{u}'\boldsymbol{u}} \quad \text{if} \quad \boldsymbol{u} \neq \boldsymbol{u}',$$

but $|\boldsymbol{v}_{\boldsymbol{u}\boldsymbol{u}'}| = |\boldsymbol{v}_{\boldsymbol{u}'\boldsymbol{u}}| < 1$ and, as nonrelativistically, the corresponding boosts map them into the opposite of each other (see (5.30)):

$$\boldsymbol{B}_{\boldsymbol{u}'\boldsymbol{u}} \cdot \boldsymbol{v}_{\boldsymbol{u}'\boldsymbol{u}} = -\boldsymbol{v}_{\boldsymbol{u}\boldsymbol{u}'}, \quad \boldsymbol{B}_{\boldsymbol{u}\boldsymbol{u}'} \cdot \boldsymbol{v}_{\boldsymbol{u}\boldsymbol{u}'} = -\boldsymbol{v}_{\boldsymbol{u}'\boldsymbol{u}}. \tag{10.35}$$

If $\boldsymbol{v} \in \frac{\mathbf{S}_{\boldsymbol{u}}}{\mathbb{T}}$ and $|\boldsymbol{v}| < 1$, then

$$\boldsymbol{u}_{\boldsymbol{v}} := \frac{\boldsymbol{u} + \boldsymbol{v}}{\sqrt{1 - |\boldsymbol{v}|^2}} \tag{10.36}$$

is an absolute velocity and $\boldsymbol{v}$ is the standard relative velocity of $\boldsymbol{u}_{\boldsymbol{v}}$ with respect to $\boldsymbol{u}$.

Similar to the previous formulas, the standard relative velocity with respect to the standard inertial frame $\boldsymbol{u}$, of a light signal having the light direction $\boldsymbol{w}$ is

$$\boldsymbol{v}_{\boldsymbol{w}\boldsymbol{u}} := \frac{\boldsymbol{w}}{-\boldsymbol{u} \cdot \boldsymbol{w}} - \boldsymbol{u};$$

for that we have $|\boldsymbol{v}_{\boldsymbol{w}\boldsymbol{u}}| = 1$, as it must be, according to how the standard synchronization is established.

If $\boldsymbol{v} \in \frac{\mathbf{S}_{\boldsymbol{u}}}{\mathbb{T}}$ and $|\boldsymbol{v}| = 1$, then

$$\boldsymbol{w}_{\boldsymbol{v}} := \boldsymbol{u} + \boldsymbol{v}$$

is an absolute light direction and $\boldsymbol{v}$ is the standard relative velocity of $\boldsymbol{w}_{\boldsymbol{v}}$ with respect to $\boldsymbol{u}$.

Finally, it is worth emphasizing again that the relative velocities here concern standard synchronizations. Other synchronizations result in other relative velocities (see Subsection 13.2.3).

Arithmetic formulas

The Lorentz product of the absolute velocities

$$\boldsymbol{u} := \gamma(1, v, 0, 0) \quad \text{and} \quad \boldsymbol{u}' := \gamma'(1, 0, v', 0)$$

is $-\gamma\gamma'$, so the standard relative velocities are

$$\mathbf{v}_{u'u} = \frac{1}{\gamma}\begin{pmatrix}1\\0\\v'\\0\end{pmatrix} - \gamma\begin{pmatrix}1\\v\\0\\0\end{pmatrix} = \begin{pmatrix}-\gamma v^2\\-\gamma v\\v'/\gamma\\0\end{pmatrix},$$

$$\mathbf{v}_{uu'} = \frac{1}{\gamma'}\begin{pmatrix}1\\v\\0\\0\end{pmatrix} - \gamma'\begin{pmatrix}1\\0\\v'\\0\end{pmatrix} = \begin{pmatrix}-\gamma'(v')^2\\v/\gamma'\\-\gamma v'\\0\end{pmatrix}.$$

It is well seen that they are not opposite to each other. In particular (for $v' = 0$), the standard relative velocity of

$$\gamma\begin{pmatrix}1\\v\\0\\0\end{pmatrix} \text{ with respect to } \begin{pmatrix}1\\0\\0\\0\end{pmatrix} \text{ is } \begin{pmatrix}0\\v\\0\\0\end{pmatrix},$$

$$\begin{pmatrix}1\\0\\0\\0\end{pmatrix} \text{ with respect to } \gamma\begin{pmatrix}1\\v\\0\\0\end{pmatrix} \text{ is } \gamma\begin{pmatrix}-v^2\\-v\\0\\0\end{pmatrix}.$$

They are not opposite to each other but the corresponding boosts map them into the opposite to each other; omitting the last two components, (10.21) and (10.22) give

$$\gamma\begin{pmatrix}1 & -v\\-v & 1\end{pmatrix}\begin{pmatrix}-\gamma v^2\\-\gamma v\end{pmatrix} = \begin{pmatrix}0\\-v\end{pmatrix}, \quad \gamma\begin{pmatrix}1 & v\\v & 1\end{pmatrix}\begin{pmatrix}0\\v\end{pmatrix} = \begin{pmatrix}\gamma v^2\\\gamma v\end{pmatrix}.$$

10.6.4 Addition of relative velocities

It is highly important that, contrary to the nonrelativistic case, the addition of standard relative velocities does not hold, i. e., except the trivial case $\mathbf{u}'' = \mathbf{u}'$ and $\mathbf{u}' = \mathbf{u}$,

$$\mathbf{v}_{u''u} \neq \mathbf{v}_{u''u'} + \mathbf{v}_{u'u}.$$

It is evident that an equality cannot hold, in general, because $\mathbf{v}_{u''u'}$ is in the three-dimensional vector space $\frac{\mathbf{S}_{u'}}{\mathbb{T}}$ and the two others are in the different three- dimensional vector space $\frac{\mathbf{S}_u}{\mathbb{T}}$. The best we can assert is that $\mathbf{v}_{u''u}$ is a linear combination of $\mathbf{v}_{u''u'}$ boosted to $\frac{\mathbf{S}_u}{\mathbb{T}}$ and $\mathbf{v}_{u'u}$:

$$\mathbf{v}_{u''u} = \lambda \mathbf{B}_{uu'} \cdot \mathbf{v}_{u''u'} + \mu \mathbf{v}_{u'u}.$$

For the sake of brevity, introducing the notation $\alpha := -\boldsymbol{u}'\cdot\boldsymbol{u}, \beta := -\boldsymbol{u}''\cdot\boldsymbol{u}', \gamma := -\boldsymbol{u}''\cdot\boldsymbol{u}$, we have $\boldsymbol{v}_{u''u} = \frac{\boldsymbol{u}''}{\gamma} - \boldsymbol{u}$, $\boldsymbol{v}_{u'',u'} = \frac{\boldsymbol{u}''}{\beta} - \boldsymbol{u}'$, $\boldsymbol{v}_{u'u} = \frac{\boldsymbol{u}'}{\alpha} - \boldsymbol{u}$, and

$$\boldsymbol{B}_{uu'} \cdot \boldsymbol{v}_{u''u'} = \boldsymbol{u} + \frac{\boldsymbol{u}''}{\beta} - \frac{(\boldsymbol{u} + \boldsymbol{u}')(\beta + \gamma)}{\beta(1 + \alpha)}. \tag{10.37}$$

Then the coefficients λ and μ can be determined and we get the **formula of addition of relative velocities**:

$$\boldsymbol{v}_{u''u} = \frac{\beta}{\gamma}\boldsymbol{B}_{uu'} \cdot \boldsymbol{v}_{u''u'} + \frac{\alpha(\beta + \gamma)}{\gamma(1 + \alpha)}\boldsymbol{v}_{u'u}. \tag{10.38}$$

With the notation $\boldsymbol{v}_1 := \boldsymbol{v}_{u'u}$, $\boldsymbol{v}_2 := \boldsymbol{B}_{uu'} \cdot \boldsymbol{v}_{u''u'}$, $\boldsymbol{v}_3 := \boldsymbol{v}_{u''u}$, (10.38) can be transformed in

$$\boldsymbol{v}_3 = \frac{\boldsymbol{v}_2 + \boldsymbol{v}_1 - \frac{\alpha}{1+\alpha}(\boldsymbol{v}_2 \times \boldsymbol{v}_1) \times \boldsymbol{v}_1}{1 + \boldsymbol{v}_1 \cdot \boldsymbol{v}_2} \tag{10.39}$$

where $\times$ is the vectorial product. In fact, it is simpler to obtain (10.38) from (10.39) using the equalities $(\boldsymbol{v}_2 \times \boldsymbol{v}_1) \times \boldsymbol{v}_1 = (\boldsymbol{v}_2 \cdot \boldsymbol{v}_1)\boldsymbol{v}_1 - |\boldsymbol{v}_1|^2\boldsymbol{v}_2$, $\boldsymbol{v}_1 \cdot \boldsymbol{v}_2 = \boldsymbol{v}_{u'u} \cdot \boldsymbol{B}_{uu'} \cdot \boldsymbol{v}_{u'',u'} = \frac{\gamma}{\alpha\beta} - 1$ and $|\boldsymbol{v}_1|^2 = \frac{\alpha^2-1}{\alpha^2}$.

10.6.5 Measuring relative velocity

In practice, standard relative velocities are measured by radar apparatus. A light signal launched at the standard instant t is reflected from a vehicle and returns after a time period $2\boldsymbol{t}_1$. A second light signal launched $\boldsymbol{s}$ time interval later returns after a time period $2\boldsymbol{t}_2$. Then (now, from an everyday practical point of view, we write c for the light speed) at the instant $t + \boldsymbol{t}_1$ the distance between the apparatus and the vehicle is $c\boldsymbol{t}_1$ and at the instant $t + \boldsymbol{s} + \boldsymbol{t}_2$ the distance is $c\boldsymbol{t}_2$. So, the magnitude of the relative velocity is

$$v := \left|\frac{c\boldsymbol{t}_2 - c\boldsymbol{t}_1}{(t + \boldsymbol{s} + \boldsymbol{t}_2) - (t + \boldsymbol{t}_1)}\right| = \frac{\boldsymbol{t}_1 - \boldsymbol{t}_2}{\boldsymbol{s} - (\boldsymbol{t}_1 - \boldsymbol{t}_2)}c.$$

This formula contains only time periods measured by the radar apparatus.

Knowing the magnitude of the standard relative velocity, the boost of a vector parallel to the relative velocity can be realized according to Section 9.6. Namely, $\boldsymbol{t}' = \frac{2\boldsymbol{d}'}{c}$ is the time period between the direct and the reflected light signals starting towards $\boldsymbol{u}$ from the end point of the $\boldsymbol{u}'$-space vector, and $\boldsymbol{t}_-$ is the time period between their arrivals at a $\boldsymbol{u}$-space point. Then the length $\boldsymbol{d}'$ can be computed from equality (10.29).

10.6.6 Exercises

1. Prove equalities (10.38) and (10.39).
2. The transitivity of boosts holds (see Subsection 10.1.4) if and only if $\boldsymbol{u}$, $\boldsymbol{u}'$ and $\boldsymbol{u}''$ are coplanar. Show that this is equivalent to that $\boldsymbol{v}_{u''u}$ is parallel to $\boldsymbol{v}_{u'u}$ or $\boldsymbol{v}_{u''u'}$ is parallel to $\boldsymbol{v}_{uu'}$.
3. $(1,1,0,0)$ is an absolute light direction in the arithmetic spacetime model. What are its standard relative velocities with respect to $\gamma(1,v,0,0)$ and $\gamma'(1,0,v',0)$?

10.7 Standard vectorial splitting and transformation rules

10.7.1 Preliminaries

For an absolute velocity $\boldsymbol{u}$, we list some formulas, similar to the nonrelativistic ones in Subsection 5.6.1:
(i) $\tau_u : \mathbf{M} \to \mathbb{T}$, $\tau_u \cdot \boldsymbol{x} = -\boldsymbol{u} \cdot \boldsymbol{x}$, its transpose $\tau_u^* : \mathbb{T}^* \to \mathbf{M}^*$ is a linear injection.
(ii) $\boldsymbol{i}_u : \mathbf{S}_u \to \mathbf{M}$ is the embedding, its transpose $\boldsymbol{i}_u^* : \mathbf{M}^* \to \mathbf{S}_u^*$ is a linear surjection.
(iii) $\pi_u = \mathbf{1} + \boldsymbol{u} \otimes \boldsymbol{u} : \mathbf{M} \to \mathbf{S}_u$ is a linear surjection, its transpose $\pi_u^* : \mathbf{S}_u^* \to \mathbf{M}^*$ is a linear injection.

Now because of the identifications $\mathbf{M}^* \equiv \frac{\mathbf{M}}{\mathbb{T}\otimes\mathbb{T}}$ and $\mathbf{S}_u^* \equiv \frac{\mathbf{S}_u}{\mathbb{T}\otimes\mathbb{T}}$, we find the following simple relations:
(i) $\tau_u^* = \tau_u = -\boldsymbol{u}$:

$$(\tau_u^* e) \cdot \boldsymbol{x} = e(\tau_u \cdot \boldsymbol{x}) = e(-\boldsymbol{u} \cdot \boldsymbol{x}) = (-e\boldsymbol{u}) \cdot \boldsymbol{x}.$$

(ii) $\boldsymbol{i}_u^* = \pi_u$, and $\pi_u^* = \boldsymbol{i}_u$:

$$(\boldsymbol{i}_u^* \cdot \boldsymbol{x}) \cdot \boldsymbol{p} = \boldsymbol{x} \cdot (\boldsymbol{i}_u \cdot \boldsymbol{p}) = \boldsymbol{x} \cdot \boldsymbol{p} = (\pi_u \cdot \boldsymbol{x})\boldsymbol{p}.$$

10.7.2 Splitting

For an absolute velocity $\boldsymbol{u}$, $\mathbb{T}\boldsymbol{u}$ and $\mathbf{S}_u$ are complementary subspaces, so every world vector $\boldsymbol{x}$ can be uniquely given as the sum of vectors in these subspaces: $\boldsymbol{x} = (-\boldsymbol{u} \cdot \boldsymbol{x})\boldsymbol{u} + \pi_u \cdot \boldsymbol{x}$ (see (10.5) and (10.10)). Accordingly, (see Figure 10.9)

$$\xi_u := \begin{pmatrix} -\boldsymbol{u} \\ \pi_u \end{pmatrix} : \mathbf{M} \to \begin{pmatrix} \mathbb{T} \\ \mathbf{S}_u \end{pmatrix}, \quad \boldsymbol{x} \mapsto \begin{pmatrix} \tau_u \cdot \boldsymbol{x} \\ \pi_u \cdot \boldsymbol{x} \end{pmatrix} = \begin{pmatrix} -\boldsymbol{u} \cdot \boldsymbol{x} \\ \pi_u \cdot \boldsymbol{x} \end{pmatrix} \tag{10.40}$$

is a linear bijection, called the **standard splitting of world vectors** according to $\boldsymbol{u}$, or simply **vectorial $\boldsymbol{u}$-splitting**;
- $-\boldsymbol{u} \cdot \boldsymbol{x} = \tau_u \cdot \boldsymbol{x}$ is the **$\boldsymbol{u}$-time-like component** of $\boldsymbol{x}$.
- $\pi_u \cdot \boldsymbol{x} = \boldsymbol{x} + \boldsymbol{u}(\boldsymbol{u} \cdot \boldsymbol{x})$ is the **$\boldsymbol{u}$-space-like component** of $\boldsymbol{x}$.

The notation $\tau_u \cdot x$ is useful for comparison with the nonrelativistic case but in applications $-u \cdot x$ is more convenient.

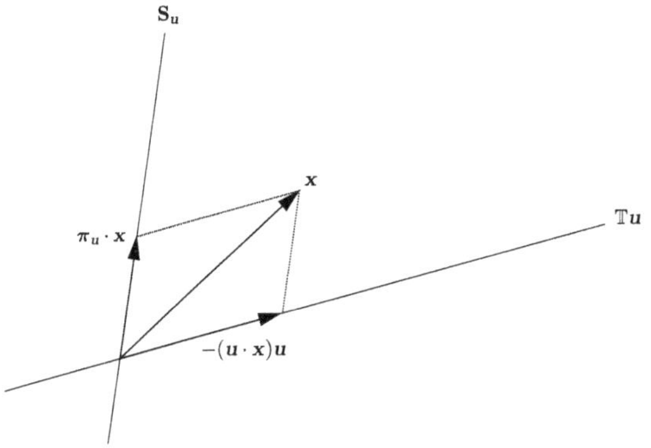

Figure 10.9: Splitting of world vectors.

Note the essential difference from the nonrelativistic formula: there the u-space-like components are in the same vector space $\mathbf{S}$ for all u, whereas here the u-space-like components are in different vector spaces for different u-s.

It is trivial that

$$\xi_u^{-1} \cdot \begin{pmatrix} t \\ q \end{pmatrix} = tu + q = tu + i_u \cdot q$$

for $(t, q) \in \mathbb{T} \times \mathbf{S}_u$.

Giving the Lorentz form

$$g_u\left(\begin{pmatrix} s \\ p \end{pmatrix}, \begin{pmatrix} t \\ q \end{pmatrix}\right) := -st + p \cdot q \tag{10.41}$$

on $\mathbb{T} \times \mathbf{S}_u$, we have

$$g(y, x) = -(-u \cdot y)(-u \cdot x) + (\pi_u \cdot y) \cdot (\pi_u \cdot x) = g_u(\xi_u \cdot y, \xi_u \cdot x).$$

Of course, the tensor products and quotients of $\mathbf{M}$ by measure lines such as $\frac{\mathbf{M}}{\mathbb{T}}$ or $\frac{\mathbf{M}}{\mathbb{T} \otimes \mathbb{T}}$ are split in a similar way, because the multiplication and division by element of measure lines can be interchanged with linear maps. For instance,

– The u-time-like component of an absolute velocity u',

$$\tau_u \cdot u' = -u \cdot u',$$

is the relativistic factor between u and u'.

– The $\boldsymbol{u}$-space-like component of $\boldsymbol{u}'$,

$$\pi_{\boldsymbol{u}} \cdot \boldsymbol{u}' = \boldsymbol{u}' + (\boldsymbol{u} \cdot \boldsymbol{u}')\boldsymbol{u}$$

is the standard relative velocity of $\boldsymbol{u}'$ with respect to $\boldsymbol{u}$, multiplied by the relativistic factor; thus,

$$\xi_{\boldsymbol{u}} \cdot \boldsymbol{u}' = \frac{1}{\sqrt{1 - |\boldsymbol{v}_{\boldsymbol{u}'\boldsymbol{u}}|^2}} \begin{pmatrix} 1 \\ \boldsymbol{v}_{\boldsymbol{u}'\boldsymbol{u}} \end{pmatrix}.$$

The splitting of vectors determines the splitting of covectors by the formula

$$(\xi_{\boldsymbol{u}}{}^{-1})^* : \mathbf{M}^* \to \mathbb{T}^* \times \mathbf{S}_{\boldsymbol{u}}^*, \tag{10.42}$$

called the **splitting of world covectors** according to $\boldsymbol{u}$, or simply **covectorial $\boldsymbol{u}$-splitting.**

For a covector $\boldsymbol{k}$, similar to the nonrelativistic formula (see in Subsection 5.6.2),

$$(\xi_{\boldsymbol{u}}{}^{-1})^* \cdot \boldsymbol{k} = (\boldsymbol{k} \cdot \boldsymbol{u}, \boldsymbol{k} \cdot \boldsymbol{i}_{\boldsymbol{u}}) = (\boldsymbol{u} \cdot \boldsymbol{k}, \pi_{\boldsymbol{u}} \cdot \boldsymbol{k}),$$

so

$$(\xi_{\boldsymbol{u}}{}^{-1})^* = (\boldsymbol{u}, \pi_{\boldsymbol{u}}).$$

According to the identification $\mathbf{M}^* \equiv \frac{\mathbf{M}}{\mathbb{T}\otimes\mathbb{T}}$, a covector can also be split as a vector; then the formula above and (10.40) show that:

(i) The **$\boldsymbol{u}$-time-like component** in the covectorial splitting is the negative of the $\boldsymbol{u}$-time-like component in the vectorial splitting.

(ii) The **$\boldsymbol{u}$-space-like component** in the covectorial splitting equals the $\boldsymbol{u}$-space-like component in the vectorial splitting.

The inverse of the covectorial splitting is $\xi_{\boldsymbol{u}}{}^* : (\mathbb{T}^* \quad \mathbf{S}_{\boldsymbol{u}}^*) \to \mathbf{M}^*$,

$$\xi_{\boldsymbol{u}}{}^* \cdot (\boldsymbol{e}, \boldsymbol{p}) = -\boldsymbol{e}\boldsymbol{u} + \boldsymbol{p}$$

for $(\boldsymbol{e}.\boldsymbol{p}) \in \mathbb{T}^* \times \mathbf{S}_{\boldsymbol{u}}^*$.

Splitting of vectors and covectors has the physical meaning that a standard inertial frame $\boldsymbol{u}$ perceives the components of vectors and covectors, not the vectors and covectors themselves; for instance, we have seen already that the $\boldsymbol{u}$-space-like component of $\boldsymbol{u}'$ divided by its $\boldsymbol{u}$-time-like component is just the standard relative velocity of $\boldsymbol{u}'$ with respect to $\boldsymbol{u}$.

Arithmetic formulas

Split forms of vectors may be highly complicated in the arithmetic model because the space-like components depend on the absolute velocity making the splitting. The single exception is the absolute velocity $(1,0,0,0)$, which establishes the splitting

$$\mathbb{R}^4 \to \begin{pmatrix} \mathbb{R} \\ \mathbb{R}^3 \end{pmatrix}, \quad \begin{pmatrix} x^0 \\ x^1 \\ x^2 \\ x^3 \end{pmatrix} \to \begin{pmatrix} x_0 \\ \begin{pmatrix} x^1 \\ x^2 \\ x^3 \end{pmatrix} \end{pmatrix},$$

similar to the arithmetic formula in Subsection 5.6.2.

In an other splitting, the time-like components are real numbers, the space-like components are in a three-dimensional subspace which, however, can only be given by four components. For instance, the vector $\begin{pmatrix} x^0 \\ x^1 \\ x^2 \\ x^3 \end{pmatrix}$, split by $\gamma \begin{pmatrix} 1 \\ v \\ 0 \\ 0 \end{pmatrix}$, has

- The time-like component $\gamma(x^0 - vx^1)$
- The space-like component $\begin{pmatrix} -\gamma^2 v(vx^0 - x^1) \\ \gamma^2(-vx^0 + x^1) \\ x^2 \\ x^3 \end{pmatrix}$

so the split form of the vector could be written in the terrible expression

$$\begin{pmatrix} \gamma(x^0 - vx^1) \\ \begin{pmatrix} -\gamma^2 v(vx^0 - x^1) \\ \gamma^2(-vx^0 + x^1) \\ x^2 \\ x^3 \end{pmatrix} \end{pmatrix}. \tag{10.43}$$

We see again, how arithmetic formulas can be complicated.

10.7.3 Transformation rules

Different standard inertial frames split world vectors differently. The comparison of different splits is not so simple as in the nonrelativistic case because the split forms are elements of different vector spaces. More closely, the split forms $(\boldsymbol{t},\boldsymbol{q})$ and $(\boldsymbol{t}',\boldsymbol{q}')$ of the same vector according to $\boldsymbol{u}$ and $\boldsymbol{u}'$ are elements of $\mathbb{T}\times\mathbf{S}_u$ and $\mathbb{T}\times\mathbf{S}_{u'}$, respectively. That is why $\xi_{u'}\cdot\xi_u^{-1} : \mathbb{T}\times\mathbf{S}_u \to \mathbb{T}\times\mathbf{S}_{u'}$ is not suitable for the comparison.

Now we can utilize the boost, which makes a natural correspondence between $\mathbf{S}_{u'}$ and $\mathbf{S}_u$. Namely, the **vectorial transformation rule** from the $\boldsymbol{u}$-splitting to the $\boldsymbol{u}'$-splitting is defined to be

$$\xi_{u'u} := (1, \boldsymbol{B}_{uu'})(\xi_{u'}\cdot\xi_u^{-1}) : \mathbb{T}\times\mathbf{S}_u \to \mathbb{T}\times\mathbf{S}_u \tag{10.44}$$

where $(1, \boldsymbol{B}_{uu'})$ means that the first component (an element of $\mathbb{T}$) is multiplied by 1 and $\boldsymbol{B}_{uu'}$ is applied to the second component (an element of $\mathbf{S}_{u'}$).

With the standard relative velocity $\boldsymbol{v}_{u'u} := \frac{\boldsymbol{u}'}{-\boldsymbol{u}\cdot\boldsymbol{u}'} - \boldsymbol{u}$ and the notation $v := |\boldsymbol{v}_{u'u}|$,

$$\gamma(v) := \frac{1}{\sqrt{1-|v|^2}} = -\boldsymbol{u}\cdot\boldsymbol{u}',$$

$$J(\boldsymbol{v}_{u'u}) := \frac{1}{\gamma(v)}\left(\mathrm{id}_{\mathbf{S}_u} + \frac{\gamma(v)^2}{1+\gamma(v)}\boldsymbol{v}_{u'u}\otimes\boldsymbol{v}_{u'u}\right) \tag{10.45}$$

we obtain the vectorial transformation rule: if $(\boldsymbol{t}, \boldsymbol{q})$ is the split form of a vector due to $\boldsymbol{u}$ and $(\boldsymbol{t}', \boldsymbol{q}')$ is its split form due to $\boldsymbol{u}'$, **boosted** to $\boldsymbol{u}$, then

$$\begin{pmatrix}\boldsymbol{t}'\\ \boldsymbol{q}'\end{pmatrix} = \gamma(v)\begin{pmatrix}\boldsymbol{t} - \boldsymbol{v}_{u'u}\cdot\boldsymbol{q}\\ J(\boldsymbol{v}_{u'u})\cdot\boldsymbol{q} - \boldsymbol{t}\boldsymbol{v}_{u'u}\end{pmatrix}.$$

In a block matrix form,

$$\xi_{u'u} = \gamma(v)\begin{pmatrix}1 & -\boldsymbol{v}_{u'u}\\ -\boldsymbol{v}_{u'u} & J(\boldsymbol{v}_{u'u})\end{pmatrix} \tag{10.46}$$

The equality

$$\boldsymbol{u}'\cdot\boldsymbol{q} = (-\boldsymbol{u}\cdot\boldsymbol{u}')\left(\frac{\boldsymbol{u}'}{-\boldsymbol{u}\cdot\boldsymbol{u}'} - \boldsymbol{u}\right)\cdot\boldsymbol{q} = \gamma(v)\boldsymbol{v}_{u'u}\cdot\boldsymbol{q}$$

will be used repeatedly.

The $\boldsymbol{u}'$-time-like component of $\xi_u^{-1}\cdot\left(\begin{smallmatrix}t\\ q\end{smallmatrix}\right)$ is

$$\xi_{u'}\cdot(\boldsymbol{u}\boldsymbol{t}+\boldsymbol{q}) = (-\boldsymbol{u}'\cdot\boldsymbol{u})\boldsymbol{t} - \boldsymbol{u}'\cdot\boldsymbol{q} = \gamma(v)(\boldsymbol{t} - \boldsymbol{v}_{u'u}\cdot\boldsymbol{q}) = \boldsymbol{t}'.$$

The $\boldsymbol{u}'$-space-like component of $\xi_u^{-1}\cdot\left(\begin{smallmatrix}t\\ q\end{smallmatrix}\right)$, boosted to $\boldsymbol{u}$ is

$$\begin{aligned}\boldsymbol{B}_{uu'}\cdot\pi_{u'}\cdot(\boldsymbol{u}\boldsymbol{t}+\boldsymbol{q}) &= \boldsymbol{B}_{uu'}\cdot((\gamma(v)\boldsymbol{v}_{uu'}\boldsymbol{t} + \boldsymbol{B}_{uu'}\cdot\boldsymbol{q} + \boldsymbol{u}'(\boldsymbol{u}'\cdot\boldsymbol{q}))\\ &= -\gamma(v)\boldsymbol{v}_{u'u}\boldsymbol{t} + \boldsymbol{B}_{uu'}\cdot\boldsymbol{q} + \boldsymbol{u}(\boldsymbol{u}'\cdot\boldsymbol{q}) = \boldsymbol{q}'.\end{aligned}$$

Since $\boldsymbol{B}_{uu'}$ maps $\mathbf{S}_{u'}$ onto $\mathbf{S}_u$, we can apply π_u to the last expression without changing its value; the first term remains as it is, the third term disappears, and

$$\pi_u\cdot\boldsymbol{B}_{uu'}\cdot\boldsymbol{q} = \boldsymbol{q} + \frac{(\pi_u\cdot\boldsymbol{u}')(\boldsymbol{u}'\cdot\boldsymbol{q})}{1-\boldsymbol{u}\cdot\boldsymbol{u}'} = \frac{\gamma(v)^2\boldsymbol{v}_{u'u}(\boldsymbol{v}_{u'u}\cdot\boldsymbol{q})}{1+\gamma(v)}.$$

We get a simpler and more familiar form if the element $\boldsymbol{q}$ of $\mathbf{S}_u$ is given as a sum of two components, $\boldsymbol{q}_\parallel$ being parallel to $\boldsymbol{v}_{u'u}$ and $\boldsymbol{q}_\perp$ being perpendicular (orthogonal) to $\boldsymbol{v}_{u'u}$. Then $(\boldsymbol{t}, \boldsymbol{q}) = (\boldsymbol{t}, \boldsymbol{q}_\parallel) + (\boldsymbol{0}, \boldsymbol{q}_\perp)$ and

$$\boldsymbol{q}'_\perp = \boldsymbol{q}_\perp, \tag{10.47}$$

$$\boldsymbol{t}' = \frac{1}{\sqrt{1-v^2}}(\boldsymbol{t} - \boldsymbol{v}_{u'u}\cdot\boldsymbol{q}_\parallel), \quad \boldsymbol{q}'_\parallel = \frac{1}{\sqrt{1-v^2}}(\boldsymbol{q}_\parallel - \boldsymbol{t}\boldsymbol{v}_{u'u}). \tag{10.48}$$

This is just the **Lorentz transformation rule**, well known usually with $\mathbb{R}$ and $\mathbb{R}^3$ instead of $\mathbb{T}$ and $\mathbf{S}_u$, respectively.

Note that

a Lorentz transformation is a linear map $\mathbf{M} \to \mathbf{M}$,
a Lorentz transformation rule is a linear map $\mathbb{T} \times \mathbf{S}_u \to \mathbb{T} \times \mathbf{S}_u$.

Because of the Lorentz identification of covectors and vectors, the splitting of covectors is essentially (apart from a minus sign) the same as the splitting of vectors; as a consequence, the **covectorial transformation rule** is essentially (apart from a minus sign) the same as the vectorial transformation rule. More closely, if $(\boldsymbol{e}, \boldsymbol{p})$ is the $\boldsymbol{u}$-split form of a covector, then its $\boldsymbol{u}'$-split form, **boosted** to $\boldsymbol{u}$ is

$$\boldsymbol{p}'_\perp = \boldsymbol{p}_\perp,$$
$$\boldsymbol{e}' = \frac{1}{\sqrt{1-v^2}}(\boldsymbol{e} + \boldsymbol{v}_{u'u} \cdot \boldsymbol{p}_\|), \quad \boldsymbol{p}'_\| = \frac{1}{\sqrt{1-v^2}}(\boldsymbol{p}_\| + \boldsymbol{e}\boldsymbol{v}_{u'u}).$$

Arithmetic formulas

(10.43) and (10.23) convince us not to even try to deal with general transformation rules, We only consider the transformation rule from $\boldsymbol{u} := (1,0,0,0)$ to $\boldsymbol{u}' := \gamma(1,v,0,0)$.

Let $\left(\begin{pmatrix} t \\ x \\ y \\ z \end{pmatrix}\right)$ be the $\boldsymbol{u}$-split form of a vector. Then

$$\xi_u^{-1} \cdot \left(\begin{pmatrix} t \\ x \\ y \\ z \end{pmatrix}\right) = \begin{pmatrix} t \\ x \\ y \\ z \end{pmatrix}$$

whose $\boldsymbol{u}'$-split form is (see (10.43))

$$\left(\begin{pmatrix} \gamma(t - vx) \\ -\gamma^2 v(vt - x) \\ \gamma^2(x - vt) \\ y \\ z \end{pmatrix}\right).$$

The transformation rule from the $\boldsymbol{u}$-splitting to the $\boldsymbol{u}'$-splitting is obtained

- By the first component in the outer parentheses above.
- By applying the boost (10.22) to the components in the inner parentheses.

Then we have

$$\gamma \begin{pmatrix} 1 & -v & 0 & 0 \\ -v & 1 & 0 & 0 \\ 0 & 0 & 1 & 0 \\ 0 & 0 & & 1 \end{pmatrix} \gamma^2 \begin{pmatrix} v(x - vt) \\ x - vt \\ y \\ z \end{pmatrix} = \begin{pmatrix} 0 \\ \gamma(x - vt) \\ y \\ z \end{pmatrix}.$$

Omitting the zero from the right side, the transformation rule in question becomes

$$\left(\begin{pmatrix} t \\ x \\ y \\ z \end{pmatrix}\right) \mapsto \left(\begin{pmatrix} \gamma(t - vx) \\ \gamma(x - vt) \\ y \\ z \end{pmatrix}\right). \tag{10.49}$$

Ignoring the inner parentheses, we get

$$\begin{pmatrix} t \\ x \\ y \\ z \end{pmatrix} \mapsto \begin{pmatrix} \gamma(t - vx) \\ \gamma(x - vt) \\ y \\ z \end{pmatrix}, \tag{10.50}$$

which is just the action of the boost (10.22) on (t, x, y, z).

1. A boost in usual treatments of special relativity is a Lorentz transformation rule ("passive transformation") when "the corresponding axes of the moving spatial coordinate systems are parallel," which is considered a fact not to be explained (in our precise notation it is (10.49)). Then, ignoring the difference between $\mathbb{R} \times \mathbb{R}^3$ and $\mathbb{R}^4$, it is considered the Lorentz transformation ("active transformation") "without rotation" (10.50).
2. Boosts in our treatment are Lorentz transformations, which among others, make sense to transformation rules.
3. A boost in usual treatments is always a transformation rule between the "rest frame" $(1, 0, 0, 0)$ and a "moving frame" $\gamma(1, v, 0, 0)$. The question does not even arise: what is the transformation rule between two "moving frames," $\gamma(1, v, 0, 0)$ and $\gamma'(1, 0, v', 0)$? The asymmetry of "resting" and "moving" suggests a false point of view causing the mistaken length contraction, time dilation, and mass increase (see Section 10.12 and Section 11.3).

10.8 Standard tensorial splitting and transformation rules

10.8.1 Splitting

In a number of physical theories—e. g., in electromagnetism and continuum mechanics—not only vectors and covectors but various tensors are also included. The mathematical supplement helps the reader to be familiar with tensors.

The various tensors, i. e., the elements of $\mathbf{M} \otimes \mathbf{M}$, $\mathbf{M} \otimes \mathbf{M}^*$, $\mathbf{M}^* \otimes \mathbf{M}$, $\mathbf{M}^* \otimes \mathbf{M}^*$, too, can be split by absolute velocities. Their splitting can be obtained as nonrelativistically, with the modification that:

(i) τ and τ^* are replaced by $-\boldsymbol{u}$.
(ii) $\boldsymbol{i}$, $\boldsymbol{i}^*$, and $\pi_{\boldsymbol{u}}^*$ are replaced by $\pi_{\boldsymbol{u}}$.

(1) The tensor $\boldsymbol{G} \in \mathbf{M} \otimes \mathbf{M}$ has the $\boldsymbol{u}$-split form

$$\begin{pmatrix} \boldsymbol{u} \cdot \boldsymbol{G} \cdot \boldsymbol{u} & -\boldsymbol{u} \cdot \boldsymbol{G} \cdot \pi_{\boldsymbol{u}} \\ -\pi_{\boldsymbol{u}} \cdot \boldsymbol{G} \cdot \boldsymbol{u} & \pi_{\boldsymbol{u}} \cdot \boldsymbol{G} \cdot \pi_{\boldsymbol{u}} \end{pmatrix}. \tag{10.51}$$

The antisymmetric tensors play a fundamental role in electromagnetism.
If $\boldsymbol{G}$ is an antisymmetric tensor, i. e., $\boldsymbol{G} = -\boldsymbol{G}^*$, then $\boldsymbol{u} \cdot \boldsymbol{G} = -\boldsymbol{G} \cdot \boldsymbol{u}$. So, $\boldsymbol{u} \cdot \boldsymbol{G} \cdot \boldsymbol{u} = 0$, and that is why it has the $\boldsymbol{u}$-split form

$$\begin{pmatrix} 0 & -\boldsymbol{u} \cdot \boldsymbol{G} \\ -\boldsymbol{G} \cdot \boldsymbol{u} & \boldsymbol{G} - \boldsymbol{u} \wedge (\boldsymbol{u} \cdot \boldsymbol{G}) \end{pmatrix}.$$

Here, the two "lower" components determine the other ones; therefore, we refer to the $\boldsymbol{u}$-split form as

$$((-\boldsymbol{G} \cdot \boldsymbol{u}, \boldsymbol{G} - \boldsymbol{u} \wedge (\boldsymbol{u} \cdot \boldsymbol{G}))) \in (\mathbf{S}_{\boldsymbol{u}} \otimes \mathbb{T}) \times (\mathbf{S}_{\boldsymbol{u}} \wedge \mathbf{S}_{\boldsymbol{u}}). \tag{10.52}$$

They are called the **$\boldsymbol{u}$-time-like component** and the **$\boldsymbol{u}$-space-like component** of $\boldsymbol{G}$, respectively.
It is a simple fact that if the $\boldsymbol{u}$-split form of $\boldsymbol{G}$ is $((\boldsymbol{D}_{\boldsymbol{u}}, \boldsymbol{H}_{\boldsymbol{u}}))$, then

$$\boldsymbol{G} = \boldsymbol{H}_{\boldsymbol{u}} - \boldsymbol{u} \wedge \boldsymbol{D}_{\boldsymbol{u}}. \tag{10.53}$$

(2) The cotensor $\boldsymbol{F} \in \mathbf{M}^* \otimes \mathbf{M}^*$ has the $\boldsymbol{u}$-split form

$$\begin{pmatrix} \boldsymbol{u} \cdot \boldsymbol{F} \cdot \boldsymbol{u} & \boldsymbol{u} \cdot \boldsymbol{F} \cdot \pi_{\boldsymbol{u}} \\ \pi_{\boldsymbol{u}} \cdot \boldsymbol{F} \cdot \boldsymbol{u} & \pi_{\boldsymbol{u}} \cdot \boldsymbol{F} \cdot \pi_{\boldsymbol{u}} \end{pmatrix}. \tag{10.54}$$

The antisymmetric cotensors of $\mathbf{M}^* \otimes \mathbf{M}^*$ play a fundamental role in electromagnetism.
If $\boldsymbol{F}$ is an antisymmetric cotensor, i. e., $\boldsymbol{F} = -\boldsymbol{F}^*$ then $\boldsymbol{u} \cdot \boldsymbol{F} = -\boldsymbol{F} \cdot \boldsymbol{u}$. So, $\boldsymbol{u} \cdot \boldsymbol{F} \cdot \boldsymbol{u} = 0$, and it has the $\boldsymbol{u}$-split form

$$\begin{pmatrix} 0 & -\boldsymbol{F} \cdot \boldsymbol{u} \\ \boldsymbol{F} \cdot \boldsymbol{u} & \boldsymbol{F} + \boldsymbol{u} \wedge (\boldsymbol{u} \cdot \boldsymbol{F}) \end{pmatrix}.$$

As previously, only the "lower" components will be considered for referring to the $\boldsymbol{u}$-split form:

$$((\boldsymbol{F} \cdot \boldsymbol{u}, \boldsymbol{F} + \boldsymbol{u} \wedge (\boldsymbol{u} \cdot \boldsymbol{F}))) \in (\mathbf{S}_{\boldsymbol{u}}^* \otimes \mathbb{T}^*) \times (\mathbf{S}_{\boldsymbol{u}}^* \wedge \mathbf{S}_{\boldsymbol{u}}^*). \tag{10.55}$$

They are called **$\boldsymbol{u}$-time-like component** and the **$\boldsymbol{u}$-space-like component** of $\boldsymbol{F}$, respectively.
It is a simple fact that if the $\boldsymbol{u}$-split form of $\boldsymbol{F}$ is $((\boldsymbol{E}_{\boldsymbol{u}}, \boldsymbol{Bu}))$, then

$$\boldsymbol{F} = \boldsymbol{B}_{\boldsymbol{u}} + \boldsymbol{u} \wedge \boldsymbol{E}_{\boldsymbol{u}}. \tag{10.56}$$

(3) The $\boldsymbol{u}$-split form of $\boldsymbol{P} \in \mathbf{M}^* \otimes \mathbf{M}$ is

$$\begin{pmatrix} -\boldsymbol{u}\cdot\boldsymbol{P}\cdot\boldsymbol{u} & \boldsymbol{u}\cdot\boldsymbol{P}\cdot\pi_u \\ -\pi_u\cdot\boldsymbol{P}\cdot\boldsymbol{u} & \pi_u\cdot\boldsymbol{P}\cdot\pi_u \end{pmatrix}.$$

(4) The $\boldsymbol{u}$-split form of a Lorentz transformation $\boldsymbol{L} \in \mathbf{M} \otimes \mathbf{M}^*$ is

$$\begin{pmatrix} -\boldsymbol{u}\cdot\boldsymbol{L}\cdot\boldsymbol{u} & -\boldsymbol{u}\cdot\boldsymbol{L}\cdot\pi_u \\ \pi_u\cdot\boldsymbol{L}\cdot\boldsymbol{u} & \pi_u\cdot\boldsymbol{L}\cdot\pi_u \end{pmatrix}.$$

In particular, the components of the $\boldsymbol{u}$-split form of the boost $\boldsymbol{B}_{u'u}$ from $\boldsymbol{u}$ to $\boldsymbol{u}'$ are easily obtained: with the notation $v := |\boldsymbol{v}_{u'u}|$,

$$-\boldsymbol{u}\cdot\boldsymbol{B}_{u'u}\cdot\boldsymbol{u} = -\boldsymbol{u}\cdot\boldsymbol{u}' = \frac{1}{\sqrt{1-v^2}},$$

$$-\boldsymbol{u}\cdot\boldsymbol{B}_{u'u}\cdot\pi_u = \pi_u\cdot\boldsymbol{B}_{u'u}\cdot\boldsymbol{u} = \frac{1}{\sqrt{1-v^2}}\boldsymbol{v}_{u'u},$$

$$\pi_u\cdot\boldsymbol{B}_{u'u}\cdot\pi_u = \pi_u + \frac{\pi_u\cdot\boldsymbol{u}'\otimes\pi_u\cdot\boldsymbol{u}'}{1-\boldsymbol{u}\cdot\boldsymbol{u}'} = \mathrm{id}_{\mathbf{S}_u} + \frac{\gamma(v)^2\boldsymbol{v}_{u'u}\otimes\boldsymbol{v}_{u'u}}{1+\gamma(v)}.$$

After all, in view of (10.45), the $\boldsymbol{u}$-split form of the boost from $\boldsymbol{u}$ to $\boldsymbol{u}'$ is

$$\gamma(v)\begin{pmatrix} 1 & \boldsymbol{v}_{u'u} \\ \boldsymbol{v}_{u'u} & J(\boldsymbol{v}_{u'u}) \end{pmatrix}. \tag{10.57}$$

The $\boldsymbol{u}'$-time reversal (10.24) has the $\boldsymbol{u}$-split form

$$\frac{1}{1-v^2}\begin{pmatrix} -(1+v^2) & 2\boldsymbol{v}_{u'u} \\ -2\boldsymbol{v}_{u'u} & (1-v^2)\mathbf{1} + 2\boldsymbol{v}_{u'u}\otimes\boldsymbol{v}_{u'u} \end{pmatrix}, \tag{10.58}$$

which equals the linear map $(\boldsymbol{t},\boldsymbol{q}) \mapsto (-\boldsymbol{t},\boldsymbol{q})$ if and only if $\boldsymbol{u}' = \boldsymbol{u}$.

(5) Third-order tensors, $\mathbf{M}\otimes\mathbf{M}\otimes\mathbf{M}$, $\mathbf{M}\otimes\mathbf{M}\otimes\mathbf{M}^*$, $\mathbf{M}^*\otimes\mathbf{M}^*\otimes\mathbf{M}$, etc. also appear in some applications. They can be split by considering them linear maps $\mathbf{M}^* \to \mathbf{M}\otimes\mathbf{M}$, $\mathbf{M} \to \mathbf{M}\otimes\mathbf{M}$, $\mathbf{M}^* \to \mathbf{M}^*\otimes\mathbf{M}^*$, etc. In particular, we encounter antisymmetric elements of $\mathbf{M}^*\otimes\mathbf{M}^*\otimes\mathbf{M}^*$ in electromagnetism.

(6) **Tensors multiplied or divided** by some measure line also appear in practice; their splits have the same forms.

Arithmetic formulas

In view of (10.43), it is not worth wasting time for terrible $\otimes$ terrible formulas of split tensors.

10.8.2 Transformation rules

Comparing different splits of various tensorial splits, we get tensorial transformation rules. We treat only the formulas concerning antisymmetric tensors and cotensors.

If $((\boldsymbol{D},\boldsymbol{H}))$ and $((\boldsymbol{D}',\boldsymbol{H}'))$ are the $\boldsymbol{u}$-split form and the $\boldsymbol{u}'$-split form **boosted** to $\boldsymbol{u}$, respectively, of an antisymmetric tensor, then $((\boldsymbol{D}',\boldsymbol{H}')) = \xi_{u'u} \cdot ((\boldsymbol{D},\boldsymbol{H})) \cdot \xi^*_{u'u}$; using the notation $\boldsymbol{v} := \boldsymbol{v}_{u'u}$ and (10.45) we have

$$((\boldsymbol{D}',\boldsymbol{H}')) = \gamma(v)^2 \begin{pmatrix} 1 & -\boldsymbol{v} \\ -\boldsymbol{v} & J(\boldsymbol{v}) \end{pmatrix} \begin{pmatrix} 0 & -\boldsymbol{D} \\ \boldsymbol{D} & \boldsymbol{H} \end{pmatrix} \begin{pmatrix} 1 & -\boldsymbol{v} \\ -\boldsymbol{v} & J(\boldsymbol{v}) \end{pmatrix}.$$

Here, too, the decomposition

$$\boldsymbol{D} = \boldsymbol{D}_\perp + \boldsymbol{D}_\parallel, \quad \boldsymbol{H} = \boldsymbol{H}_\perp + \boldsymbol{H}_\parallel$$

gives a more tractable form where $\boldsymbol{D}_\perp$ and $\boldsymbol{D}_\parallel$ are the $\boldsymbol{u}$-space-like vectors perpendicular and parallel to $\boldsymbol{v}$, respectively, $\boldsymbol{H}_\perp$ and $\boldsymbol{H}_\parallel$ are antisymmetric $\boldsymbol{u}$-space-like tensors whose kernel is perpendicular and parallel to $\boldsymbol{v}$, respectively. Then

$$\boldsymbol{D}'_\perp = \boldsymbol{D}_\perp, \quad \boldsymbol{H}'_\perp = \boldsymbol{H}_\perp,$$
$$\boldsymbol{D}'_\parallel = \frac{1}{\sqrt{1-v^2}}(\boldsymbol{D}_\parallel + \boldsymbol{H}_\parallel \cdot \boldsymbol{v}), \quad \boldsymbol{H}'_\parallel = \frac{1}{\sqrt{1-v^2}}(\boldsymbol{v} \wedge \boldsymbol{D}_\parallel + \boldsymbol{H}_\parallel).$$

Apart from two minus signs, similar transformation rule holds for antisymmetric cotensors, too. More closely, if $((\boldsymbol{E},\boldsymbol{B}))$ and $((\boldsymbol{E}',\boldsymbol{B}'))$ are the $\boldsymbol{u}$-split form and the $\boldsymbol{u}'$-split form **boosted** to $\boldsymbol{u}$, respectively, of an antisymmetric cotensor then

$$\boldsymbol{E}'_\perp = \boldsymbol{E}_\perp, \quad \boldsymbol{B}'_\perp = \boldsymbol{B}_\perp,$$
$$\boldsymbol{E}'_\parallel = \frac{1}{\sqrt{1-v^2}}(\boldsymbol{E}_\parallel - \boldsymbol{B}_\parallel \cdot \boldsymbol{v}), \quad \boldsymbol{B}'_\parallel = \frac{1}{\sqrt{1-v^2}}(-\boldsymbol{v} \wedge \boldsymbol{E}_\parallel + \boldsymbol{B}_\parallel).$$

10.8.3 Exercises

1. Show that the $\boldsymbol{u}$-split components of a tensor have the following detailed forms:

$$-\boldsymbol{u} \cdot \boldsymbol{G} \cdot \pi_u = -\boldsymbol{u} \cdot \boldsymbol{G} - \boldsymbol{u}(\boldsymbol{u} \cdot \boldsymbol{G} \cdot \boldsymbol{u}), \quad -\pi_u \cdot \boldsymbol{G} \cdot \boldsymbol{u} = -\boldsymbol{G} \cdot \boldsymbol{u} - \boldsymbol{u}(\boldsymbol{u} \cdot \boldsymbol{G} \cdot \boldsymbol{u}),$$
$$\pi_u \cdot \boldsymbol{G} \cdot \pi_u = \boldsymbol{G} + \boldsymbol{u} \otimes (\boldsymbol{u} \cdot \boldsymbol{G}) + (\boldsymbol{G} \cdot \boldsymbol{u}) \otimes \boldsymbol{u} + \boldsymbol{u} \otimes \boldsymbol{u}(\boldsymbol{u} \cdot \boldsymbol{G} \cdot \boldsymbol{u}).$$

2. Prove equalities (10.53) and (10.56).
3. Demonstrate (10.58).

10.9 Standard spacetime splitting and transformation rules

10.9.1 Splitting

A standard reference frame $\boldsymbol{u}$ characterizes occurrences by giving when and where they happen: it **splits** spacetime into $\boldsymbol{u}$-time and $\boldsymbol{u}$-space; to a world point x it assigns the corresponding $\boldsymbol{u}$-time point $\tau_u(x) = x + \mathbf{S}_u$ and the corresponding $\boldsymbol{u}$-space point $\pi_u(x) = x + \mathbb{T}\boldsymbol{u}$. The splitting (see Figure 10.10)

$$\xi_u : \mathrm{M} \to \mathrm{T}_u \times \mathrm{S}_u, \quad x \mapsto (\tau_u(x), \pi_u(x)), \tag{10.59}$$

is an affine bijection over the vectorial splitting $\boldsymbol{\xi}_u$ as it is well seen from equalities (10.33), (10.11), and (10.40). The inverse of this splitting—which gives the world point corresponding to a $\boldsymbol{u}$-time point and a $\boldsymbol{u}$-space point—is

$$\xi_u^{-1} : \mathrm{T}_u \times \mathrm{S}_u \to \mathrm{M}, \quad (t, q) \mapsto t \cap q.$$

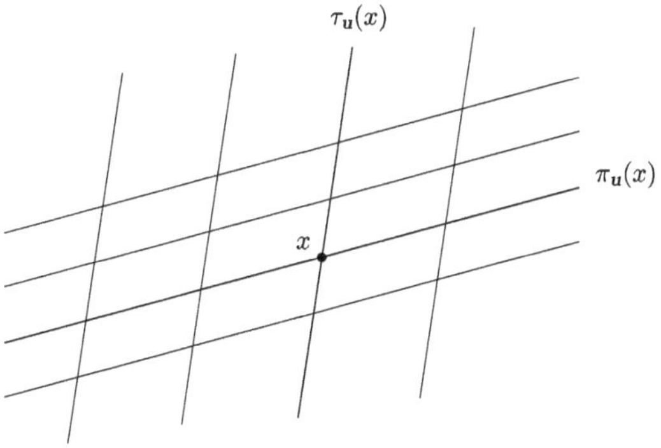

Figure 10.10: Standard splitting of spacetime.

Instead of affine spaces it is often more suitable to deal with the underlying vector spaces; therefore, a standard reference frame—corresponding to the everyday usage when time points are represented by time intervals that passed from a given time point and space points are represented by vectors from an origin—choosing an "initial" $\boldsymbol{u}$-time point t_o and a $\boldsymbol{u}$-"origin" q_o, vectorizes $\boldsymbol{u}$-time and $\boldsymbol{u}$-space by the assignment

$$\mathrm{T}_u \times \mathrm{S}_u \to \mathbb{T} \times \mathbf{S}_u, \quad (t, q) \mapsto (t - t_o, q - q_o). \tag{10.60}$$

Choosing a t_o and a q_o is equivalent to choosing a "spacetime origin" $o := t_o \cap q_o$; then $t_o = \tau(o) = o + \mathbf{S}_u$, $q_o = \pi_u(o) = o + \mathbb{T}\boldsymbol{u}$. The pair $(\boldsymbol{u}, o)$ is called the **standard inertial frame $\boldsymbol{u}$ with origin** o.

It is a simple fact that the $\boldsymbol{u}$-splitting of spacetime followed by the vectorization of $\boldsymbol{u}$-time and $\boldsymbol{u}$-space is the same as the vectorization of spacetime followed by the vectorial splitting according to $\boldsymbol{u}$; thus, the **vectorized splitting of spacetime** by o and $\boldsymbol{u}$ is

$$\xi_{u,o} : \mathrm{M} \to \mathbb{T} \times \mathbf{S}_{u}, \quad x \mapsto \boldsymbol{\xi}_{u} \cdot (x - o) = \big(-\boldsymbol{u} \cdot (x - o), \boldsymbol{\pi}_{u} \cdot (x - o)\big). \tag{10.61}$$

It has the inverse

$$\mathbb{T} \times \mathbf{S}_{u} \to \mathrm{M}, \quad (\boldsymbol{t}, \boldsymbol{q}) \mapsto o + \boldsymbol{t}\boldsymbol{u} + \boldsymbol{q}. \tag{10.62}$$

Briefly,

$$\xi_{u,o}^{-1} = o + \boldsymbol{\xi}_{u}^{-1}.$$

10.9.2 Transformation rules

The transformation rule between two spacetime splits would be $\xi_{u',o'}\xi_{u,o}^{-1}$ from $\mathrm{T}_u \times \mathrm{S}_u \to \mathrm{T}_{u'} \times \mathrm{S}_{u'}$. We face the problem again that the domain and the range of this map are different. To overcome this problem, we consider vectorized splitting combined with boosts, getting a transformation rule $\mathbb{T} \times \mathbf{S}_u \to \mathbb{T} \times \mathbf{S}_u$.

Let $(\boldsymbol{t}, \boldsymbol{q})$ and $(\boldsymbol{t}', \boldsymbol{q}')$ be the vectorized split forms of a world point, due to $(\boldsymbol{u}, o)$ and $(\boldsymbol{u}', o')$, respectively, the latter one boosted to $\boldsymbol{u}$. Then

$$(\boldsymbol{t}', \boldsymbol{q}') = (1, \boldsymbol{B}_{uu'}) \cdot \big(\xi_{u',o'}(\xi_{u,o}^{-1}(\boldsymbol{t}, \boldsymbol{q}))\big).$$

Omitting a trivial step coming from the previous formulas and putting $\boldsymbol{t}_0 := -\boldsymbol{u}' \cdot (o' - o)$ and $\boldsymbol{q}_0 = \boldsymbol{B}_{uu'} \cdot \boldsymbol{\pi}_{u'} \cdot (o' - o)$, we obtain

$$\boldsymbol{t}' = \gamma(v)(\boldsymbol{t} - \boldsymbol{v}_{u'u} \cdot \boldsymbol{q}) - \boldsymbol{t}_0, \quad \boldsymbol{q}' = \gamma(v)(\boldsymbol{J}(\boldsymbol{v}_{u'u}) \cdot \boldsymbol{q} - \boldsymbol{t}\boldsymbol{v}_{u'u}) - \boldsymbol{q}_0, \tag{10.63}$$

which is called the **inhomogeneous Lorentz transformation rule** (well known usually with $\mathbb{R}$ and $\mathbb{R}^3$ instead of $\mathbb{T}$ and $\mathbf{S}_u$, respectively).

This equals the vectorial Lorentz transformation rule if the standard inertial frames choose the same spacetime origin ($o' = o$). In other words, the standard inertial frames choose $\boldsymbol{u}$-time point t_o and $\boldsymbol{u}$-space point q_o, $\boldsymbol{u}'$-time point t_o' and $\boldsymbol{u}'$-space points q_o' in such a way that the meeting of q_o and q_o' occurs at the $\boldsymbol{u}$-instant t_o and at the $\boldsymbol{u}'$-instant t_o' ($t_o \cap q_o = t_o' \cap q_o'$).

Note that

an inhomogeneous Lorentz transformation rule is an affine map, which serves for comparing splits of spacetime,

a Lorentz transformation rule is a linear map, which serves for comparing the splits of world vectors.

10.9.3 Exercises

1. Prove that (10.59) followed by (10.60) equals (10.61).
2. Prove that $\xi_{u,o}$ is an affine map over $\boldsymbol{\xi}_{\boldsymbol{u}}$.

10.10 Transformations and transformation rules

It is worth repeating the main points of transformations and transformation rules.

1. Spacetime vectors

The $\boldsymbol{u}$-split forms of Lorentz transformations are more complicated, in general, than the $\boldsymbol{u}$-split forms of Galilei transformations in the nonrelativistic case. Exception is the $\boldsymbol{u}$-split form of the boost from $\boldsymbol{u}$ to $\boldsymbol{u}'$ (see (10.57)),

$$\gamma(v)\begin{pmatrix} 1 & \boldsymbol{v}_{u'u} \\ \boldsymbol{v}_{u'u} & J(\boldsymbol{v}_{u'u}) \end{pmatrix},$$

which is similar in form to the Lorentz transformation rule from $\boldsymbol{u}$ to $\boldsymbol{u}'$ (see (10.46))

$$\gamma(v)\begin{pmatrix} 1 & -\boldsymbol{v}_{u'u} \\ -\boldsymbol{v}_{u'u} & J(\boldsymbol{v}_{u'u}) \end{pmatrix}.$$

Keep in mind that
Lorentz transformations are linear bijections $\mathbf{M} \to \mathbf{M}$:

(1) Split forms of Lorentz transformations are linear bijections $\mathbb{T} \times \mathbf{S} \to \mathbb{T} \times \mathbf{S}$.
(2) Lorentz transformation rules, too, are linear bijections $\mathbb{T} \times \mathbf{S} \to \mathbb{T} \times \mathbf{S}$,

which may have similar mathematical forms but are completely different from a physical point of view.

2. Spacetime

Let L be a Poincaré transformation over the Lorentz transformation $\boldsymbol{L}$. Since $L(x) = L(o) + \boldsymbol{L}(x - o)$, the $(\boldsymbol{u}, o)$-split form of L is

$$\begin{aligned} \xi_{u,o} L \xi_{u,o}^{-1} &= \xi_{u,o} L(o + \boldsymbol{\xi}_{\boldsymbol{u}}^{-1}) = \xi_{u,o}(L(o) + \boldsymbol{L} \cdot \boldsymbol{\xi}_{\boldsymbol{u}}^{-1}) \\ &= \boldsymbol{\xi}_{\boldsymbol{u}} \cdot (L(o) - o + \boldsymbol{L} \cdot \boldsymbol{\xi}_{\boldsymbol{u}}^{-1}) = \boldsymbol{\xi}_{\boldsymbol{u}} \cdot (L(o) - o) + \boldsymbol{\xi}_{\boldsymbol{u}} \cdot \boldsymbol{L} \cdot \boldsymbol{\xi}_{\boldsymbol{u}}^{-1}. \end{aligned} \tag{10.64}$$

Formally, this is the same as (5.60), only its detailed expression is different because the $\boldsymbol{u}$-split form of a Lorentz transformation is much more complicated than the $\boldsymbol{u}$-split form of a Galilei transformation, even for boosts. Nevertheless, we can repeat, according to the sense, what we said in Section 5.9.

Keep in mind that

Poincaré transformations are $\mathrm{M} \to \mathrm{M}$ affine bijections:

(1) Split forms of Poincaré transformations are affine bijections $\mathbb{T} \times \mathbf{S} \to \mathbb{T} \times \mathbf{S}$.
(2) Inhomogeneous Lorentz transformation rules are also affine bijections $\mathbb{T} \times \mathbf{S} \to \mathbb{T} \times \mathbf{S}$,

which may have similar mathematical forms but are completely different from a physical point of view.

Remarks.

1. Formula (10.64) enlightens why the Poincaré group is usually called the inhomogeneous Lorentz group.
2. If spacetime is treated in split form (and even in coordinates) then transformations (spacetime symmetries) and transformation rules have similar form.

A distinction is made usually by saying [20, 2]
- **Active** transformations (which correspond to spacetime symmetries).
- **Passive** transformations (which correspond to transformation rules).

Nevertheless, they are often confused, which causes conceptual errors as we pointed out in the nonrelativistic case, too (see Section 5.9). An excellent example is the following [19, volume I, p. 50]: "A symmetry transformation is a change of our point of view that does not change the result of possible experiments. If the observer O sees a system in a state ..., then an equivalent observer O' who looks at the same system will observe it in a different state ..., but the two observers must find the same transition probabilities."

3. A quantity is usually called "covariant" if it is "well transformed" when changing from a reference frame to another and one requires to use covariant quantities in physical theories. We require **absolute** quantities; the split forms of absolute quantities are well transformed automatically.

4. For a clear distinction, we summarize:
(1) split forms of transformation arise from **one** symmetry due to **one** absolute velocity,
(2) transformation rules compare **two** split forms of **one** symmetry that are due to **two** absolute velocities.

5. The (inhomogeneous) Lorentz transformation rule is a **single** axiom in usual treatments, while it is a consequence of **two** different ways of splitting and **one** boost. Among others this exhibits the

efficacy of our treatment similar to the efficacy of a magnifier glass, which shows that one smear consists of three spots.

6. Transformation rules are the cornerstone of usual treatments, while they have only practical importance in the spacetime model. On the contrary, spacetime transformations have an essential importance.

10.11 Coordinate systems

In practice, to make computations, a convenient coordinate system is introduced in which time points are represented by numbers and space points are represented by triplets of numbers; thus, world points are represented by quartets of numbers.

This is done as follows. A standard inertial frame $\boldsymbol{u}$ chooses:

(i) A time unit s
(ii) An "origin" t_o in time
(iii) An "origin" q_o in $\boldsymbol{u}$-space
(iv) Three orthogonal straight lines (coordinate axes) in $\boldsymbol{u}$-space passing through the origin, directed by the "right-handed" orthogonal space vectors $(\boldsymbol{e}_1, \boldsymbol{e}_2, \boldsymbol{e}_3)$ of length s

and represents the world points by time coordinates and space coordinates, elements of $\mathbb{R} \times \mathbb{R}^3$ in such a way (in our setting) that

1. Splits spacetime M into $\boldsymbol{u}$-time $\mathrm{T}_{\boldsymbol{u}}$ and $\boldsymbol{u}$-space $S_{\boldsymbol{u}}$, $x \mapsto (t, q)$ where $t := \tau(x) = x + \mathbf{S}_{\boldsymbol{u}}$ and $q := \pi_{\boldsymbol{u}}(x) = x + \mathbb{T}\boldsymbol{u}$ (thus, $x = t \cap q$).
2. Represents the $\boldsymbol{u}$-time point t by the real number $\frac{t-t_o}{\mathrm{s}}$.
3. Represents the $\boldsymbol{u}$-space point q by the three real numbers $\frac{\boldsymbol{e}_1\cdot(q-q_o)}{\mathrm{s}^2}, \frac{\boldsymbol{e}_2\cdot(q-q_o)}{\mathrm{s}^2}, \frac{\boldsymbol{e}_3\cdot(q-q_o)}{\mathrm{s}^2}$.

This procedure having a clear physical meaning is equivalent to the following mathematical procedure: with the world "origin" $o := t_o \cap q_o$ (i. e., $t_o = o + \mathbf{S}_{\boldsymbol{u}}$, $q_o = o + \mathbb{T}\boldsymbol{u}$), the world point x is represented by the four coordinates

$$\xi^0 := \frac{-\boldsymbol{u}\cdot(x-o)}{\mathrm{s}}, \quad \xi^k := \frac{\boldsymbol{e}_k \cdot \pi_{\boldsymbol{u}} \cdot (x-o)}{\mathrm{s}^2} \quad (k = 1, 2, 3).$$

Then, if the coordinates are known, the world point is

$$x = o + \xi^0 \mathrm{s}\boldsymbol{u} + \xi^1 \boldsymbol{e}_1 + \xi^2 \boldsymbol{e}_2 + \xi^3 \boldsymbol{e}_3.$$

It is convenient to introduce $\boldsymbol{e}_0 := \mathrm{s}\boldsymbol{u}$ and according to the formulas above, a **standard inertial coordinate system** is $(o, \mathrm{s}, \boldsymbol{e}_0, \boldsymbol{e}_1, \boldsymbol{e}_2, \boldsymbol{e}_3)$, where o is a world point, s is a time unit, $(\boldsymbol{e}_0, \boldsymbol{e}_1, \boldsymbol{e}_2, \boldsymbol{e}_3)$ is a positively oriented Lorentz orthogonal basis, normalized to s.

Returning to Section 10.2, we can state that such a standard coordinate system establishes an isomorphism to the arithmetic spacetime model. We see very well, how many arbitrary objects are hidden in the use of the arithmetic spacetime model: a spacetime origin, a time unit, a standard inertial frame, and three orthogonal vectors in the space of the inertial frame.

In a concise form: a **coordination** of spacetime by a standard coordinate system is

$$C : \mathrm{M} \to \mathbb{R}^4, \quad x \mapsto (\xi^0, \xi^1, \xi^2, \xi^3). \tag{10.65}$$

C is an affine map, the underlying linear map $\boldsymbol{C}$ represents vectors, covectors, and various tensors by coordinates, too. The coordinates $\boldsymbol{C} \cdot \boldsymbol{x}$ of a vector $\boldsymbol{x}$ are

$$x^i := \frac{\boldsymbol{e}_i \cdot \cdot \boldsymbol{x}}{\mathrm{s}^2} \quad (i = 0, 1, 2, 3),$$

the coordinates $(\boldsymbol{C}^{-1})^* \cdot \boldsymbol{k}$ of a covector $\boldsymbol{k}$ are

$$(\boldsymbol{C}^{-1})^* \boldsymbol{k} = k_i := \boldsymbol{k} \cdot \boldsymbol{e}_i \quad (i = 0, 1, 2, 3).$$

The indexes of the coordinates of a covector are usually written as subscripts, for distinguishing them from the coordinates of vectors.

Now, because of the Lorentz identification $\mathbf{M}^* \equiv \frac{\mathbf{M}}{\mathbb{T} \otimes \mathbb{T}}$, coordinates of covectors are obtained essentially in the same way as coordinates of vectors: the difference is a minus sign in the zeroth (time-like) coordinate. Accordingly, one can consider the vectorial coordinates $(x^i : i = 0, 1, 2, 3)$ of a vector $\boldsymbol{x}$ (which are the coordinates of $\boldsymbol{x}$ itself) and the covectorial coordinates $(x_i : i = 0, 1, 2, 3)$ of $\boldsymbol{x}$ (which are the coordinates of $\frac{\boldsymbol{x}}{\mathrm{s}^2}$ as a covector); then $x_0 = -x^0$, $x_k = x^k$ $(k = 1, 2, 3)$.

The Lorentz form $\boldsymbol{g}$ in coordinates has the form

$$g_{ik} := \frac{\boldsymbol{g}(\boldsymbol{e}_i, \boldsymbol{e}_k)}{\mathrm{s}^2};$$

thus, $g_{00} = -1$, $g_{kk} = 1$ for $k = 1, 2, 3$, and $g_{ik} = 0$ if $i \neq k$.

Introducing $g^{ik} := g_{ik}$, we can formulate the identification of vectors and covectors by $\boldsymbol{g}$, i. e., the pulling of indices up and down as follows:

$$x_i = g_{ik} x^k, \quad x^i = g^{ik} x_k \quad \text{(Einstein summation)}.$$

The coordinates of a tensor $\boldsymbol{G} \in \mathbf{M} \otimes \mathbf{M}$ are $(i, k = 1, 2, 3)$:

$$G^{00} := \frac{\boldsymbol{e}_0 \cdot \boldsymbol{G} \cdot \boldsymbol{e}_0}{\mathrm{s}^2}, \quad G^{0k} = -\frac{\boldsymbol{e}_0 \cdot \boldsymbol{G} \cdot \boldsymbol{e}_k}{\mathrm{s}^2}, \quad G^{k0} = -\frac{\boldsymbol{e}_k \cdot \boldsymbol{G} \cdot \boldsymbol{e}_0}{\mathrm{s}^2}, \quad G^{ik} = \frac{\boldsymbol{e}_i \cdot \boldsymbol{G} \cdot \boldsymbol{e}_k}{\mathrm{s}^2}.$$

The coordinates of a cotensor $\boldsymbol{F} \in \mathbf{M}^* \otimes \mathbf{M}^*$ are $(i, k = 0, 1, 2, 3)$:

$$F_{ik} = \boldsymbol{e}_i \cdot \boldsymbol{F} \cdot \boldsymbol{e}_k.$$

Remark. It is emphasized that all the formulas above are valid only for standard coordinate systems. There are other types of coordinate systems, too, e. g., in which the basis of spacetime vectors consists of four future-like vectors.

Now more complications may arise if spacetime is treated in coordinates than in the nonrelativistic case. For instance, it gets lost that the space vectors of different observers are different, and that is why it is taken for granted that the relative velocities of two observers are opposite to each other; this resulted in the paradox which will be treated in Section 15.3.

Exercises

1. Two coordinate systems are equivalent by the general definition if there is a proper Poincaré transformation that maps one of them to the other. Prove that the standard coordinate systems $(o, \mathrm{s}, \boldsymbol{e}_0, \boldsymbol{e}_1, \boldsymbol{e}_2, \boldsymbol{e}_3)$ and $(o', \mathrm{s}', \boldsymbol{e}_0', \boldsymbol{e}_1', \boldsymbol{e}_2', \boldsymbol{e}_3')$ are equivalent if and only if $\mathrm{s}' = \mathrm{s}$.
2. Give the coordinates of the world point x in the coordinate system defined by a world point o and an arbitrary basis $\boldsymbol{n}_1, \boldsymbol{n}_2, \boldsymbol{n}_3, \boldsymbol{n}_4$ of $\mathbf{M}$.
3. Define a coordinate system in which a standard inertial frame splits spacetime and uses spherical coordinates in its space.

10.12 Comparison of lengths and comparison of time intervals

10.12.1 Length contraction?

Though boosts are already defined by light signals, it is worth examining instantaneous prints that gave the boosts in the nonrelativistic case.

Let us consider a vector $\boldsymbol{q}'$, a "rod with tip" in the space of an inertial observer $\boldsymbol{u}'$. The print of $\boldsymbol{q}'$ made by a standard inertial frame $\boldsymbol{u}$ is obtained by marking those $\boldsymbol{u}$-space points that meet the rod at an arbitrary standard $\boldsymbol{u}$-instant.

To formulate this procedure, recall that the spaces of the inertial observers $\boldsymbol{u}'$ and $\boldsymbol{u}$ are the sets of lines in spacetime directed by $\boldsymbol{u}'$ and $\boldsymbol{u}$, respectively.

Let $q' := x + \mathbb{T}\boldsymbol{u}'$ and $p' := x + \boldsymbol{q}' + \mathbb{T}\boldsymbol{u}'$ be the end points of the rod in the $\boldsymbol{u}'$-space.

The instantaneous print of p' and q' in $\boldsymbol{u}$-space are p and q in such a way that the meeting point of p' and p is $\boldsymbol{u}$-simultaneous with the meeting point of q' and q. Then $q := x + \mathbb{T}\boldsymbol{u}$ and $p := x + \boldsymbol{q}' + \boldsymbol{t}'\boldsymbol{u}' + \mathbb{T}\boldsymbol{u}$ where $\boldsymbol{t}'$ is determined by the condition that the difference of $x + \boldsymbol{q}' + \boldsymbol{t}'\boldsymbol{u}'$ and x are $\boldsymbol{u}$-simultaneous (see Figure 10.11):

$$\boldsymbol{q}' + \boldsymbol{t}'\boldsymbol{u}' =: \boldsymbol{q} \quad \text{is in} \quad \mathbf{S}_{\boldsymbol{u}}.$$

Taking the Lorentz product of this equality by $\boldsymbol{u}$ and then eliminating $\boldsymbol{t}'$, we get

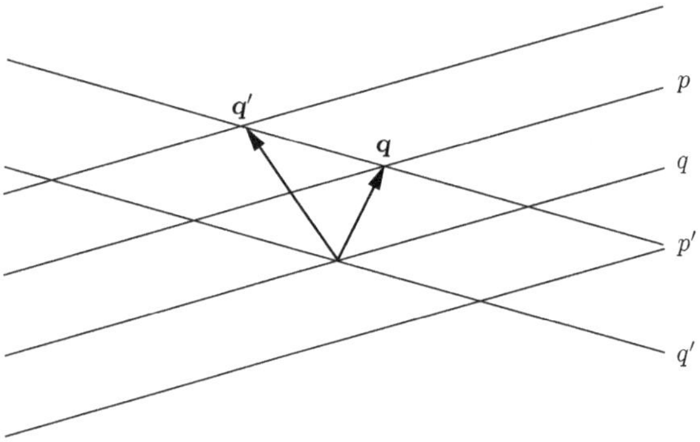

Figure 10.11: $p - q = \boldsymbol{q}$ is the $\boldsymbol{u}$-simultaneous print of $p' - q' = \boldsymbol{q}'$.

$$\boldsymbol{q} = \boldsymbol{q}' + \frac{\boldsymbol{u} \cdot \boldsymbol{q}'}{-\boldsymbol{u} \cdot \boldsymbol{u}'}\boldsymbol{u}' = \boldsymbol{q}' + \left(\left(\frac{\boldsymbol{u}}{-\boldsymbol{u} \cdot \boldsymbol{u}'} - \boldsymbol{u}'\right) \cdot \boldsymbol{q}'\right)\boldsymbol{u}' = \boldsymbol{q}' + (\boldsymbol{v}_{\boldsymbol{u}\boldsymbol{u}'} \cdot \boldsymbol{q}')\boldsymbol{u}';$$

thus, the square length of the print is

$$|\boldsymbol{q}|^2 = |\boldsymbol{q}'|^2 - (\boldsymbol{v}_{\boldsymbol{u}\boldsymbol{u}'} \cdot \boldsymbol{q}')^2.$$

This is the famous **Lorentz contraction**: If the vector is perpendicular to the relative velocity of $\boldsymbol{u}$ with respect to $\boldsymbol{u}'$, then the length of the vector and the length of its print are equal. Otherwise the print is shorter; it is the shortest if the vector is parallel to the relative velocity: then its length is $\sqrt{1 - |\boldsymbol{v}_{\boldsymbol{u}\boldsymbol{u}'}|^2}$ times the length of the vector.

The Lorentz contraction is often declared so that a "moving rod is contracted." This is mistaken:

The rod moves with respect to a lot of inertial frames, so it would shorten in a lot of different manners, which is nonsense.

Simple example: If $\boldsymbol{n}' := \frac{\boldsymbol{q}'}{|\boldsymbol{q}'|}$, then $\boldsymbol{u}_\alpha := \frac{\boldsymbol{u}'+\alpha\boldsymbol{n}'}{\sqrt{1-\alpha^2}}$ is an absolute velocity for all $0 < \alpha < 1$, $\boldsymbol{v}_{\boldsymbol{u}_\alpha\boldsymbol{u}'} = \alpha\boldsymbol{n}'$, and the rod would shorten $\sqrt{1-\alpha^2}$ times according to $\boldsymbol{u}_\alpha$.

Moreover, using a synchronization different from the standard one, we get a different result (see Section 13.1).

The Lorentz contraction is not a physical fact, it is an illusion connected with a synchronization.

We emphasize that physical relations "demonstrated" by the Lorentz contraction can be accepted only if a demonstration without synchronization can be found.

A typical example for the incorrect use of Lorentz contraction is the following [15]. Let us consider a disk of radius r rotating with angular velocity of magnitude ω. Consider a radial rod: it moves perpendicularly to itself, so its "moving length" is r. Pick up a tiny segment of the circumference. It moves parallel to itself, so it would be Lorentz contracted if it "was possible." Namely, the disk is "rigid," so the tiny segments hinder each other to be shorter. On the other hand, a tiny measuring rod perpendicular to the radius is contracted, so more of them is necessary to cover the circumference as if the disk were at rest. Taking into account the Lorentz contraction of the measuring rod, the length of the circumference becomes $\frac{2r\pi}{\sqrt{1-(\omega r)^2}}$. So, the Euclidean geometry does not hold for the rotating disk.

The following questions show the absurdity of this reasoning. Why is the disk rigid? Why are the measuring rods not rigid? If the disk is rigid, then it is of the same size as if it did not rotate, is not it? Its circumference is larger because it is measured with defective rods?

The reasoning has two essential errors. The more important one is that the contraction is taken to be a physical fact without any synchronization. The second one is that the Lorentz contraction refers to inertial objects but the measuring rods on the rotating disk are not inertial.

The rotating rigid disk and the distances—measured by light signals—on it are not simple but having their precise definition, we can establish the ratio of the radius and the circumference. This can be found in the book *Spacetime without Reference Frames.*

10.12.2 Time dilation?

An inertial observer wants to measure how a moving chronometer works, i. e., to make a comparison between the frequency of ticks of a moving chronometer and the frequency of ticks of its own chronometers.

The chronometer moves with respect to the observer, so its different ticks occur in different space points of the observer. As a consequence, the observer cannot make the comparison using a single chronometer of its (resting in some space point of the observer). To fulfil the task, a synchronization is necessary.

Let us consider a standard inertial frame $\boldsymbol{u}$ measuring the frequency of ticks of a chronometer with absolute velocity $\boldsymbol{u}'$.

Let the chronometer meet the $\boldsymbol{u}$-space points $x + \mathbb{T}\boldsymbol{u}$ and $x + \boldsymbol{t}'\boldsymbol{u}' + \mathbb{T}\boldsymbol{u}$. Then the $\boldsymbol{u}$-time period $\boldsymbol{t}$ between the meetings is determined by the condition that $x + \boldsymbol{t}'\boldsymbol{u}'$ and $x + \boldsymbol{t}\boldsymbol{u}$ are $\boldsymbol{u}$-simultaneous (see Figure 10.12), i. e., their difference is Lorentz orthogonal to $\boldsymbol{u}$: $\boldsymbol{u} \cdot (\boldsymbol{t}'\boldsymbol{u}' - \boldsymbol{t}\boldsymbol{u}) = 0$; thus,

$$\boldsymbol{t} := \boldsymbol{t}'(-\boldsymbol{u} \cdot \boldsymbol{u}') = \frac{\boldsymbol{t}'}{\sqrt{1 - |\boldsymbol{v}_{\boldsymbol{u}\boldsymbol{u}'}|^2}}.$$

This is the famous **time dilation**: the measured time period is longer than the proper time period.

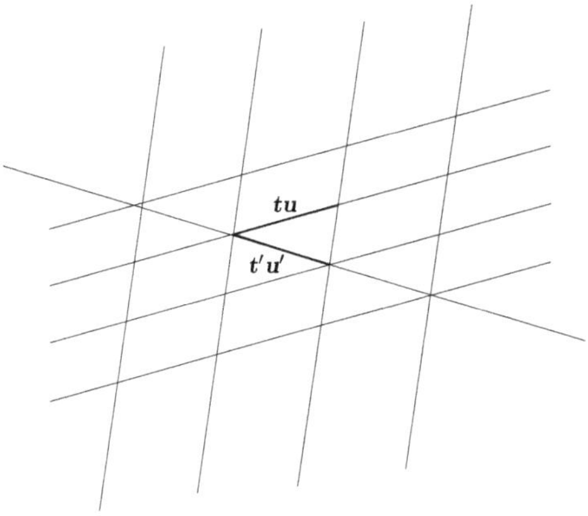

Figure 10.12: The standard inertial frame measures time period $\boldsymbol{t}$ corresponding to the proper time period $\boldsymbol{t}'$ of the chronometer.

The time dilation is often declared so that a "moving clock (chronometer) works more slowly than a resting one." This is mistaken:

The chronometer moves with respect to a lot of inertial frames, so it would slow down in a lot of different manners, which is nonsense.

Simple example: $\boldsymbol{u}_\alpha := \frac{\alpha\boldsymbol{u}'+\boldsymbol{n}'}{\sqrt{\alpha^2-1}}$ is an absolute velocity for all $\alpha > 1$ and $\boldsymbol{n}' \in \mathbf{S}_{\boldsymbol{u}'}$, $|\boldsymbol{n}'| = 1$; $-\boldsymbol{u}_\alpha \cdot \boldsymbol{u}' = \frac{\alpha}{\sqrt{\alpha^2-1}}$, so the chronometer in the $\boldsymbol{u}'$-space would slow down $\frac{\alpha}{\sqrt{\alpha^2-1}}$ times according to $\boldsymbol{u}_\alpha$.

Moreover, using a synchronization different from the standard one, we get a different result (see Section 13.1).

The time dilation is not a physical fact, it is an illusion connected with a synchronization.

We emphasize that physical relations "demonstrated" by the time dilation can be accepted only if a demonstration without synchronization can be found.

11 Elements of point mechanics in the spacetime model

Classical mechanics is a well-known and well-elaborated theory (in coordinates, as usual), and that is why it offers a good possibility to deepen our knowledge on spacetime and it very well shows the important physical difference between the nonrelativistic theory and the relativistic theory.

11.1 World line functions

A nonrelativistic world line can be parameterized by absolute time in a natural way. Here, there is no such a possibility; we take the generally defined proper time parametrization and world line function (see Subsection 2.3.4).

If r is a world line function, then $\dot{r}(\boldsymbol{s})$ is an absolute velocity, $\dot{r}(\boldsymbol{s}) \cdot \dot{r}(\boldsymbol{s}) = -1$, so it is immediate that

$$\dot{r}(\boldsymbol{s}) \cdot \ddot{r}(\boldsymbol{s}) = 0, \quad \text{i. e.,} \quad \ddot{r}(\boldsymbol{s}) \in \frac{\mathbf{S}_{\dot{r}(\boldsymbol{s})}}{\mathbb{T} \otimes \mathbb{T}}.$$

The momentary absolute acceleration is Lorentz orthogonal to the actual absolute velocity, $\ddot{r}(\boldsymbol{s})$ *is* $\dot{r}(\boldsymbol{s})$*-space-like.*

We find it convenient to use the symbol $\dot{x}$ for denoting absolute velocities regarding world lines (see in Section 11.3 and later). So, we assert that the set of **absolute accelerations belonging to the absolute velocity** $\dot{x}$ is the three- dimensional Euclidean vector space $\frac{\mathbf{S}_{\dot{x}}}{\mathbb{T} \otimes \mathbb{T}}$. In contrast to the absolute velocities,

- There is a zero absolute acceleration.
- An absolute acceleration has a magnitude.
- The angle between two absolute accelerations belonging to the same absolute velocity is meaningful (but it is not necessarily meaningful if they belong to different absolute velocities).

Exercises

1. Show that the transform of a world line function r by an arrow orientation preserving Poincaré transformation L,

$$\boldsymbol{s} \mapsto L(r(\boldsymbol{s}))$$

   is a world line function.
2. Recall (10.26) and show that if r is a world line function then

$$\boldsymbol{s} \mapsto T_{\boldsymbol{u},o} r(\boldsymbol{s}_0 + (\boldsymbol{s}_0 - \boldsymbol{s}))$$

   is also a world line function where $\boldsymbol{s}_0$ is an arbitrary given proper time value.

https://doi.org/10.1515/9783112219553-014

11.2 Movements

11.2.1 Proper time and the synchronization time

Recall that the description of motion with respect to an observer requires a synchronization; in other words, a motion can be described only with respect to a reference frame (consisting of an observer and a synchronization) by giving when and where a material point is.

A standard inertial frame $\boldsymbol{u}$ perceives the history of a material point (a world line) as a motion and describes it by assigning to a $\boldsymbol{u}$-instant t (a hyperplane directed by $\mathbf{S}_u$) the $\boldsymbol{u}$-space point (straight line directed by $\boldsymbol{u}$), which meets the world line in question at t. Then the $\boldsymbol{u}$-motion corresponding to the world line function r is (see Figure 11.1)

$$\mathrm{T}_u \to \mathrm{S}_u, \quad t \mapsto r_u(t) := \pi_u(t \cap \mathrm{Ran}(r)).$$

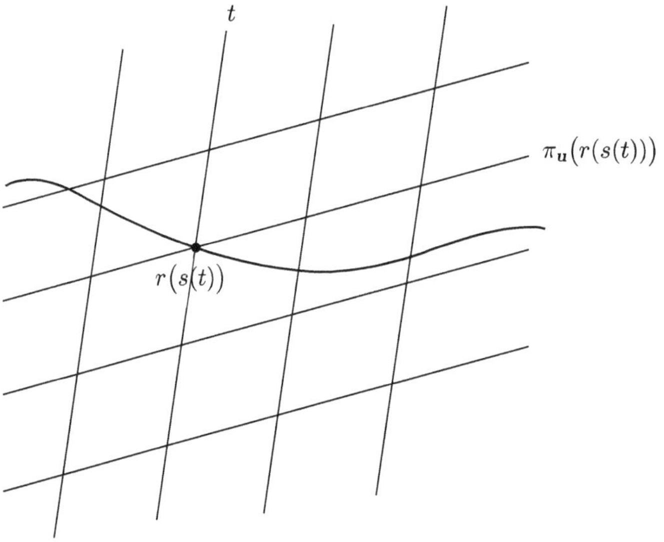

Figure 11.1: Description of motion.

This function can be well handled if we give the relation between the proper time of the world line (the variable of r) and the synchronization time (the variable of r_u).

This relation is described by the function that assigns to the proper time value $\boldsymbol{s}$ the $\boldsymbol{u}$-time instant t (hyperplane directed by $\mathbf{S}_u$) containing $r(\boldsymbol{s})$:

$$t(\boldsymbol{s}) := r(\boldsymbol{s}) + \mathbf{S}_u = \tau_u(r(\boldsymbol{s})).$$

Then

$$\frac{\mathrm{d}t(\boldsymbol{s})}{\mathrm{d}\boldsymbol{s}} = \tau_{\boldsymbol{u}}\dot{r}(\boldsymbol{s}) = -\boldsymbol{u}\cdot\dot{r}(\boldsymbol{s}) \geq 1; \tag{11.1}$$

therefore, the function $\boldsymbol{s} \mapsto t(\boldsymbol{s})$ is strictly monotone increasing (injective); for its inverse, denoted by $t \mapsto \boldsymbol{s}(t)$,

$$\frac{\mathrm{d}\boldsymbol{s}(t)}{\mathrm{d}t} = \frac{1}{-\boldsymbol{u}\cdot\dot{r}(\boldsymbol{s}(t))} \tag{11.2}$$

holds. As a consequence, $t \cap \mathrm{Ran}(r) = r(\boldsymbol{s}(t))$, so

$$r_{\boldsymbol{u}}(t) = \pi_{\boldsymbol{u}}(r(\boldsymbol{s}(t))).$$

11.2.2 Relative velocities and relative accelerations

The velocity of the material point relative to the standard inertial frame $\boldsymbol{u}$ is the time derivative of the $\boldsymbol{u}$-motion:

$$\begin{aligned}\frac{\mathrm{d}r_{\boldsymbol{u}}(t)}{\mathrm{d}t} &= \frac{\mathrm{d}\pi_{\boldsymbol{u}}(r(\boldsymbol{s}(t)))}{\mathrm{d}t} = \pi_{\boldsymbol{u}}\cdot\frac{\mathrm{d}r(\boldsymbol{s}(t))}{\mathrm{d}t} = \pi_{\boldsymbol{u}}\cdot\frac{\dot{r}(\boldsymbol{s}(t))}{-\boldsymbol{u}\cdot\dot{r}(\boldsymbol{s}(t))}\\ &= \left(\frac{\dot{r}}{-\boldsymbol{u}\cdot\dot{r}} - \boldsymbol{u}\right)(\boldsymbol{s}(t)) = \boldsymbol{v}_{\dot{r}(\boldsymbol{s}(t))\boldsymbol{u}}.\end{aligned} \tag{11.3}$$

This is a generalization of our result concerning inertial world lines, obtained in Subsection 10.6.3. Accordingly, by (10.16) we have

$$-\boldsymbol{u}\cdot\dot{r}(\boldsymbol{s}) = \frac{1}{\sqrt{1-|\boldsymbol{v}_{\dot{r}(\boldsymbol{s})\boldsymbol{u}}|^2}}, \tag{11.4}$$

and if $t \mapsto \boldsymbol{v}(t)$ is the $\boldsymbol{u}$-relative velocity function, then

$$\dot{r}(\boldsymbol{s}) = \frac{\boldsymbol{u}+\boldsymbol{v}(t(\boldsymbol{s}))}{\sqrt{1-|\boldsymbol{v}(t(\boldsymbol{s}))|^2}} \tag{11.5}$$

by (10.36).

In the sequel, we often meet functions depending on the frame time via the proper time. For the sake of avoiding involved formulas, we refer to that by the symbol $\bullet$. If φ is a function of proper time, then $\varphi\bullet$ depends on frame time: $(\varphi\bullet)(t) := \varphi(\boldsymbol{s}(t))$. The differentiation with respect to proper time is denoted by a dot, the differentiation with respect to the frame time is denoted by an inverted comma. Accordingly, on the base of equality (11.2) we have

$$(\varphi\bullet)' = \left(\frac{\dot{\varphi}}{-\boldsymbol{u}\cdot\dot{r}}\right)\bullet.$$

Using this notation, the $\boldsymbol{u}$-relative velocity corresponding to the world line function r is

$$r'_u = \left(\frac{\dot{r}}{-\boldsymbol{u}\cdot\dot{r}} - \boldsymbol{u}\right)\bullet = \boldsymbol{v}_{\dot{r}\boldsymbol{u}}\bullet .$$

Accordingly, the $\boldsymbol{u}$-relative acceleration is

$$r''_u = \left(\left(\ddot{r} + \frac{\dot{r}(\boldsymbol{u}\cdot\ddot{r})}{-\boldsymbol{u}\cdot\dot{r}}\right)\frac{1}{(\boldsymbol{u}\cdot\dot{r})^2}\right)\bullet .$$

Let us add $(\boldsymbol{u}\cdot\ddot{r})\boldsymbol{u}$ to $\ddot{r}$ in the inner parenthesis and let us subtract it from the second member: $\ddot{r} + (\boldsymbol{u}\cdot\ddot{r})\boldsymbol{u} + \frac{\dot{r}(\boldsymbol{u}\cdot\ddot{r})}{-\boldsymbol{u}\cdot\dot{r}} - (\boldsymbol{u}\cdot\ddot{r})\boldsymbol{u}$. Because of $\dot{r}\cdot\ddot{r} = 0$, we get $\boldsymbol{u}\cdot\ddot{r} = -\boldsymbol{v}_{\dot{r}\boldsymbol{u}}\cdot(\boldsymbol{\pi}_{\boldsymbol{u}}\cdot\ddot{r})$. Then applying equality (11.4), we obtain a nicer formula for the relative acceleration:

$$r''_u = (((1 - |\boldsymbol{v}_{\dot{r}\boldsymbol{u}}|^2))\bullet)((\mathbf{1} - \boldsymbol{v}_{\dot{r}\boldsymbol{u}}\otimes\boldsymbol{v}_{\dot{r}\boldsymbol{u}})\boldsymbol{\pi}_{\boldsymbol{u}}\cdot\ddot{r})\bullet .$$

It is not hard to see that this gives

$$(\boldsymbol{\pi}_{\boldsymbol{u}}\cdot\ddot{r})\bullet = \frac{1}{1 - |r'_u|^2}\left(\mathbf{1} + \frac{r'_u\otimes r'_u}{1 - |r'_u|^2}\right)r''_u .$$

Note that, contrary to the nonrelativistic case,

the $\boldsymbol{u}$-relative acceleration is far from being equal to the absolute acceleration.

This is not surprising since the $\boldsymbol{u}$-relative acceleration has all its values in $\frac{\mathbf{S}_u}{\mathbb{T}\otimes\mathbb{T}}$ while the absolute acceleration at the proper time instant $\boldsymbol{s}$ is in $\frac{\mathbf{S}_{\dot{r}(s)}}{\mathbb{T}\otimes\mathbb{T}}$. Even

the $\boldsymbol{u}$-relative acceleration is not equal to the $\boldsymbol{u}$-space-like component of the absolute acceleration.

11.2.3 Exercises

1. A vector $\boldsymbol{x}$ is said to be running parallel along a world line (world line function r) if it depends on the proper time in such a way that its values are related by the corresponding boost, $\boldsymbol{x}(\boldsymbol{t}) = \boldsymbol{B}_{\dot{r}(t),\dot{r}(s)}\cdot\boldsymbol{x}(\boldsymbol{s})$ for all $\boldsymbol{t}$, $\boldsymbol{s}$.
   (i) Prove that $\dot{r}$ is running parallel along the world line.
   (ii) Using that $(\dot{r}(\boldsymbol{t})+\dot{r}(\boldsymbol{s}))\cdot\boldsymbol{x}(\boldsymbol{s}) = (\dot{r}(\boldsymbol{t})-\dot{r}(\boldsymbol{s}))\cdot\boldsymbol{x}(\boldsymbol{s})$ if $\dot{r}(\boldsymbol{s})\cdot\boldsymbol{x}(\boldsymbol{s}) = 0$, prove that a vector Lorentz orthogonal to $\dot{r}$ and running parallel along the world line satisfies the differential equation $\dot{\boldsymbol{x}} = \dot{r}(\ddot{r}\cdot\boldsymbol{x})$.
   (iii) A world line is uniformly accelerated by definition if its absolute acceleration is running parallel along the world line. Show that such a world line function r has the form $r(\boldsymbol{s}) = x_0 + \frac{u_0}{a}\sinh(a\boldsymbol{s}) + \frac{a_0}{a^2}(\cosh(a\boldsymbol{s}) - 1)$.
2. A material point moving on a circle path of radius ρ in the space of the inertial frame $(1, 0, 0, 0)$ (in the arithmetic spacetime model) is described by the function $t \mapsto (0, \rho\cos\frac{vt}{\rho}, \rho\sin\frac{vt}{\rho}, 2)$. What is the corresponding world line?

3. A light signal moving on a circle path of radius ρ in the space of the inertial frame $(1, 0, 0, 0)$ (in the arithmetic spacetime model) is described by the function $t \mapsto (0, \rho \cos \frac{t}{\rho}, \sin \frac{t}{\rho}, 2)$. What is the corresponding light line?

11.3 Absolute and relative momentum

As in the nonrelativistic case—taking the Planck constant to be 1—the measure line of mass is $\frac{\mathbb{T}}{\mathbb{L}\otimes\mathbb{L}}$; now, with the identification $\mathbb{L} = \mathbb{T}$, $(2.99792458\ldots)10^8\mathrm{m} := \mathrm{s}$ the measure line of mass becomes $\frac{\mathbb{R}}{\mathbb{T}}$; then

$$\mathrm{kg} := \frac{2\pi(2.99792458\ldots)^2 10^{16}}{(6.62607015\ldots)10^{-34}} \frac{1}{\mathrm{s}} = (8.52288558\ldots)10^{50} \frac{1}{\mathrm{s}}.$$

This choice is unusual in practical applications but it makes the theoretical exposition easier.

The reader is asked for recalling Section 6.3 in order to better understand the following.

Let us consider a material point having constant mass m having absolute velocity $\dot{x}$. Then we accept that:

1. Absolute momentum = mass × absolute velocity:

$$m\dot{x},$$

which involves that
proper time derivative of absolute momentum = mass × absolute acceleration:

$$(m\dot{x})^{\cdot} = m\ddot{x}.$$

Further, we have that:

2. $\boldsymbol{u}$-time-like component of absolute momentum ≠ mass,

$$\tau_{\boldsymbol{u}} \cdot (m\dot{x}) = m(-\boldsymbol{u} \cdot \dot{x}) = \frac{m}{\sqrt{1 - |\boldsymbol{v}_{\dot{x}\boldsymbol{u}}|^2}}.$$

3. $\boldsymbol{u}$-space-like component of absolute momentum ≠ mass × ($\boldsymbol{u}$-relative velocity):

$$\pi_{\boldsymbol{u}} \cdot (m\dot{x}) = \frac{m}{\sqrt{1 - |\boldsymbol{v}_{\dot{x}\boldsymbol{u}}|^2}} \boldsymbol{v}_{\dot{x}\boldsymbol{u}}.$$

4. $\boldsymbol{u}$-time derivative of $\boldsymbol{u}$-space-like component of absolute momentum ≠ $\boldsymbol{u}$-time derivative of ≠ mass × ($\boldsymbol{u}$-relative velocity) ≠ mass × ($\boldsymbol{u}$-relative acceleration):

$$(\pi_{\boldsymbol{u}} \cdot (m\dot{x})_{\bullet})' \neq m(\boldsymbol{v}_{\dot{x}\boldsymbol{u}\bullet})'$$

because

$$(\pi_u \cdot (m\dot{x})\bullet)' = \left(\left(\frac{m}{\sqrt{1-|\mathbf{v}_{\dot{x}u}|^2}}\mathbf{v}_{\dot{x}u}\right)\bullet\right)'$$
$$= \left(\frac{m}{\sqrt{1-|\mathbf{v}_{\dot{x}u}|^2}}\left(1+\frac{\mathbf{v}_{\dot{x}u}\otimes\mathbf{v}_{\dot{x}u}}{1-|\mathbf{v}_{\dot{x}u}|^2}\right)\bullet\right)(\mathbf{v}_{\dot{x}u}\bullet)'.$$

Call attention to that in the nonrelativistic case the same quantity appears in three different roles:

(i) m is the multiplier of absolute velocity and the multiplier of absolute acceleration.
(ii) m is the multiplier of relative velocity.
(iii) m is the multiplier of relative acceleration.

In the relativistic case, however, three different quantities appear in three different roles:

(i) m is the multiplier of absolute velocity and the multiplier of absolute acceleration.
(ii) $\frac{m}{\sqrt{1-|\mathbf{v}_{\dot{x}u}|^2}}$ is the multiplier of relative velocity.
(iii) $\frac{m}{\sqrt{1-|\mathbf{v}_{\dot{x}u}|^2}}(\mathbf{1}+\frac{\mathbf{v}_{\dot{x}u}\otimes\mathbf{v}_{\dot{x}u}}{1-|\mathbf{v}_{\dot{x}u}|^2})$ is the "multiplier" of relative acceleration.

The three different relativistic quantities above are usually called masses with various signifiers. Quantity (iii), the amalgamation of "longitudinal mass," parallel to the relative velocity, and "transverse mass," perpendicular to the relative velocity is rarely mentioned.

On the other hand, "rest mass" for quantity (i) and "moving mass" for quantity (ii) are common names.

Quantity (ii) has another (by the way, usual) name: **relative energy** (see Section 11.6).

The best is to call **mass** *only quantity (i) and telling relative energy instead of moving mass.*

Since relative energy is larger then mass, the unfortunate usual names caused the false claim that moving mass increases.

A material point moves with respect to a lot of inertial frames, so its mass would increase in a lot of different manners, which is nonsense.

11.4 Newton equation

11.4.1 Absolute Newton equation

We accept, as in the nonrelativistic case, that the **absolute Newton equation** is *proper time derivative of absolute momentum = absolute force,*

$$(x : \mathrm{T}_a \to \mathrm{M})? \quad (m\dot{x})^{\cdot} = m\ddot{x} = \boldsymbol{f}(x, \dot{x}), \tag{11.6}$$

which is a second-order ordinary differential equation to determine the world line functions of a material point with **given mass** m under the action of a **given** absolute force $\boldsymbol{f}$.

To have a unique solution of that differential equation, the initial spacetime position and absolute velocity of the mass point must be given. That is why, if $\boldsymbol{s} \mapsto r(\boldsymbol{s})$ is a solution, it is suitable to consider the pair $(r, \dot{r})$ the **process** of the mass point: its value at an arbitrary instant determines the whole function.

The **evolution space** of a mass point is the set in which the processes take values: $\mathrm{M} \times \mathrm{V}(1)$.

In what follows, we accept the notation (widely used in physics) that:

(i) The elements of the evolution space are written in the form $(x, \dot{x})$ (note that then $\dot{x}$ is independent of x, it can denote an arbitrary absolute velocity).
(ii) An arbitrary ("abstract") process—i. e., a time function—is denoted by $(x, \dot{x})$ as well.
(iii) An actual process is denoted by $(r, \dot{r})$.

An absolute force $\boldsymbol{f}$ is a function defined in the evolution space, its values are determined by the fact that $m\ddot{x}$, which has values in $\mathbb{T}^* \otimes \frac{\mathbf{M}}{\mathbb{T}\otimes\mathbb{T}} \equiv \mathbb{T}^* \otimes \mathbf{M}^* \equiv \frac{\mathbf{M}^*}{\mathbb{T}}$, and that is why

$$\boldsymbol{f} : \mathrm{M} \times \mathrm{V}(1) \longrightarrow \frac{\mathbf{M}^*}{\mathbb{T}}.$$

Call attention to that a nonrelativistic force maps in the three-dimensional vector space $\frac{\mathbf{S}^*}{\mathbb{T}}$, its values are absolute space-like. A relativistic force maps in the four-dimensional vector space $\frac{\mathbf{M}^*}{\mathbb{T}}$, its values, however, are "three-dimensional" and space-like. Namely, the absolute acceleration is Lorentz orthogonal to the absolute velocity, so the force must satisfy the equality

$$\boldsymbol{f}(x, \dot{x}) \cdot \dot{x} = 0:$$

at $\dot{x}$ the force has $\dot{x}$-space-like values, i. e., it has values in $\frac{\mathbf{S}_{\dot{x}}^*}{\mathbb{T}}$.

11.4.2 Relative Newton equation

$\boldsymbol{u}$-motion is obtained from the **$\boldsymbol{u}$-relative Newton equation**

$\boldsymbol{u}$-time derivative of $\boldsymbol{u}$-momentum = action of the $\boldsymbol{u}$-relative force;

the question is, however, what is $\boldsymbol{u}$-relative momentum and what is the $\boldsymbol{u}$-relative force.

The answer is not so simple as in the nonrelativistic case because according to item 2 in Section 11.3 the $\boldsymbol{u}$-space-like component of absolute momentum is not equal to mass times $\boldsymbol{u}$-relative velocity: which of them is acceptable for $\boldsymbol{u}$-relative momentum?

Applications of relativity theory has made the decision:

$\boldsymbol{u}$-relative momentum = $\boldsymbol{u}$-space-like component of absolute momentum.

Expressing the elements of the absolute evolution space by the relative variables $\boldsymbol{u}$-time points (t), $\boldsymbol{u}$-space points (q), and $\boldsymbol{u}$-relative velocities ($\dot{q}$), we have the $\boldsymbol{u}$-relative Newton equation according to item 3 of Section 11.3:

$$(q : \mathrm{T}_{\boldsymbol{u}} \to \mathrm{S}_{\boldsymbol{u}})? \quad \left(\frac{mq'}{\sqrt{1-|q'|^2}}\right)' = \boldsymbol{f}_{\boldsymbol{u}}(t, q, q').$$

This form is well understandable though is not precise because $\boldsymbol{t}$ is involved only in the right side. We allow this little inaccuracy for the sake of simpler formulas.

Executing the differentiation, we get for the left side:

$$\frac{m}{\sqrt{1-|q'|^2}}\left(\mathbf{1} + \frac{q' \otimes q'}{1-|q'|^2}\right)q''.$$

This shows the important fact that, contrary to the nonrelativistic case,

the relative acceleration is not parallel to the relative force.

The $\boldsymbol{u}$-relative force is not simply the $\boldsymbol{u}$-space-like component of the absolute force, expressed in relative variables because differentiation with respect to $\boldsymbol{u}$-time differs from differentiation with respect to proper time; their relation is given by (11.2). That is why the $\boldsymbol{u}$-relative force is

$$\frac{\pi_{\boldsymbol{u}} \cdot \boldsymbol{f}(x, \dot{x})}{-\boldsymbol{u} \cdot \dot{x}} \tag{11.7}$$

expressed by relative variables, i. e.,

$$\boldsymbol{f}_{\boldsymbol{u}}(t, q, q') = \pi_{\boldsymbol{u}} \cdot \boldsymbol{f}\left(t \cap q, \frac{\boldsymbol{u} + q'}{\sqrt{1-|q'|^2}}\right)\sqrt{1-|q'|^2}. \tag{11.8}$$

Finally, we show how the absolute force can be restored from the relative one.

For the sake of simplicity, we omit denoting variables provisionally. The $\boldsymbol{u}$-relative force from the absolute one is obtained in the form

$$\boldsymbol{f}_{\boldsymbol{u}} = \frac{\mathbf{1} + \boldsymbol{u} \otimes \boldsymbol{u}}{-\boldsymbol{u} \cdot \dot{x}} \cdot \boldsymbol{f}$$

Since

$$\left(\mathbf{1} + \frac{\boldsymbol{u} \otimes \dot{x}}{-\boldsymbol{u} \cdot \dot{x}}\right) \cdot (\mathbf{1} + \boldsymbol{u} \otimes \boldsymbol{u}) = \mathbf{1} + \frac{\boldsymbol{u} \otimes \dot{x}}{-\boldsymbol{u} \cdot \dot{x}},$$

and $\dot{x} \cdot \boldsymbol{f} = 0$, we get

$$\boldsymbol{f} = (-\boldsymbol{u} \cdot \dot{x})\left(\mathbf{1} + \frac{\boldsymbol{u} \otimes \dot{x}}{-\boldsymbol{u} \cdot \dot{x}}\right) \cdot \boldsymbol{f}_{\boldsymbol{u}} = (-\boldsymbol{u} \cdot \dot{x})\boldsymbol{f}_{\boldsymbol{u}} + \boldsymbol{u}(\dot{x} \cdot \boldsymbol{f}_{\boldsymbol{u}}) = (\boldsymbol{u} \wedge \boldsymbol{f}_{\boldsymbol{u}}) \cdot \dot{x}.$$

We have the explicit result, denoting the variables, too,

$$\boldsymbol{f}(x, \dot{x}) = \big(\boldsymbol{u} \wedge \boldsymbol{f}_{\boldsymbol{u}}(\tau_{\boldsymbol{u}}(x), \pi_{\boldsymbol{u}}(x), \boldsymbol{v}_{\dot{x}\boldsymbol{u}})\big) \cdot \dot{x}. \tag{11.9}$$

11.5 Some special absolute forces

11.5.1 Forces having a potential

A special type of absolute forces comes from **absolute field strength** $\boldsymbol{F} : \mathrm{M} \to \mathbf{M}^* \wedge \mathbf{M}^*$:

$$\boldsymbol{f}(x, \dot{x}) := \boldsymbol{F}(x) \cdot \dot{x}.$$

According to (11.7) the $\boldsymbol{u}$-relative force is

$$\pi_{\boldsymbol{u}} \cdot \boldsymbol{F}(x) \cdot \dot{x} \frac{1}{-\boldsymbol{u} \cdot \dot{x}} = \pi_{\boldsymbol{u}} \cdot \boldsymbol{F}(x) \cdot \boldsymbol{u} + \pi_{\boldsymbol{u}} \cdot \boldsymbol{F}(x) \cdot \left(\frac{\dot{x}}{-\boldsymbol{u} \cdot \dot{x}} - \boldsymbol{u}\right);$$

the first member is just the $\boldsymbol{u}$-time-like component of $\boldsymbol{F}$, in the second member $\boldsymbol{v}_{\dot{x}\boldsymbol{u}} = \pi_{\boldsymbol{u}} \cdot \boldsymbol{v}_{\dot{x}\boldsymbol{u}}$ stands behind $\boldsymbol{F}(x)$, so there the $\boldsymbol{u}$-space-like component of $\boldsymbol{F}$ and the $\boldsymbol{u}$-relative velocity are included; thus,

$$\pi_{\boldsymbol{u}} \cdot \boldsymbol{F}(x) \cdot \dot{x} \frac{1}{-\boldsymbol{u} \cdot \dot{x}} \quad \text{in the } \boldsymbol{u}\text{-split form is} \quad \boldsymbol{E}_{\boldsymbol{u}} + \boldsymbol{B}_{\boldsymbol{u}} \cdot \boldsymbol{v}_{\dot{x}\boldsymbol{u}}.$$

This is the well-known form of an electromagnetic force (note that here the magnetic field is an antisymmetric $\boldsymbol{u}$-space-like tensor instead of which the corresponding axial vector is used in practice). Of course, the above formulas are valid not only for electromagnetic forces

It frequently occurs—as in electromagnetism—that the field strength is the antisymmetric derivative of an **absolute potential** $\boldsymbol{K} : \mathrm{M} \to \mathbf{M}^*$,

$$\boldsymbol{F} = \mathcal{D} \wedge \boldsymbol{K},$$

and the force is

$$\boldsymbol{f}(x, \dot{x}) := \boldsymbol{F}(x) \cdot \dot{x}.$$

The $\boldsymbol{u}$-split form of $\boldsymbol{K}$ is denoted by $(-V_{\boldsymbol{u}}, \boldsymbol{A}_{\boldsymbol{u}})$ where $V_{\boldsymbol{u}}$ is the $\boldsymbol{u}$-scalar potential and $\boldsymbol{A}_{\boldsymbol{u}}$ is the $\boldsymbol{u}$-vector potential; then the $\boldsymbol{u}$-split form $\mathcal{D} \wedge \boldsymbol{K}$ (see Section 12.2) is

$$((-\nabla_{\boldsymbol{u}} V_{\boldsymbol{u}} - \mathcal{D}_{\boldsymbol{u}} \boldsymbol{A}_{\boldsymbol{u}}, \nabla_{\boldsymbol{u}} \wedge \boldsymbol{A}_{\boldsymbol{u}})).$$

In contrast to the nonrelativistic case,

there are no absolute scalar potentials (absolute time-like covector fields) in the relativistic spacetime model.

11.5.2 The simplest special forces

(1) Force **depending only on absolute velocity**: there is function $\boldsymbol{h} : \mathrm{V}(1) \to \frac{\mathbf{M}^*}{\mathbb{T}}$ such that

$$\boldsymbol{f}(x, \dot{x}) = \boldsymbol{h}(\dot{x}).$$

The corresponding $\boldsymbol{u}$-relative force is

$$\boldsymbol{f}_{\boldsymbol{u}}(t, q, q') = \boldsymbol{\pi}_{\boldsymbol{u}} \cdot \boldsymbol{h}\left(\frac{\boldsymbol{u} + q'}{\sqrt{1 - |q'|^2}}\right)\sqrt{1 - |q'|^2}.$$

(2) **There is no** force independent absolute velocity. It may happen, however, that a $\boldsymbol{u}$-relative force is independent of $\boldsymbol{u}$-relative velocity: there is a function $\boldsymbol{h} : \mathrm{T}_{\boldsymbol{u}} \times \mathrm{S}_{\boldsymbol{u}} \to \frac{\mathbf{S}_{\boldsymbol{u}}^*}{\mathbb{T}}$ such that

$$\boldsymbol{f}_{\boldsymbol{u}}(t, q, q') = \boldsymbol{h}(t, q).$$

Then the absolute force is recovered by (11.9):

$$\boldsymbol{f}(x, \dot{x}) = (\boldsymbol{u} \wedge \boldsymbol{h}(\tau_{\boldsymbol{u}}(x), \pi_{\boldsymbol{u}}(x))) \cdot \dot{x}. \tag{11.10}$$

(3) **There is no** force depending only on time. It may happen, however, that a $\boldsymbol{u}$-relative force depends only on $\boldsymbol{u}$-time: there is a function $\boldsymbol{h} : \mathrm{T}_{\boldsymbol{u}} \to \frac{\mathbf{S}^*}{\mathbb{T}}$ such that

$$\boldsymbol{f}_{\boldsymbol{u}}(t, q, q') = \boldsymbol{h}(t).$$

The absolute force is

$$\boldsymbol{f}(x, \dot{x}) = (\boldsymbol{u} \wedge \boldsymbol{h}(\tau_{\boldsymbol{u}}(x))) \cdot \dot{x}.$$

(4) **There is no constant** force. It may happen, however, that a $\boldsymbol{u}$-relative force is constant: there is a $\boldsymbol{h} \in \frac{\mathbf{S}^*}{\mathbb{T}}$ such that

$$\boldsymbol{f}_{\boldsymbol{u}}(t, q, q') = \boldsymbol{h}.$$

Then the absolute force is

$$\boldsymbol{f}(x,\dot{x}) = (\boldsymbol{u}\wedge\boldsymbol{h})\cdot\dot{x}.$$

This force has a potential: for an arbitrary world point o,

$$\boldsymbol{K}(x) = -\boldsymbol{h}\cdot(x-o)\boldsymbol{u},$$

which corresponds to the nonrelativistic (6.6) because now $-\boldsymbol{u} = \tau_{\boldsymbol{u}}$.

(5) **The $\boldsymbol{u}_s$-static** absolute force can be defined as nonrelativistically:

$$\boldsymbol{f}(x,\dot{x}) = \tilde{\boldsymbol{f}}(\pi_{\boldsymbol{u}_s}(x),\dot{x}).$$

The explicit expression for the $\boldsymbol{u}$-relative force (11.8) is highly complicated.

(6) **In general, there are no central** forces because there is no world point of the center that is uniquely simultaneous with a world point x.
Exceptions are the inertial centers when we can consider standard simultaneity according to the center. Let $\boldsymbol{u}_c$ be the absolute velocity of the center. Then we accept that the $\boldsymbol{u}_c$-relative force has the form known in the nonrelativistic case

$$\boldsymbol{f}_{\boldsymbol{u}_c}(t,q,q') = a(|q-q_c|)(q-q_c)$$

where q_c is the position of the center in the space of the inertial observer $\boldsymbol{u}_c$ and $a:\mathbb{T}\to\frac{\mathbb{R}}{\mathbb{T}\otimes\mathbb{T}\otimes\mathbb{T}}$ is a function.
q_c and q are straight lines in spacetime, directed by $\boldsymbol{u}_c$. Let o be an arbitrary occurrence of q_c, i. e., $q_c = o+\mathbb{T}\boldsymbol{u}_c$ and let x be an arbitrary occurrence of q, i. e., $q = x+\mathbb{T}\boldsymbol{u}_c$. Then $q-q_c = (x+\mathbb{T}\boldsymbol{u}_c)-(o+\mathbb{T}\boldsymbol{u}_c) = \pi_{\boldsymbol{u}_c}\cdot(x-o)$, so the absolute force is

$$\boldsymbol{f}(x,\dot{x}) = a(|\pi_{\boldsymbol{u}_c}\cdot(x-o)|)(\boldsymbol{u}\wedge\pi_{\boldsymbol{u}_c}\cdot(x-o))\cdot\dot{x}$$

where o is an arbitrary world point of the world line of the center.
This $\boldsymbol{u}_c$-static force has a potential,

$$\boldsymbol{K}(x) = -b(|\pi_{\boldsymbol{u}_c}\cdot(x-o)|)\boldsymbol{u}_c,$$

where b is a primitive function of $\boldsymbol{s}\mapsto a(\boldsymbol{s})\boldsymbol{s}$. Taking into account that $-\boldsymbol{u}_c = \tau_{\boldsymbol{u}_c}$, this formula, too, agrees with the nonrelativistic (6.8).
In particular,

- **The inertial electric** force is

$$\boldsymbol{f}(x,\dot{x}) = -e_c e\frac{(\boldsymbol{u}_c\wedge(x-o))\cdot\dot{x}}{|\pi_{\boldsymbol{u}_c}(x-o)|^3},$$

  having the potential

$$\boldsymbol{K}(x) = -e_c e \frac{\boldsymbol{u}_c}{|\pi_{\boldsymbol{u}_c}(x - o)|}$$

- **The inertial elastic** force is

$$\boldsymbol{f}(x, \dot{x}) = -k(\boldsymbol{u}_c \wedge (x - o)) \cdot \dot{x}$$

having the potential

$$\boldsymbol{K}(x) = -k|\pi_{\boldsymbol{u}_c}(x - o)|^2 \boldsymbol{u}_c.$$

In these formulas, o is an arbitrary world point of the world line of the center.

Arithmetic formula

Let us consider the $\boldsymbol{u}_c := \gamma(1, v, 0, 0)$-static elastic force. Then $x - o = (x^0, x^1, x^2, x^3)$, $\dot{x} = (\dot{x}^0, \dot{x}^1, \dot{x}^2, \dot{x}^2)$ and

$$\boldsymbol{u}_c((x - o) \cdot \dot{x}) = \gamma(-x^0\dot{x}^0 + x^1\dot{x}^1 + x^2\dot{x}^2 + x^3\dot{x}^3)(1, v, 0, 0), \tag{11.11}$$

$$(x - o)(\boldsymbol{u}_c \cdot \dot{x}) = \gamma(x^0, x^1, x^2, x^3)(-\dot{x}^0 + v\dot{x}^1). \tag{11.12}$$

The corresponding absolute force is k times the difference (11.11)–(11.12).

This is an excellent example of how complicated the arithmetic formulas may be.

11.5.3 Exercises

1. The absolute forces $\boldsymbol{f}$ and $\boldsymbol{f}'$ are physically equivalent by the definition in Section 3.3 if there is a proper Poincaré transformation (over the proper Lorentz transformation $\boldsymbol{L}$) such that $\boldsymbol{f}'(L(x), \boldsymbol{L} \cdot \dot{x}) = \boldsymbol{L} \cdot \boldsymbol{f}(x, \dot{x})$.
   Accordingly, characterize the physically equivalent central forces given by $\boldsymbol{u}_c$, k and $\boldsymbol{u}_{c'}$, k', respectively
2. An absolute force $\boldsymbol{f}$ is invariant for the $(\boldsymbol{u}, o)$-time reversal if and only if $\boldsymbol{f}(T_{\boldsymbol{u},o}(x), -\boldsymbol{T}_{\boldsymbol{u}} \cdot \dot{x}) = \boldsymbol{T}_{\boldsymbol{u}} \cdot \boldsymbol{f}(x, \dot{x})$. Prove that if $\boldsymbol{f}$ is invariant:
   (i) For the $(\boldsymbol{u}, o)$-time reversal with a given $\boldsymbol{u}$ and for all o, then it is constant on every straight line directed by $\boldsymbol{u}$, i. e., it depends only on $\boldsymbol{u}$-space points and $\boldsymbol{u}$-relative velocities.
   (ii) For all $(\boldsymbol{u}, o)$-time reversal, then it is zero.
   Compare this result with Exercise 3 in 6.5.3.

11.6 Kinetic energy and power

We had to decide which possibility is to be chosen for relative momentum. Now we make a similar decision: corresponding to the nonrelativistic case, we accept that the **relative power is the product of relative force and relative velocity.**

For the sake of brevity, in the sequel the variables of forces (both absolute and relative ones) are omitted from the notation.

Using the notation introduced in Subsection 11.2.2, the $\boldsymbol{u}$-relative power of the force f is

$$(\boldsymbol{f}_{\boldsymbol{u}} \cdot \boldsymbol{v}_{\dot{x}\boldsymbol{u}})\bullet = \left(\frac{\boldsymbol{f} + \boldsymbol{u}(\boldsymbol{u} \cdot \boldsymbol{f})}{-\boldsymbol{u} \cdot \dot{x}} \cdot \left(\frac{\dot{x}}{-\boldsymbol{u} \cdot \dot{x}} - \boldsymbol{u}\right)\right)\bullet = \left(\frac{-\boldsymbol{u} \cdot \boldsymbol{f}}{-\boldsymbol{u} \cdot \dot{x}}\right)\bullet,$$

which is just the negative of the $\boldsymbol{u}$-time-like component of the force considered a covector, multiplied by the reciprocal of the relativistic factor. The reader is asked to recall what was previously said in Section 6.6.

Further, we have

$$\left(\frac{-\boldsymbol{u} \cdot \boldsymbol{f}}{-\boldsymbol{u} \cdot \dot{x}}\right)\bullet = \left(\frac{-\boldsymbol{u} \cdot m\ddot{x}}{-\boldsymbol{u} \cdot \dot{x}}\right)\bullet = (m(-\boldsymbol{u} \cdot \dot{x})\bullet)' = \left(\frac{m}{\sqrt{1 - |\boldsymbol{v}_{\dot{x}\boldsymbol{u}}|^2}}\bullet\right)'.$$

Based on this formula, the quantity in the parenthesis of the right side of the last (or last but one) equality is usually called energy because its time derivative is the power. This is, however, not correct for two reasons:

(1) Adding an arbitrary constant to it, the derivative remains the same.
(2) In the nonrelativistic case, it is the relative *kinetic energy* whose time derivative equals the product of relative force and relative velocity.

Kinetic energy is zero if relative velocity is zero. We will be in accordance with the nonrelativistic case if we add a constant to the quantity in the parenthesis to obtain a quantity, which is zero for zero relative velocity.

That is why we accept that

$$m(-\boldsymbol{u} \cdot \dot{x}) - m = \frac{m}{\sqrt{1 - |\boldsymbol{v}_{\dot{x}\boldsymbol{u}}|^2}} - m$$

is the ***u*-kinetic energy**.

This choice is supported by the fact, too, that for relative velocities whose magnitude is much less than 1 (the light speed), we have—from the power series of the square root—that this equals approximately the nonrelativistic kinetic energy $\frac{1}{2}m|\boldsymbol{v}_{\dot{x}\boldsymbol{u}}|^2$.

Nevertheless,

$$m(-\boldsymbol{u} \cdot \dot{x}) = \frac{m}{\sqrt{1 - |\boldsymbol{v}_{\dot{x}\boldsymbol{u}}|^2}}$$

can be accepted as the **u-relative energy** but this does not follow from the previous considerations; this will be supported by other processes in which material points join and separate or emit light particles (photons), which will be treated in Section 11.9.

11.7 Conservation laws

11.7.1 There is no instantaneous interaction

Contrary to the nonrelativistic case, the interaction of two material points—briefly particles—cannot be described by forces because there are no absolute instants:

Newton's law of action–reaction does not hold in relativity theory.

Let us consider an electron and a proton moving with respect to each other. They interact by emitting and absorbing electromagnetic radiation. This means that the "particle–particle" interaction is replaced by "particle–field–particle–field" interaction. It is a fundamental problem, however [8]:

At present, we have no differential equation for describing processes of mechanical-electromagnetic interaction.

11.7.2 Collisions

The possibility of direct particle-particle interaction occurs only in collisions for which we accept as a fundamental physical fact that *absolute momentum is conserved.*

Let two particles meet and join together (inelastic collision). Let the particles have masses m_1 and m_2 and absolute velocities $\boldsymbol{u}_1$ and $\boldsymbol{u}_2$, respectively. Let the arising new particle have mass m_3 and absolute velocity $\boldsymbol{u}_3$. The conservation of total momentum gives

$$m_1\boldsymbol{u}_1 + m_2\boldsymbol{u}_2 = m_3\boldsymbol{u}_3. \tag{11.13}$$

The $\boldsymbol{u}$-space-like component of this equality gives the **conservation of $\boldsymbol{u}$-relative momentum** for arbitrary inertial frame $\boldsymbol{u}$:

$$m_1(\boldsymbol{u}_1 + \boldsymbol{u}(\boldsymbol{u}\cdot\boldsymbol{u}_1)) + m_2(\boldsymbol{u}_2 + \boldsymbol{u}(\boldsymbol{u}\cdot\boldsymbol{u}_2)) = m_3(\boldsymbol{u}_3 + \boldsymbol{u}(\boldsymbol{u}\cdot\boldsymbol{u}_3)),$$

which expressed by relative velocities, is

$$\frac{m_1\boldsymbol{v}_{u_1u}}{\sqrt{1-|\boldsymbol{v}_{u_1u}|^2}} + \frac{m_2\boldsymbol{v}_{u_2u}}{\sqrt{1-|\boldsymbol{v}_{u_2u}|^2}} = \frac{m_3\boldsymbol{v}_{u_3u}}{\sqrt{1-|\boldsymbol{v}_{u_3u}|^2}}.$$

Of course, the $\boldsymbol{u}$-time-like component gives

$$m_1(-\boldsymbol{u}\cdot\boldsymbol{u}_1) + m_2(-\boldsymbol{u}\cdot\boldsymbol{u}_2) = m_3(-\boldsymbol{u}\cdot\boldsymbol{u}_3), \tag{11.14}$$

which supported by later arguments, is the **conservation of $\boldsymbol{u}$-relative energy**. Expressed by relative velocities,

$$\frac{m_1}{\sqrt{1-|\boldsymbol{v}_{\boldsymbol{u}_1\boldsymbol{u}}|^2}} + \frac{m_2}{\sqrt{1-|\boldsymbol{v}_{\boldsymbol{u}_2\boldsymbol{u}}|^2}} = \frac{m_3}{\sqrt{1-|\boldsymbol{v}_{\boldsymbol{u}_3\boldsymbol{u}}|^2}}.$$

Let us take $\boldsymbol{u}_3$ for the role of $\boldsymbol{u}$ (let us consider the standard inertial frame in which the new particle is at rest). Then

$$\frac{m_1}{\sqrt{1-|\boldsymbol{v}_{\boldsymbol{u}_1\boldsymbol{u}_3}|^2}} + \frac{m_2}{\sqrt{1-|\boldsymbol{v}_{\boldsymbol{u}_2\boldsymbol{u}_3}|^2}} = m_3;$$

the multipliers on the left side are greater than 1—if $\boldsymbol{u}_1$ and $\boldsymbol{u}_2$ does not equal $\boldsymbol{u}_3$ (this is the case of a real collision)—consequently,

$$m_1 + m_2 < m_3.$$

Mass is not conserved when two particles join in a collision.

Experience suggested us introducing in the nonrelativistic case the notion of internal energy, which admitted us to formulate energy conservation: the internal energy of the particle after collision is greater then the sum of the internal energies of the particles before collision, the difference is just the difference between the kinetic energies after and before collision, respectively.

Now the difference between the $\boldsymbol{u}$-kinetic energies after and before collision is

$$\big(m_1(-\boldsymbol{u}\cdot\boldsymbol{u}_1) - m_1\big) + \big(m_2(-\boldsymbol{u}\cdot\boldsymbol{u}_2) - m_2\big) - \big(m_3(-\boldsymbol{u}\cdot\boldsymbol{u}_3) - m_3\big) = m_3 - (m_1 + m_2).$$

The difference between the $\boldsymbol{u}$-kinetic energies is just the difference between the masses after and before collision. In analogy with the nonrelativistic case, we can say that here mass takes the role of internal energy, and we accept

$$\begin{aligned}\boldsymbol{u}\text{-relative energy} &:= \boldsymbol{u}\text{-kinetic energy} + \text{mass}\\ &= \boldsymbol{u}\text{-time-like component of absolute momentum.}\end{aligned}$$

11.8 Newton equation with varying mass

The basic principle of the relativistic absolute Newton equation for a material body, which loses or gains mass during its existence is momentum conservation, as in the nonrelativistic case.

There must be given

- The mass of the body as a function of the proper time, $m : \mathbb{T} \to \mathbb{T}^* = \frac{\mathbb{R}}{\mathbb{T}}$.
- The relative velocity of the leaving or joining mass with respect to the body as a function of proper time, $\boldsymbol{v} : \mathbb{T} \to \frac{\mathbf{M}}{\mathbb{T}}$ in such a way that if r is the world line function of the body then $\boldsymbol{v}(\boldsymbol{s})$ is in $\frac{\mathbf{S}_{\dot{r}(\boldsymbol{s})}}{\mathbb{T}}$, i. e., $\dot{r}(\boldsymbol{s}) \cdot \boldsymbol{v}(\boldsymbol{s}) = 0$.

If r is the world line function of the body, then the change of absolute momentum in a "short" proper time interval $\boldsymbol{t}$ is

$$m(\boldsymbol{s}+\boldsymbol{t})\dot{r}(\boldsymbol{s}+\boldsymbol{t}) - m(\boldsymbol{s})\dot{r}(\boldsymbol{s}) = \big(m(\boldsymbol{s})\ddot{r}(\boldsymbol{s}) + \dot{m}(\boldsymbol{s})\dot{r}(\boldsymbol{s})\big)\boldsymbol{t} + \operatorname{ordo}(\boldsymbol{t}), \tag{11.15}$$

which is due to the force acting on the body and to the momentum of the leaving or joining mass. (11.15) formally is the same as the nonrelativistic (6.11), only t is replaced by $\boldsymbol{s}$. For the acting force, we have (6.12), too, in such a way that t is replaced by $\boldsymbol{s}$:

$$\boldsymbol{f}\big(r(\boldsymbol{s}) + \operatorname{Ordo}(\boldsymbol{t}), \dot{r}(\boldsymbol{s}) + \operatorname{Ordo}(\boldsymbol{t})\big)\boldsymbol{t}. \tag{11.16}$$

On the contrary, (6.13) cannot be copied simply; the less problematic is that $\dot{r}(t) + \boldsymbol{v}(t)$ must be replaced by $\frac{\dot{r}(\boldsymbol{s})+\boldsymbol{v}(\boldsymbol{s})}{\sqrt{1-|\boldsymbol{v}(\boldsymbol{s})|^2}}$ but $m(t+\boldsymbol{t}) - m(t)$ cannot be replaced by $m(\boldsymbol{s}+\boldsymbol{t}) - m(\boldsymbol{s})$ because mass need not be conserved (see Subsection 11.7.2). For the time being, we can say only that it is of the form $(\mu(\boldsymbol{s})\boldsymbol{t} + \operatorname{ordo}(\boldsymbol{t}))$ for some μ. Then the momentum change due to the leaving or joining mass is

$$\big(\mu(\boldsymbol{s})\boldsymbol{t} + \operatorname{ordo}(\boldsymbol{t})\big)\left(\frac{\dot{r}(\boldsymbol{s}) + \boldsymbol{v}(\boldsymbol{s})}{\sqrt{1-|\boldsymbol{v}(\boldsymbol{s})|^2}}\right). \tag{11.17}$$

Dividing by $\boldsymbol{t}$ the equality (11.15) = (11.16) + (11.17) and then taking the limit $\boldsymbol{t} \to 0$, we get

$$m(\boldsymbol{s})\ddot{r}(\boldsymbol{s}) + \dot{m}(\boldsymbol{s})\dot{r}(\boldsymbol{s}) = \boldsymbol{f}\big(r(\boldsymbol{s}), \dot{r}(\boldsymbol{s})\big) + \mu(\boldsymbol{s})\frac{\dot{r}(\boldsymbol{s}) + \boldsymbol{v}(\boldsymbol{s})}{\sqrt{1-|\boldsymbol{v}(\boldsymbol{s})|^2}}.$$

Lorentz multiplying by $\dot{r}(\boldsymbol{s})$, we obtain

$$\mu = \dot{m}\sqrt{1-|\boldsymbol{v}|^2}.$$

Finally, we have the equation

$$(x : \mathbb{T} \to \mathrm{M})? \quad m\ddot{x} = \boldsymbol{f}(x, \dot{x}) + \dot{m}\boldsymbol{v},$$

which is formally the same as the nonrelativistic one.

There is a trouble, however. Namely, this equation would determine the world line of the body but the function $\boldsymbol{v}$ can be given only if the world line function of the body is known because $v(s) \cdot \dot{r}(s) = 0$ must hold. We can avoid this trouble if we imagine the body in the space of the inertial observer $u := \dot{r}(0)$ and we give the relative velocity of the leaving or joining mass in that space, i. e., a function $\hat{v} : \mathbb{T} \to \frac{\mathrm{S}_u}{\mathbb{T}}$, which will be boosted to the actual absolute velocity, $v(s) := B_{\dot{r}(s),u}\hat{v}(s)$. Then the equation

$$(x : \mathbb{T} \mapsto \mathrm{M})? \quad m\ddot{x} = f(x, \dot{x}) + \dot{m}B_{\dot{x},u} \cdot \hat{v}$$

is well-defined.

Remark. The relativistic Newton equation in the form $(m\dot{x})^{\cdot} = \dot{m}\dot{x} + m\ddot{x} = f(x, \dot{x})$ for varying mass has the same trouble as in the nonrelativistic case (see Section 6.8). Lorentz-multiplying by $\dot{x}$ we obtain $\dot{m} = 0$. This trouble gets lost if the exact form of functions is neglected. Namely, as a seemingly meaningful result one has [1] $f_\mu u^\mu = \dot{m}$ (u^μ being the absolute velocity) without realizing that the left side is zero; so, the right side is zero, too.

11.9 Particles and photons

11.9.1 Collisions

As said, we have no differential equation for describing processes of mechanical-electromagnetic interaction. Some properties of this interaction—emission, absorption and reflections of light—can be obtained by considering collisions of mechanical particles and electromagnetic particles, called photons. It is a simple empirical fact that a body emitting or absorbing light will be colder or warmer, respectively, so the idea in Section 11.6 regarding energy will be more convincing in this way.

Note that the absolute momentum p of a particle with mass m is a future-like vector for which $p \cdot p = -m^2$ holds.

The absolute momentum k a photon is future light-like, so $k \cdot k = 0$, which is conceived that a photon has no mass.

Let a particle absorb a photon; the balance of momentum is

$$m_1 u_1 + k = m_2 u_2. \tag{11.18}$$

Its u-time-like component is

$$m_1(-u \cdot u_1) - u \cdot k = m_2(-u \cdot u_2);$$

taking into account the meaning of the terms containing the masses, this equality expresses the conservation of u-relative energy in such a way that $-u \cdot k$ is the u-energy of the photon.

In particular, for $\boldsymbol{u} = \boldsymbol{u}_2$, we obtain

$$m_1(-\boldsymbol{u}_2 \cdot \boldsymbol{u}_1) - \boldsymbol{u}_2 \cdot \boldsymbol{k} = m_2.$$

Since $-\boldsymbol{u}_2 \cdot \boldsymbol{u}_1 > 1$ and $-\boldsymbol{u}_2 \cdot \boldsymbol{k} > 0$, we see that $m_2 > m_1$. We have got that light absorption—which means internal energy increase in the nonrelativistic case—results in mass increase in the relativistic case. The increase of mass is

$$m_2 - m_1 = (m_1(-\boldsymbol{u}_2 \cdot \boldsymbol{u}_1) - m_1) - \boldsymbol{u}_2 \cdot \boldsymbol{k}.$$

The $\boldsymbol{u}_2$-kinetic energy before absorption is $m_1(-\boldsymbol{u}_2 \cdot \boldsymbol{u}_1) - m_1$, and after absorption is zero (the particle after absorption is at rest with respect to $\boldsymbol{u}_2$). According to the nonrelativistic reasoning, we would say that the kinetic energy before collision and the energy of the photon is transformed into internal energy, i. e., the sum of the kinetic energy of the particle and the energy of the photon before collision equals the inner energy increase of the particle. In the right side of the above equality, we find the total energy before collision while the left side is the mass increase. This supports our idea that in the relativistic case mass takes the role of internal energy.

$\boldsymbol{u}_2$ in (11.18) cannot equal $\boldsymbol{u}_1$ because the future-like vector $(m_2 - m_1)\boldsymbol{u}_1$ cannot be equal the future light-like vector $\boldsymbol{k}$. The absolute velocity of the particle changes due to the absorption of a photon.

If the particle with absolute velocity $\boldsymbol{u}_1$ absorbs two photons having opposite $\boldsymbol{u}_1$-relative momentum, $\pi_{\boldsymbol{u}_1} \cdot \boldsymbol{k}_2 = -\pi_{\boldsymbol{u}_1} \cdot \boldsymbol{k}_1$, then

$$m_1\boldsymbol{u}_1 + \boldsymbol{k}_1 + \boldsymbol{k}_2 = m_2\boldsymbol{u}_1,$$

the particle "remains at rest," only its mass changes:

$$m_1 - \boldsymbol{u}_1 \cdot \boldsymbol{k}_1 - \boldsymbol{u}_1 \cdot \boldsymbol{k}_2 = m_2. \tag{11.19}$$

More interesting is when a particle is annihilated by radiating two photons. It has the balance of momentum

$$m\boldsymbol{u} = \boldsymbol{k}_1 + \boldsymbol{k}_2$$

from which we get

$$m = -\boldsymbol{u} \cdot \boldsymbol{k}_1 - \boldsymbol{u} \cdot \boldsymbol{k}_2.$$

The particle is annihilated, mass disappears, and photons appear. These photons can be absorbed by another particle, which causes mass increase (internal energy increase from a nonrelativistic point of view) of that particle.

11.9.2 Equivalence of mass and internal energy

Let us examine Einstein's famous result, the "equivalence of mass and energy," in its usual form, $E = mc^2$. Summarizing our results, we can say that

In the nonrelativistic case:
(i) Relative energy = kinetic energy + internal energy
(ii) Kinetic energy $\rightarrow$ increase of internal energy
(iii) Relative momentum = mass $\times$ relative velocity

whereas in the relativistic case
(i) Relative energy = kinetic energy + mass
(ii) Kinetic energy $\rightarrow$ increase of mass
(iii) Relative momentum = relative energy $\times$ relative velocity

In the relativistic relations (i) and (ii), mass takes the role of nonrelativistic internal energy; in the relation (iii), relative energy takes the role of mass. This shows that the roles of mass and some kind of energy cannot be interchanged in every respect.

Let us make it clear: energy is a relative notion (depending on reference frames) both nonrelativistically and relativistically, whereas mass (and internal energy) are absolute (independent of reference frames). Accordingly, we can state that in the relativistic case:

- Mass unites nonrelativistic mass and internal energy,
  even, taking into account particle annihilation,
- Mass and internal energy are the same notions;

briefly,

internal energy = mass; in formula, $E_{\text{internal}} = mc^2$.

It is emphasized: Relative energy cannot be taken instead of internal energy, just because mass is absolute, energy is relative. In the formula of relative momentum, relative velocity is multiplied by

- Mass in the nonrelativistic case
- Relative energy in the relativistic case

but this is by no means an equivalence.

Einstein's first declaration about mass and energy was not sufficiently clear but later he made it definite that his formula is about internal energy and mass. Unfortunately, most works on relativity contains a misunderstanding in this respect, though there are several articles explaining the truth [3].

11.9.3 The Doppler effect

According to a standard inertial frame $\boldsymbol{u}$, light is an electromagnetic wave; the $\boldsymbol{u}$-frequency of the wave

- Is proportional to the $\boldsymbol{u}$-energy transferred by the light
- Manifests itself in the color of the light

Switching from waves to light particles, if a photon has absolute momentum $\boldsymbol{k}$ then its $\boldsymbol{u}$-energy $-\boldsymbol{u}\cdot\boldsymbol{k}$ is conceived as its $\boldsymbol{u}$-frequency.

Let a standard inertial frame $\boldsymbol{u}'$ emit an electromagnetic wave with frequency f'; an other standard inertial frame finds that this wave has an other frequency f. In other words, for a photon emitted with absolute momentum $\boldsymbol{k}$, the corresponding frequencies are

$$f' := -\boldsymbol{u}'\cdot\boldsymbol{k}, \quad f := -\boldsymbol{u}\cdot\boldsymbol{k};$$

it is trivial then the frequencies colors of the same light are different for different frames.

The relationship between the two frequencies is the Doppler effect, which is obtained as follows.

Let a source having absolute velocity $\boldsymbol{u}'$ emit a photon (an electromagnetic wave) with absolute momentum $\boldsymbol{k}$. Let $\boldsymbol{u}$ be the absolute velocity of a relay. The relative velocity $\boldsymbol{v}_{\boldsymbol{k}\boldsymbol{u}}$ of the photon with respect to $\boldsymbol{u}$ is a unit vector, $|\boldsymbol{v}_{\boldsymbol{k}\boldsymbol{u}}| = 1$. For the relative velocity $\boldsymbol{v}_{\boldsymbol{u}'\boldsymbol{u}}$, we put $v := |\boldsymbol{v}_{\boldsymbol{u}'\boldsymbol{u}}| < 1$. Let α be the angle between those relative velocities, in other words, between the direction of the $\boldsymbol{u}$-motion of the photon and the direction of the $\boldsymbol{u}$-motion of the light source. Then

$$\cos\alpha = \frac{\boldsymbol{v}_{\boldsymbol{u}'\boldsymbol{u}}}{v}\cdot\boldsymbol{v}_{\boldsymbol{k}\boldsymbol{u}} = \frac{1}{v}\left(\frac{\boldsymbol{u}'}{-\boldsymbol{u}\cdot\boldsymbol{u}'} - \boldsymbol{u}\right)\cdot\left(\frac{\boldsymbol{k}}{-\boldsymbol{u}\cdot\boldsymbol{k}} - \boldsymbol{u}\right) = \frac{\boldsymbol{u}'\cdot\boldsymbol{k}}{(-\boldsymbol{u}\cdot\boldsymbol{k})(-\boldsymbol{u}'\cdot\boldsymbol{u})} + 1,$$

which yields

$$\frac{f'}{f} = \frac{1 - v\cos\alpha}{\sqrt{1-v^2}}.$$

In particular, if the photon moves with respect to $\boldsymbol{u}$:

(i) In the same direction as $\boldsymbol{u}'$ moves then $\cos\alpha = 1$ and $\frac{f'}{f} = \sqrt{\frac{1-v}{1+v}}$

(ii) In the opposite direction as $\boldsymbol{u}'$ moves then $\cos\alpha = -1$ and $\frac{f'}{f} = \sqrt{\frac{1+v}{1-v}}$

which are the results obtained in Subsection 10.5.5, too.

12 Elements of electromagnetism in the spacetime model

The equations of electromagnetism—Maxwell equations—played a fundamental role in constructing relativity theory. Light is an electromagnetic phenomenon. We based the relativistic spacetime model for the properties of light propagation. This section will show how the relativistic spacetime model corrects the troubles of nonrelativistic electromagnetism.

12.1 Splitting of spacetime functions

We consider a spacetime splitting by a standard inertial frame $\boldsymbol{u}$ with origin o (see Subsection 5.8.1).

12.1.1 Vector fields, covector fields

The **half-split form** by $\boldsymbol{u}$ of a vector field $\boldsymbol{J} : \mathrm{M} \to \mathbf{M}$ is obtained by splitting its range,

$$\mathrm{M} \xrightarrow[\boldsymbol{J}]{} \mathbf{M} \xrightarrow[\xi_{\boldsymbol{u}}]{} \mathbb{T} \times \mathbf{S}_{\boldsymbol{u}} = (\rho_{\boldsymbol{u}}, \boldsymbol{j}_{\boldsymbol{u}}) : \mathrm{M} \to \mathbb{T} \times \mathbf{S}_{\boldsymbol{u}}, \tag{12.1}$$

$$\xi_{\boldsymbol{u}} \cdot \boldsymbol{J} = \begin{pmatrix} -\boldsymbol{u} \cdot \boldsymbol{J} \\ \boldsymbol{\pi}_{\boldsymbol{u}} \cdot \boldsymbol{J} \end{pmatrix} =: \begin{pmatrix} \rho_{\boldsymbol{u}} \\ \boldsymbol{j}_{\boldsymbol{u}} \end{pmatrix}.$$

The **completely split form** by $(\boldsymbol{u}, o)$ of $\boldsymbol{J}$ is obtained by splitting its domain, too,

$$\mathbb{T} \times \mathbf{S}_{\boldsymbol{u}} \xrightarrow[\xi_{\boldsymbol{u},o}^{-1}]{} \mathrm{M} \xrightarrow[\boldsymbol{J}]{} \mathbf{M} \xrightarrow[\xi_{\boldsymbol{u}},]{} \mathbb{T} \times \mathbf{S}_{\boldsymbol{u}}$$

$$\xi_{\boldsymbol{u}} \cdot \boldsymbol{J}(\xi_{\boldsymbol{u},o}^{-1}) =: \begin{pmatrix} \hat{\rho}_{\boldsymbol{u}} \\ \hat{\boldsymbol{j}}_{\boldsymbol{u}} \end{pmatrix}.$$

A **coordinated form** of $\boldsymbol{J}$ is obtained by a coordinate system adapted to the $(\boldsymbol{u}, o)$-splitting (see (10.65)):

$$\mathbb{R}^4 \xrightarrow[C^{-1}]{} \mathbb{T} \times \mathbf{S}_{\boldsymbol{u}} \xrightarrow[\xi_{\boldsymbol{u},o}^{-1}]{} \mathrm{M} \xrightarrow[\boldsymbol{J}]{} \mathbf{M} \xrightarrow[\xi_{\boldsymbol{u}}]{} \mathbb{T} \times \mathbf{S}_{\boldsymbol{u}} \xrightarrow[\boldsymbol{C},]{} \mathbb{R}^4$$

$$\boldsymbol{C} \cdot \xi_{\boldsymbol{u}} \cdot \boldsymbol{J}(\xi_{\boldsymbol{u},o}^{-1}(C^{-1})) =: \begin{pmatrix} \tilde{\rho}_{\boldsymbol{u}} \\ \tilde{\boldsymbol{j}}_{\boldsymbol{u}} \end{pmatrix} =: J^k \quad (k = 0, 1, 2, 3).$$

https://doi.org/10.1515/9783112219553-015

Hinting at the different forms of splitting by the symbol ~, we write

$$\boldsymbol{J} \sim (\rho_u, \boldsymbol{j}_u) \sim (\hat{\rho}_u, \hat{\boldsymbol{j}}_u) \sim (\tilde{\rho}_u, \tilde{\boldsymbol{j}}_u) \sim J^k \quad (k = 0, 1, 2, 3).$$

The **half-split form** by $\boldsymbol{u}$ of a covector field $\boldsymbol{K} : \mathrm{M} \to \mathbf{M}^*$ is obtained by splitting its range,

$$\begin{array}{ccc} \mathrm{M} \longrightarrow \mathbf{M} & \longrightarrow & \mathbb{T}^* \times \mathbf{S}_u^* \\ \boldsymbol{K} & (\xi_u{}^{-1})^*, & \end{array} \tag{12.2}$$
$$(\xi_u{}^{-1})^* \cdot \boldsymbol{K} = (\boldsymbol{u} \cdot \boldsymbol{K}\ \boldsymbol{i}^* \cdot \boldsymbol{K}) =: (-V_u\ \boldsymbol{A}_u).$$

The **completely split form** by $(\boldsymbol{u}, o)$ of $\boldsymbol{K}$ is obtained by splitting its domain, too,

$$\begin{array}{ccccc} \mathbb{T} \times \mathbf{S}_u \longrightarrow \mathrm{M} \longrightarrow \mathbf{M} & \longrightarrow & \mathbb{T}^* \times \mathbf{S}_u^* \\ \xi_{u,o}^{-1} \quad \boldsymbol{K} & (\xi_u{}^{-1})^*, & \end{array}$$
$$(\xi_u{}^{-1})^* \cdot \boldsymbol{J}(\xi_{u,o}^{-1}) =: (-\widehat{V}_u\ \widehat{\boldsymbol{A}}_u).$$

A **coordinated form** of $\boldsymbol{K}$ is obtained by a coordinate system adapted to the $(\boldsymbol{u}, o)$-splitting:

$$\begin{array}{ccccccc} \mathbb{R}^4 \longrightarrow \mathbb{T} \times \mathbf{S}_u \longrightarrow \mathrm{M} \longrightarrow \mathbf{M} & \longrightarrow & \mathbb{T}^* \times \mathbf{S}_u^* & \longrightarrow & \mathbb{R}^4 \\ C^{-1} \qquad \xi_{u,o}^{-1} \quad \boldsymbol{K} & (\xi_u{}^{-1})^* & & (C^{-1})^*, & \end{array}$$
$$(C^{-1})^* \cdot (\xi_u{}^{-1})^* \cdot \boldsymbol{K}(\xi_{u,o}^{-1}(C^{-1})) =: (-\widetilde{V}_u\ \widetilde{\boldsymbol{A}}_u) =: K_k \quad (k = 0, 1, 2, 3).$$

Concisely,

$$\boldsymbol{K} \sim (-V_u, \boldsymbol{A}) \sim (-\widehat{V}_u, \widehat{\boldsymbol{A}}_u) \sim (-\widetilde{V}_u, \widetilde{\boldsymbol{A}}_u) \sim K_k \quad (k = 0, 1, 2, 3).$$

12.1.2 Tensor fields, cotensor fields

Now we consider different split forms of an antisymmetric tensor field $\boldsymbol{G} : \mathrm{M} \to \mathbf{M} \wedge \mathbf{M}$ and an antisymmetric cotensor field $\boldsymbol{F} : \mathrm{M} \to \mathbf{M}^* \wedge \mathbf{M}^*$ applying the formulas of Subsection 5.7.1 and the notation of the previous subsection, according to the sense.

Attention!
$\boldsymbol{G}$ and $\boldsymbol{F}$ in the cited subsection are elements of $\mathbf{M} \wedge \mathbf{M}$ and $\mathbf{M}^* \wedge \mathbf{M}^*$, respectively, while now the same symbols denote tensor/cotensor valued functions defined in spacetime.

Then we have

$$\boldsymbol{G} \sim ((\boldsymbol{D}_u, \boldsymbol{H}_u)) \sim ((\widehat{\boldsymbol{D}}_u, \widehat{\boldsymbol{H}}_u)) \sim ((\widetilde{\boldsymbol{D}}_u, \widetilde{\boldsymbol{H}}_u)) \sim G^{ik} \quad (i \neq k = 0, 1, 2, 3),$$

and

$$\boldsymbol{F} \sim ((\boldsymbol{E}_u, \boldsymbol{B}_u)) \sim ((\widehat{\boldsymbol{E}}_u, \widehat{\boldsymbol{B}}_u)) \sim ((\widetilde{\boldsymbol{E}}_u, \widetilde{\boldsymbol{B}}_u)) \sim F_{ik} \quad (i \neq k = 0, 1, 2, 3),$$

12.2 Differentiation

Formulas, similar to those in Section 7.2, are valid, of course, with some modifications.

12.2.1 Scalar fields

The derivative of a scalar field $f : \mathrm{M} \to \mathbb{R}$ is the covector field $x \mapsto \mathcal{D}f[x]$.

The half-split form by u of $\mathcal{D}f$ is

$$(\boldsymbol{u} \cdot \mathcal{D}f \quad \pi_{\boldsymbol{u}} \cdot \mathcal{D}f) =: (\mathcal{D}_{\boldsymbol{u}}f \ \nabla_{\boldsymbol{u}}f);$$

$\mathcal{D}_{\boldsymbol{u}}$ is the **u-time-like derivative** of f and $\nabla_{\boldsymbol{u}}f$ is the **u-space-like derivative** of f; they have direct meanings as follows.

Let us restrict f onto the straight line passing through x and directed by $\boldsymbol{u}$, i. e., let us consider the function $\mathbb{T} \to \mathbb{R}$, $\boldsymbol{t} \mapsto f(x + \boldsymbol{t}\boldsymbol{u})$. The derivative at zero of this function—according to the chain rule of differentiation of composite functions—is $(\mathcal{D}f[x]) \cdot \boldsymbol{u} = \boldsymbol{u} \cdot (\mathcal{D}f[x])$.

Let us restrict f onto the hyperplane passing through x and directed by $\mathbf{S}_{\boldsymbol{u}}$, i. e., let us consider the function $\mathbf{S}_{\boldsymbol{u}} \to \mathbb{R}$, $\boldsymbol{q} \mapsto f(x + \boldsymbol{q}) = f(x + \boldsymbol{i}_{\boldsymbol{u}} \cdot \boldsymbol{q})$. The derivative at zero of this function—according to the chain rule of differentiation of composite functions—is $(\mathcal{D}f[x]) \cdot \boldsymbol{i} = \boldsymbol{i}^* \cdot (\mathcal{D}f[x]) = \pi_{\boldsymbol{u}} \cdot \mathcal{D}f[x]$.

The completely split form by $(\boldsymbol{u}, o)$ of $\mathcal{D}f$ is obtained by the differentiation of the function

$$\mathbb{T} \times \mathbf{S}_{\boldsymbol{u}} \to \mathbb{R}, \quad (\boldsymbol{t}, \boldsymbol{q}) \mapsto \hat{f}(\boldsymbol{t}, \boldsymbol{q}) := f(o + \boldsymbol{t}\boldsymbol{u} + \boldsymbol{q}).$$

The "partial differentiation with respect to time" and the "partial differentiation with respect to space," denoted by $\partial_{\mathbb{T}}$ and $\boldsymbol{\nabla}_{\boldsymbol{u}}$, respectively, give the completely split form by $(\boldsymbol{u}, o)$:

$$(\partial_{\mathbb{T}}\hat{f}, \boldsymbol{\nabla}_{\boldsymbol{u}}\hat{f}).$$

A coordinated form of $\mathcal{D}f$ is obtained by the partial derivatives of the function (see Section 5.10)

$$\mathbb{R}^4 \to \mathbb{R}, \quad (\xi^0, \xi^1, \xi^2, \xi^3) \mapsto \tilde{f}(\xi^0, \xi^1, \xi^2, \xi^3) := f(o + \xi^i \boldsymbol{e}_i)$$

(Einstein summation), which are

$$\partial_k \hat{f} \quad (k = 0, 1, 2, 3).$$

Concisely,

$$\mathcal{D}f \sim (\mathcal{D}_{\boldsymbol{u}}f, \nabla_{\boldsymbol{u}}f) \sim (\partial_{\mathbb{T}}\hat{f}, \boldsymbol{\nabla}_{\boldsymbol{u}}\hat{f}) \sim \partial_k \tilde{f} \quad (k = 0, 1, 2, 3).$$

12.2.2 Vector fields, tensor fields

In general,

the differentiation $\mathcal{D}$ can be considered a symbolic covector for which we have

$$\mathcal{D} \sim (\mathcal{D}_u \; \nabla_u) \sim (\partial_{\mathbb{T}} \; \nabla_u) \sim \partial_k \quad (k = 0,1,2,3).$$

Then with evident modifications of the nonrelativistic formulas, for a vector field $\boldsymbol{J} : \mathrm{M} \to \mathbf{M}$ we can write

$$\boldsymbol{J} \sim (\rho_u, \boldsymbol{j}_u) \sim (\hat{\rho}_u, \hat{\boldsymbol{j}}_u) \sim J^k \quad (k = 0,1,2,3)$$

$$\mathcal{D} \otimes \boldsymbol{J} \sim \begin{pmatrix} \mathcal{D}_u \rho_u & \mathcal{D}_u \boldsymbol{j}_u \\ \nabla_u \rho_u & \nabla_u \otimes \boldsymbol{j}_u \end{pmatrix} \sim \begin{pmatrix} \partial_{\mathbb{T}} \hat{\rho}_u & \partial_{\mathbb{T}} \hat{\boldsymbol{j}}_u \\ \boldsymbol{\nabla}_u \hat{\rho}_u & \boldsymbol{\nabla}_u \otimes \hat{\boldsymbol{j}}_u \end{pmatrix} \sim \partial_i \tilde{J}^k \quad (i, k = 0,1,2,3)$$

and the **divergence** of $\boldsymbol{J}$ is

$$\mathcal{D} \cdot \boldsymbol{J} \sim \mathcal{D}_u \rho_u + \nabla_u \cdot \boldsymbol{j}_u \sim \partial_{\mathbb{T}} \hat{\rho}_u + \boldsymbol{\nabla}_u \cdot \hat{\boldsymbol{j}}_u \sim \partial_k J^k.$$

For a covector field $\boldsymbol{K} : \mathrm{M} \to \mathbf{M}^*$, if

$$\boldsymbol{K} \sim (-V_u, \boldsymbol{A}_u) \sim (-\widehat{V}_u, \widehat{\boldsymbol{A}}_u) \sim \tilde{K}_k \quad (k = 0,1,2,3)$$

then

$$\begin{aligned} \mathcal{D} \otimes \boldsymbol{K} &\sim \begin{pmatrix} -\mathcal{D}_u V_u & \mathcal{D}_u \boldsymbol{A}_u \\ -\nabla_u V_u & \nabla_u \otimes \boldsymbol{A}_u \end{pmatrix} \\ &\sim \begin{pmatrix} -\partial_{\mathbb{T}} \widehat{V}_u & \partial_{\mathbb{T}} \widehat{\boldsymbol{A}}_u \\ -\boldsymbol{\nabla}_u \widehat{V}_u & \boldsymbol{\nabla}_u \otimes \widehat{\boldsymbol{A}}_u \end{pmatrix} \sim \partial_i \tilde{K}_k \quad (i, k = 0,1,2,3). \end{aligned}$$

The **exterior derivative** of $\boldsymbol{K}$ is

$$\begin{aligned} \mathcal{D} \wedge \boldsymbol{K} &\sim ((-\nabla_u V_u - \mathcal{D}_u \boldsymbol{A}_u, \nabla_u \wedge \boldsymbol{A}_u)) \\ &\sim ((-\boldsymbol{\nabla}_u \widehat{V}_u - \partial_{\mathbb{T}} \widehat{\boldsymbol{A}}_u, \boldsymbol{\nabla}_u \wedge \widehat{\boldsymbol{A}}_u)) \sim \partial_i \tilde{K}_k - \partial_k \tilde{K}_i \quad (i, k = 0,1,2,3). \end{aligned}$$

Contrary to the nonrelativistic case,

because of the Lorentz identification of vectors and covectors, we can take the exterior derivative of a vector field and the divergence of a covector field.

2. For an antisymmetric tensor field $\boldsymbol{G} : \mathrm{M} \to \mathbf{M} \wedge \mathbf{M}$, if

$$\boldsymbol{G} \sim ((\boldsymbol{D}_u, \boldsymbol{H}_u)) \sim ((\widehat{\boldsymbol{D}}_u, \widehat{\boldsymbol{H}}_u)) \sim \tilde{G}^{ik}$$

then its divergence is

$$\mathcal{D}\cdot\boldsymbol{G} \sim (\nabla_{\boldsymbol{u}}\cdot\boldsymbol{D}_{\boldsymbol{u}}, -\mathcal{D}_{\boldsymbol{u}}\boldsymbol{D}_{\boldsymbol{u}} + \nabla_{\boldsymbol{u}}\cdot\boldsymbol{H}_{\boldsymbol{u}})$$
$$\sim (\nabla_{\boldsymbol{u}}\cdot\widehat{\boldsymbol{D}}_{\boldsymbol{u}}, -\partial_{\mathbb{T}}\widehat{\boldsymbol{D}}_{\boldsymbol{u}} + \nabla_{\boldsymbol{u}}\cdot\widehat{\boldsymbol{H}}_{\boldsymbol{u}}) \sim \partial_i G^{ik}.$$

For an antisymmetric cotensor field $\boldsymbol{F} : \mathrm{M} \to \mathbf{M}^* \wedge \mathbf{M}^*$, if

$$\boldsymbol{F} \sim ((\boldsymbol{E}_{\boldsymbol{u}}, \boldsymbol{B}_{\boldsymbol{u}})) \sim \tilde{F}_{ik}$$

then its exterior derivative is

$$\mathcal{D}\wedge\boldsymbol{F} \sim (((\nabla_{\boldsymbol{u}}\wedge\boldsymbol{E}_{\boldsymbol{u}} + \mathcal{D}_{\boldsymbol{u}}\boldsymbol{B}_{\boldsymbol{u}}, \nabla_{\boldsymbol{u}}\wedge\boldsymbol{B}_{\boldsymbol{u}})))$$
$$\sim (((\nabla_{\boldsymbol{u}}\wedge\widehat{\boldsymbol{E}}_{\boldsymbol{u}} + \partial_{\mathbb{T}}\widehat{\boldsymbol{B}}_{\boldsymbol{u}}, \nabla_{\boldsymbol{u}}\wedge\widehat{\boldsymbol{B}}_{\boldsymbol{u}}))) \sim \partial_j F_{ik} + \partial_k F_{ji} + \partial_i F_{kj}$$

where the three parentheses means that the two quantities inside determine the whole antisymmetric tensor in $\mathbf{M}^* \wedge \mathbf{M}^* \wedge \mathbf{M}^*$.

Contrary to the nonrelativistic case,

because of the Lorentz identification of vectors and covectors, we can take the exterior derivative of a tensor field and the divergence of a cotensor field.

12.3 Maxwell equations

The usual relative Maxwell equations (7.7) and (7.8) hold relativistically, too. Four steps, similar to those in Subsection 7.3.1, with replacing $\boldsymbol{E}$ by $\boldsymbol{E}_{\boldsymbol{u}}$, $\boldsymbol{B}$ by $\boldsymbol{B}_{\boldsymbol{u}}$, etc. lead us to the equations

$$\nabla_{\boldsymbol{u}}\cdot\boldsymbol{D}_{\boldsymbol{u}} = \rho_{\boldsymbol{u}}, \quad -\mathcal{D}_{\boldsymbol{u}}\boldsymbol{D}_{\boldsymbol{u}} + \nabla_{\boldsymbol{u}}\cdot\boldsymbol{H}_{\boldsymbol{u}} = \boldsymbol{j}_{\boldsymbol{u}}, \tag{12.3}$$
$$\mathcal{D}_{\boldsymbol{u}}\boldsymbol{B}_{\boldsymbol{u}} + \nabla_{\boldsymbol{u}}\wedge\boldsymbol{E}_{\boldsymbol{u}} = 0, \quad \nabla_{\boldsymbol{u}}\wedge\boldsymbol{B}_{\boldsymbol{u}} = 0. \tag{12.4}$$

They are the half-split form by $\boldsymbol{u}$ of the absolute Maxwell equations

$$\mathcal{D}\cdot\boldsymbol{G} = \boldsymbol{J}, \quad \mathcal{D}\wedge\boldsymbol{F} = 0 \tag{12.5}$$

having the same form as nonrelativistically.

Under some “regularity” conditions $\boldsymbol{F}$ is the exterior derivative of a covector field, called **absolute potential**: $\boldsymbol{F} = \mathcal{D}\wedge\boldsymbol{K}$.

$\boldsymbol{J}$, $\boldsymbol{G}$, and $\boldsymbol{F}$ have the same physical meaning as in the nonrelativistic case and their measure lines are the same as well if $\mathbb{T}$ is taken instead of $\mathbb{L}$. In particular,

$$\boldsymbol{J} : \mathrm{M} \to \frac{\mathbf{M}}{\mathbb{T}^{(4)}}, \quad \boldsymbol{G} : \mathrm{M} \to \frac{\mathbf{M}\wedge\mathbf{M}}{\mathbb{T}^{(4)}}, \quad \boldsymbol{F} : \mathrm{M} \to \mathbf{M}^*\wedge\mathbf{M}^* \equiv \frac{\mathbf{M}\wedge\mathbf{M}}{\mathbb{T}^{(4)}}.$$

12.4 Vacuum constitutive relation

As in the nonrelativistic case, we have to give a **constitutive relation**

$$G = \Gamma(\boldsymbol{F}),$$

which reflects how a material medium in spacetime influences the electromagnetic phenomena.

The constitutive relation due to a real medium can be formulated here in the same way as in the nonrelativistic case. There is an essential difference for vacuum because now we have a convenient constitutive relation without introducing the fictitious ether.

Namely, because of the identification $\frac{\mathbf{M}\wedge\mathbf{M}}{\mathbb{T}^{(4)}} \equiv \mathbf{M}^* \wedge \mathbf{M}^*$, the electromagnetic displacement $\boldsymbol{G}$ and the electromagnetic field become quantities of the same kind.

As a consequence, the vacuum constitutive relation can be formulated as

$$\boldsymbol{G} = \boldsymbol{F}$$

and the absolute Maxwell equations in vacuum read

$$\mathcal{D} \cdot \boldsymbol{F} = \boldsymbol{J}, \quad \mathcal{D} \wedge \boldsymbol{F} = 0.$$

12.5 The problem of particle-field interaction

The Newton equation describes the processes of a particle under the action of a **given** force.

The Maxwell equations describe the processes of the electromagnetic field generated by a **given** charge current.

The Newton equation and the Maxwell equations cannot be coupled to describe the processes of a particle and the electromagnetic field together, which is surveyed as follows.

The Newton equation for a particle having mass m and charge e in an electromagnetic field $\boldsymbol{F}$ is

$$m\ddot{x} = e\boldsymbol{F}(x) \cdot \dot{x}. \tag{12.6}$$

The charge current of the particle is $e\dot{x}$, so the Maxwell equations for the electromagnetic field $\boldsymbol{F}$ generated by the particle are

$$(\boldsymbol{D} \cdot \boldsymbol{F})[x] = e\dot{x}, \quad \boldsymbol{D} \wedge \boldsymbol{F} = 0. \tag{12.7}$$

It turns out that (12.6) and (12.7) together do not form a meaningful system of equations determining both the world line of the particle and the electromagnetic field [8].

Remark. $\boldsymbol{J}$ in the Maxwell equations is a charge current density. The charge current of a single particle has no density; in a precise formulation the corresponding distribution ("generalized function") must be taken instead of $e\dot{x}$.

13 Nonstandard formulas

Coordinates in usual treatments implicitly always concern standard synchronization, so the importance of synchronizations is lost, which leads to paradoxes like the one discussed in Section 15.4. That is why it is worth dealing with nonstandard synchronizations; we shall get instructive results.

13.1 Uniform synchronizations

A uniform synchronization is given by a three-dimensional linear subspace $\mathbf{S}_s$, which is transverse to all absolute velocities. There are two possibilities:

1. $\mathbf{S}_s$ contains only space-like vectors; then the synchronization is called **space-like.** In this case, there is a unique absolute velocity $\boldsymbol{u}_s$ such that $\mathbf{S}_s = \mathbf{S}_{\boldsymbol{u}_s}$, i. e.,

$$\mathbf{S}_s = \{x \in \mathbf{M} \mid \boldsymbol{u}_s \cdot x = 0\}. \tag{13.1}$$

2. $\mathbf{S}_s$ contains space-like vectors and light-like vectors; then the synchronization is called **light-like.** In this case, there is a light direction $\boldsymbol{w}_s$ such that

$$\mathbf{S}_s = \{x \in \mathbf{M} \mid \boldsymbol{w}_s \cdot x = 0\}.$$

   Light-like vectors in $\mathbf{S}_s$ can only be the multiples of $\boldsymbol{w}_s$.

For an arbitrary absolute velocity $\boldsymbol{u}$, there are $\boldsymbol{u}$-space-like vectors $\boldsymbol{v}_1$ and $\boldsymbol{v}_2$ such that $\boldsymbol{u}, \pi_{\boldsymbol{u}} \cdot \boldsymbol{u}, \boldsymbol{v}_1, \boldsymbol{v}_2$ is a Lorentz orthogonal basis. $\boldsymbol{w}_s$ is a linear combination of $\boldsymbol{u}$ and $\pi_{\boldsymbol{u}} \cdot \boldsymbol{w}_s$, $\boldsymbol{w}_s = -(\boldsymbol{u}\boldsymbol{w}_s)\boldsymbol{u} + \pi_{\boldsymbol{u}} \cdot \boldsymbol{w}_s$, so $\boldsymbol{w}_s \cdot \boldsymbol{v}_1 = \boldsymbol{w}_s \cdot \boldsymbol{v}_2 = 0$, which means that $\boldsymbol{w}_s, \boldsymbol{v}_1, \boldsymbol{v}_2$ is a basis in $\frac{\mathbf{S}_s}{\mathbb{T}}$. $|\alpha \boldsymbol{w}_s + \alpha_1 \boldsymbol{v}_1 + \alpha \boldsymbol{v}_2|^2 = \alpha_1^2 |\boldsymbol{v}_1|^2 + \alpha_2^2 |\boldsymbol{v}_2|^2$, which is zero if and only if the linear combination in question is a multiple of $\boldsymbol{w}_s$.

In both cases, the synchronization itself, too, is denoted by $\mathbf{S}_s$; the synchronization time points are hyperplanes directed by $\mathbf{S}_s$; their set is T_s.

13.2 Inertial frame with space-like synchronization

An inertial frame is $(\boldsymbol{u}, \mathbf{S}_s)$ where $\boldsymbol{u}$ is an inertial observer and $\mathbf{S}_s$ is a uniform synchronization; such a frame is represented by Figure 3.6. If $\mathbf{S}_s \neq \mathbf{S}_{\boldsymbol{u}}$, then the synchronization of the inertial frame is **nonstandard**. In this section, only space-like synchronizations of form (13.1) are considered.

13.2.1 Splitting

The inertial frame $(\boldsymbol{u}, \mathbf{S}_s)$ splits spacetime vectors into the sum of a vector parallel to $\boldsymbol{u}$ an a vector in $\mathbf{S}_s$. It is simple that

https://doi.org/10.1515/9783112219553-016

$$\pi_{u,u_s} := \mathbf{1} + \frac{\boldsymbol{u} \otimes \boldsymbol{u}_s}{-\boldsymbol{u} \cdot \boldsymbol{u}_s} \tag{13.2}$$

is the projection onto $\mathbf{S}_s$ along $\boldsymbol{u}$.

Introducing the map

$$\tau_{u,u_s} : \mathbf{M} \to \mathbb{T}, \quad x \mapsto \frac{-\boldsymbol{u}_s \cdot x}{-\boldsymbol{u} \cdot \boldsymbol{u}_s},$$

we find that $x = (\tau_{u,u_s} \cdot x)\boldsymbol{u} + \pi_{u,u_s} \cdot x$; therefore, the splitting of vectors according to $(\boldsymbol{u}, \mathbf{S}_s)$ is

$$\xi_{u,u_s} : \mathbf{M} \to \mathbb{T} \times \mathbf{S}_s, \quad x \mapsto (\tau_{u,u_s} \cdot x, \pi_{u,u_s} \cdot x); \tag{13.3}$$

briefly,

$$\xi_{u,u_s} = (\tau_{u,u_s}, \pi_{u,u_s}).$$

The difference of two $\mathbf{S}_s$-time points is taken to be the proper time interval passing in $\boldsymbol{u}$-space points passing between the synchronization time points:

$$(y + \mathbf{S}_s) \overset{u}{-} (x + \mathbf{S}_s) := \tau_{u,u_s} \cdot (y - x). \tag{13.4}$$

$\boldsymbol{u}$-space vectors are represented by elements of $\mathbf{S}_s$ in such a way that

$$(y + \mathbb{T}\boldsymbol{u}) \overset{u_s}{-} (y + \mathbb{T}\boldsymbol{u}) := \pi_{u,u_s} \cdot (y - x). \tag{13.5}$$

Of course, the Euclidean structure of the observer does not depend on the synchronization; thus, the distance between the $\boldsymbol{u}$-space points $x + \mathbb{T}\boldsymbol{u}$ and $y + \mathbb{T}\boldsymbol{u}$ is $|\pi_{\boldsymbol{u}}(y - x)|$. Since $\pi_{\boldsymbol{u}} \cdot \pi_{u,u_s} = \pi_{\boldsymbol{u}}$, if $\boldsymbol{q}_s := \pi_{u,u_s} \cdot (y - x) \in \mathbf{S}_s$, then the square of the distance is

$$|\boldsymbol{q}_s|_{\boldsymbol{u}}^2 := |\pi_{\boldsymbol{u}} \cdot \boldsymbol{q}_s|^2 = |\boldsymbol{q}_s|^2 + (\boldsymbol{u} \cdot \boldsymbol{q}_s)^2, \tag{13.6}$$

where $|\ |$ denotes the Lorentz length and $|\ |_{\boldsymbol{u}}$ denotes the physical length of a $\boldsymbol{u}$-space vector represented by an element of $\mathbf{S}_s$; keep this formula in mind, not to commit an error: $|\boldsymbol{q}_s|_{\boldsymbol{u}} \geq |\boldsymbol{q}_s|$.

Arithmetic formulas

Let $\boldsymbol{u} := (1, 0, 0, 0)$ and $\boldsymbol{u}_s := \gamma(1, v, 0, 0)$; then $-\boldsymbol{u} \cdot \boldsymbol{u}_s = \gamma$. Let us write vectors in columns and omit the last two coordinates; then

$$\tau_{u,u_s} = \begin{pmatrix} 1 & v \\ 0 & 0 \end{pmatrix}, \quad \pi_{u,u_s} = \begin{pmatrix} 0 & -v \\ 0 & 1 \end{pmatrix}.$$

For $\boldsymbol{q}_s := \binom{vq}{q}$, we have $-\boldsymbol{u} \cdot \boldsymbol{q}_s = vq$. So,

$$|\boldsymbol{q}_s|^2 = (1 - v^2)q^2, \quad |\boldsymbol{q}_s|_{\boldsymbol{u}}^2 = q^2.$$

13.2.2 Transformation rules

The splits by $(\boldsymbol{u}, \mathbf{S}_s)$ and $(\boldsymbol{u}', \mathbf{S}_{s'})$ can be compared, as in the case of standard splitting, if $\mathbf{S}_{s'}$ is boosted to $\mathbf{S}_s$.

In general, highly complicated transformation rules are obtained in this way.

Transformation rules, of course in coordinates, different from the usual Lorentz transformation rules can be found in the literature [5, 6, 14, 15]. Not being aware that those transformation rules apply nonstandard synchronizations, they wanted to create paradoxes or to produce alternative theories. Such transformation rules, however, can be well explained in the relativistic spacetime model.

13.2.3 Relative velocities

The movement of an inertial material point with absolute velocity $\boldsymbol{u}'$ with respect to the inertial frame $(\boldsymbol{u}, \mathbf{S}_s)$ is uniform, i. e., has constant relative velocity.

Let x and y be two occurrences of the material point, i. e., $y - x = \boldsymbol{t}'\boldsymbol{u}'$ for some $\boldsymbol{t}' \in \mathbb{T}$. At the $\mathbf{S}_s$-instants $x + \mathbf{S}_s$ and $y + \mathbf{S}_s$, the material points meets the $\boldsymbol{u}$-space points $x + \mathbb{T}\boldsymbol{u}$ and $y + \mathbb{T}\boldsymbol{u}$, respectively; the $\boldsymbol{u}$-space vector between them is represented by

$$(y + \mathbb{T}\boldsymbol{u}) \overset{\boldsymbol{u}_s}{-} (x + \mathbb{T}\boldsymbol{u}) = \pi_{\boldsymbol{u},\boldsymbol{u}_s} \cdot (y - x) = \boldsymbol{t}'(\pi_{\boldsymbol{u},\boldsymbol{u}_s} \cdot \boldsymbol{u}').$$

The $\mathbf{S}_s$ duration between the two meetings is

$$(y + \mathbf{S}_s) \overset{\boldsymbol{u}}{-} (x + \mathbf{S}_s) = \tau_{\boldsymbol{u},\boldsymbol{u}_s} \cdot (y - x) = \boldsymbol{t}'(\tau_{\boldsymbol{u},\boldsymbol{u}_s} \cdot \boldsymbol{u}').$$

As a consequence, the $(\boldsymbol{u}, \mathbf{S}_s)$-relative velocity of $\boldsymbol{u}'$ is

$$\boldsymbol{v}_{\boldsymbol{u}'\boldsymbol{u},\boldsymbol{u}_s} := \frac{(y + \mathbb{T}\boldsymbol{u}) \overset{\boldsymbol{u}_s}{-} (x + \mathbb{T}\boldsymbol{u})}{(y + \mathbf{S}_s) \overset{\boldsymbol{u}}{-} (x + \mathbf{S}_s)} = \frac{\pi_{\boldsymbol{u},\boldsymbol{u}_s} \cdot \boldsymbol{u}'}{\tau_{\boldsymbol{u},\boldsymbol{u}_s} \cdot \boldsymbol{u}'} = \frac{(-\boldsymbol{u}_s \cdot \boldsymbol{u})\boldsymbol{u}'}{-\boldsymbol{u}_s \cdot \boldsymbol{u}'} - \boldsymbol{u},$$

for which

$$|\boldsymbol{v}_{\boldsymbol{u}'\boldsymbol{u},\boldsymbol{u}_s}|_{\boldsymbol{u}} = |\pi_{\boldsymbol{u}} \cdot \boldsymbol{v}_{\boldsymbol{u}'\boldsymbol{u},\boldsymbol{u}_s}| = |\boldsymbol{v}_{\boldsymbol{u}'\boldsymbol{u}}| \frac{(-\boldsymbol{u}' \cdot \boldsymbol{u})(-\boldsymbol{u}_s \cdot \boldsymbol{u})}{-\boldsymbol{u}_s \cdot \boldsymbol{u}'},$$

where $\boldsymbol{v}_{\boldsymbol{u}'\boldsymbol{u}}$ is the standard relative velocity of $\boldsymbol{u}'$ with respect to $\boldsymbol{u}$.

Similar formulas hold for light signals, i. e., for a light direction $\boldsymbol{w}$ instead of the absolute velocity $\boldsymbol{u}'$. We can see better the physical content of such a formula if we write

$$\boldsymbol{u}_s = \frac{\boldsymbol{u} + v_s \boldsymbol{n}_s}{\sqrt{1 - v_s^2}}, \qquad \boldsymbol{w} = \boldsymbol{u} + \boldsymbol{n}_w,$$

where $\boldsymbol{n}_s, \boldsymbol{n}_w \in \frac{\mathbf{S}_u}{\mathbb{T}}$ are unit vectors. $\boldsymbol{n}_s$ is a direction in $\boldsymbol{u}$-space, characterizing the synchronization; more closely, $v_s := |\boldsymbol{v}_{u_s u}|$, $\boldsymbol{n}_s = \frac{\boldsymbol{v}_{u_s u}}{v_s}$, and $\boldsymbol{n}_w$ is the direction of the light signal in $\boldsymbol{u}$-space (the relative velocity of $\boldsymbol{w}$ with respect to $\boldsymbol{u}$). Then

$$|\boldsymbol{v}_{wu,u_s}|_u = \frac{1 - v_s \boldsymbol{n}_s \cdot \boldsymbol{n}_w}{1 - v_s^2}.$$

The light speed in $\boldsymbol{u}$-space depends on the direction of light propagation (except, of course, the trivial case $v_s = 0$, which corresponds to the standard synchronization). The minimal light speed is $\frac{1}{1+v_s}$ when $\boldsymbol{n}_w = \boldsymbol{n}_s$ and the maximal light speed is $\frac{1}{1-v_s}$ when $\boldsymbol{n}_w = -\boldsymbol{n}_s$.

13.2.4 Comparison of lengths and time intervals

Let us consider the inertial frame $(\boldsymbol{u}, \mathbf{S}_s)$.

1. Let this inertial frame make instantaneous prints of $\boldsymbol{u}'$-space vectors. The print of the vector $\boldsymbol{q}' \in \mathbf{S}_{u'}$ is the vector $\boldsymbol{q}_s \in \mathbf{S}_s$ for which $\boldsymbol{q}' - \boldsymbol{q}_s$ is parallel to $\boldsymbol{u}'$: $\boldsymbol{q}' - \boldsymbol{q}_s = \boldsymbol{t}'\boldsymbol{u}'$ for some $\boldsymbol{t}'$ (see Figure 13.1). Taking the Lorentz product of this equality by $\boldsymbol{u}_s$ and then eliminating $\boldsymbol{t}'$, we get

$$\begin{aligned}\boldsymbol{q}_s &= \boldsymbol{q}' + \frac{\boldsymbol{u}_s \cdot \boldsymbol{q}'}{-\boldsymbol{u}_s \cdot \boldsymbol{u}'}\boldsymbol{u}' = \boldsymbol{q}' + \left(\left(\frac{\boldsymbol{u}_s}{-\boldsymbol{u}_s \cdot \boldsymbol{u}'} - \boldsymbol{u}'\right) \cdot \boldsymbol{q}'\right)\boldsymbol{u}' \\ &= \boldsymbol{q}' + (\boldsymbol{v}_{u_s u'} \cdot \boldsymbol{q}')\boldsymbol{u}';\end{aligned}$$

   then

$$\boldsymbol{u} \cdot \boldsymbol{q}_s = \frac{1}{\sqrt{1 - |\boldsymbol{v}_{uu'}|^2}}(\boldsymbol{v}_{uu'} - \boldsymbol{v}_{u_s u'}) \cdot \boldsymbol{q}',$$

   where we used that $\boldsymbol{u} \cdot \boldsymbol{q}' = (-\boldsymbol{u} \cdot \boldsymbol{u}')(\frac{\boldsymbol{u}}{-\boldsymbol{u} \cdot \boldsymbol{u}'} - \boldsymbol{u}') \cdot \boldsymbol{q}'$.
   In this way, (13.6) gives the following formula for the square length of the print:

$$|\boldsymbol{q}_s|_u^2 = |\boldsymbol{q}'|^2 - (\boldsymbol{v}_{u_s u'} \cdot \boldsymbol{q}')^2 + \frac{1}{1 - |\boldsymbol{v}_{uu'}|^2}((\boldsymbol{v}_{uu'} - \boldsymbol{v}_{u_s u'}) \cdot \boldsymbol{q}')^2.$$

   For $\boldsymbol{u}_s = \boldsymbol{u}$, this is reduced to the Lorentz contraction, of course. If $\boldsymbol{u}_s = \boldsymbol{u}'$ then "length dilation" occurs.

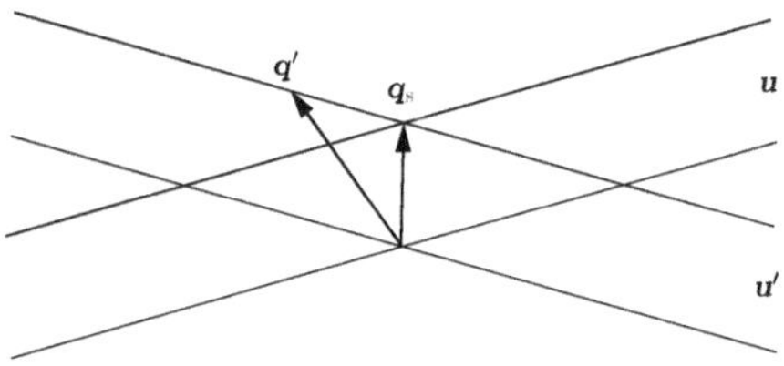

Figure 13.1: $\boldsymbol{q}_s$ is the $\mathbf{S}_s$-simultaneous $\boldsymbol{u}$-print of $\boldsymbol{q}'$.

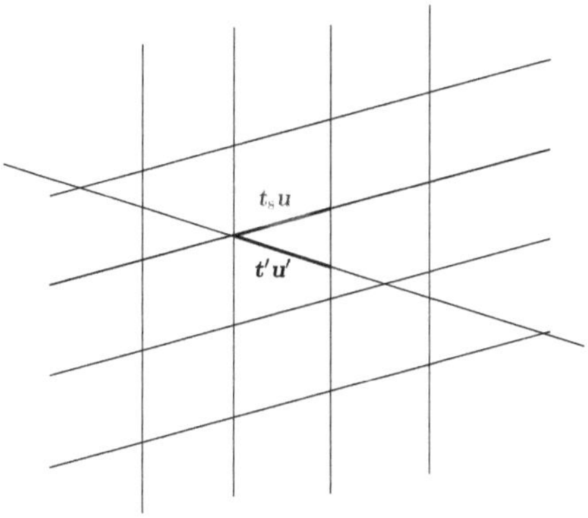

Figure 13.2: The nonstandard inertial frame measures time period $\boldsymbol{t}$ corresponding to the time period $\boldsymbol{t}'$ of the chronometer.

2. Let the inertial frame $(\boldsymbol{u}, \mathrm{S}_s)$ measure the frequency of ticks of a chronometer with absolute velocity $\boldsymbol{u}'$.
   Let $\boldsymbol{t}'$ be the proper time period of the chronometer while going from a $\boldsymbol{u}$-space point to another one. The corresponding synchronization time period $\boldsymbol{t}_s$ is defined by $\boldsymbol{u}_s \cdot (\boldsymbol{t}'\boldsymbol{u}' - \boldsymbol{t}_s\boldsymbol{u}) = 0$ (see Figure 13.2), i. e.,

$$\boldsymbol{t}_s = \frac{-\boldsymbol{u}_s \cdot \boldsymbol{u}'}{-\boldsymbol{u}_s \cdot \boldsymbol{u}}\boldsymbol{t}' = \frac{\sqrt{1 - |\boldsymbol{v}_{\boldsymbol{u}\boldsymbol{u}_s}|^2}}{\sqrt{1 - |\boldsymbol{v}_{\boldsymbol{u}'\boldsymbol{u}_s}|^2}}\boldsymbol{t}'.$$

   For $\boldsymbol{u}_s = \boldsymbol{u}$, this is reduced to the usual time dilation, of course. If $\boldsymbol{u}_s = \boldsymbol{u}'$, then "time contraction" occurs.

13.2.5 Equivalence of inertial frames

As usual, inertial frames are defined by Newton's first law: "an inertial material point moves uniformly in a straight line" and it is stated that "all inertial frames are equivalent: laws of physics assume the same form in them" (see Section 16.2).

We defined an inertial frame as an inertial observer with a uniform synchronization and we proved that Newton's first low holds in any inertial frame (see Subsection 3.5.5).

The previous results, however, show that not all the inertial frames are equivalent in the quoted sense: the one-way speed of light can be different in different inertial frames. We can state:

The inertial frames $(\boldsymbol{u}, \mathrm{S}_s)$ and $(\boldsymbol{u}', \mathrm{S}_{s'})$ are equivalent if and only if $\boldsymbol{u} \cdot \boldsymbol{u}_s = \boldsymbol{u}' \cdot \boldsymbol{u}_{s'}$.

Indeed, they are equivalent by the definition in Section 3.3 if there is a proper Lorentz transformation $\boldsymbol{L}$ such that $\boldsymbol{L} \cdot \boldsymbol{u} = \boldsymbol{u}'$ and $\boldsymbol{L}[\mathrm{S}_s] = \mathrm{S}_{s'}$; the latter condition is satisfied if and only if $\boldsymbol{L} \cdot \boldsymbol{u}_s = \boldsymbol{u}_{s'}$. Then $\boldsymbol{u} \cdot \boldsymbol{u}_s = (\boldsymbol{L} \cdot \boldsymbol{u}) \cdot (\boldsymbol{L} \cdot \boldsymbol{u}_s) = \boldsymbol{u}' \cdot \boldsymbol{u}_{s'}$.

Conversely, if $\boldsymbol{u}\cdot\boldsymbol{u}_s = \boldsymbol{u}'\cdot\boldsymbol{u}_{s'}$ then the linear map defined by

$$\boldsymbol{v}_{\boldsymbol{u}\boldsymbol{u}_s} \mapsto \boldsymbol{v}_{\boldsymbol{u}'\boldsymbol{u}_{s'}}, \quad \boldsymbol{v}_{\boldsymbol{u}_s\boldsymbol{u}} \mapsto \boldsymbol{v}_{\boldsymbol{u}_{s'}\boldsymbol{u}'},$$
$$\boldsymbol{R} : \mathbf{S}_{\boldsymbol{u}} \cap \mathbf{S}_s \to \mathbf{S}_{\boldsymbol{u}'} \cap \mathbf{S}_{s'}$$

is a proper Lorentz transformation establishing the equivalence where $\boldsymbol{R}$ is an arbitrary orientation preserving Euclidean map. Indeed, we know that $\boldsymbol{v}_{\boldsymbol{u}\boldsymbol{u}_s}$ and $\boldsymbol{v}_{\boldsymbol{u}_s\boldsymbol{u}}$ have the same magnitude and are Lorentz orthogonal to the two dimensional linear subspace $\mathbf{S}_{\boldsymbol{u}} \cap \mathbf{S}_s$ and similar holds for the "primed" quantities; thus, the linear map above is a proper Lorentz transformation $\boldsymbol{L}$. In view of $-\boldsymbol{u}\cdot\boldsymbol{u}_s = -\boldsymbol{u}'\cdot\boldsymbol{u}_{s'}$, it is a routine to show from

$$\boldsymbol{L}\cdot\left(\frac{\boldsymbol{u}}{-\boldsymbol{u}\cdot\boldsymbol{u}_s} - \boldsymbol{u}_s\right) = \frac{\boldsymbol{u}'}{-\boldsymbol{u}'\cdot\boldsymbol{u}_{s'}} - \boldsymbol{u}_{s'}, \quad \boldsymbol{L}\cdot\left(\frac{\boldsymbol{u}_s}{-\boldsymbol{u}\cdot\boldsymbol{u}_s} - \boldsymbol{u}\right) = \frac{\boldsymbol{u}_{s'}}{-\boldsymbol{u}'\cdot\boldsymbol{u}_{s'}} - \boldsymbol{u}'$$

that $\boldsymbol{L}\cdot\boldsymbol{u} = \boldsymbol{u}'$ and $\boldsymbol{L}\cdot\boldsymbol{u}_s = \boldsymbol{u}_{s'}$.

13.2.6 Exercises

1. Give the inverse of the splitting (13.3).
2. Show that the covectorial splitting is

$$\mathbf{M}^* \to \mathbb{T}^* \times \mathbf{S}_s^*, \quad \boldsymbol{k} \mapsto \left(\frac{\boldsymbol{k}\cdot\boldsymbol{u}}{-\boldsymbol{u}\cdot\boldsymbol{u}_s}, \boldsymbol{k}\cdot\boldsymbol{i}_s\right),$$

   where $\boldsymbol{i}_s : \mathbf{S}_s \to \mathbf{M}$ is the canonical embedding.
3. Let $(\boldsymbol{t}, \boldsymbol{q}) \in \mathbb{T}\times\mathbf{S}_s$ be the split components of a vector $\boldsymbol{x}$. What are the covectorial components of $\boldsymbol{x}$?
4. Take the inertial frames $(\boldsymbol{u}, \mathbf{S}_{\boldsymbol{u}})$ and $(\boldsymbol{u}', \mathbf{S}_{\boldsymbol{u}})$ (the "unprimed" frame is a standard one and the "primed" frame uses the "unprimed' synchronization). Then the split components according to both frames are in $\mathbb{T}\times\mathbf{S}_{\boldsymbol{u}}$. No boost is needed for comparing the splits. Give the transformation rule between the splits.
5. Take the absolute velocities $\boldsymbol{u}_1 := (1,0,0,0)$, $\boldsymbol{u}_2 := \gamma(1,v,0,0)$ and $\boldsymbol{u}_3 := \gamma'(1,0,v',0)$ in the arithmetic spacetime model. Give the nonstandard relative velocities in the possible cases $\boldsymbol{v}_{12,3}$, $\boldsymbol{v}_{31,2}$, etc.
6. Consider an inertial frame with light-like synchronization and examine how the statements of Subsections 13.2.1, 13.2.3, 13.2.4, and 13.2.5 change for this case.
7. For a positive $\boldsymbol{t}_0 \in \mathbb{T}$, the set $H := \{\boldsymbol{x} \in \mathrm{T}^{\to} \mid \boldsymbol{x}\cdot\boldsymbol{x} = -\boldsymbol{t}_0^2\}$ is a three- dimensional hypersurface, which defines a (nonuniform) synchronization: the world points x and y are simultaneous if and only if there are $\boldsymbol{x}, \boldsymbol{y} \in H$ in such a way that $x - y = \boldsymbol{x} - \boldsymbol{y}$. Then, given a spacetime point o and an absolute velocity $\boldsymbol{u}_0$,

$$\{o + \boldsymbol{t}\boldsymbol{u}_0 + H \mid \boldsymbol{t} \in \mathbb{T}\}$$

is the set of time points of this synchronization.
Let $\boldsymbol{u}$ be an inertial observer different from $\boldsymbol{u}_0$. Show that different proper times elapse in the $\boldsymbol{u}$-space points $o + \mathbb{T}\boldsymbol{u}$ and $o + \boldsymbol{t}_0\boldsymbol{u}_0 + \mathbb{T}\boldsymbol{u}$ between the synchronization instants labeled by $\boldsymbol{t}_0$ and $\boldsymbol{t} > \boldsymbol{t}_0$.

14 Noninertial observers

Noninertial observers can also be well treated in the special relativistic spacetime model but in a more complicated way than in the nonrelativistic case. The reason is the absence of absolute simultaneity. Now we treat only some problems. Further information, among others a detailed treatment of uniformly accelerated observers and uniformly rotating observers can be found in the book *Spacetime without Reference Frames*.

14.1 Nearly standard local synchronizations

What would be the standard synchronization of a noninertial observer?

Of course, a noninertial observer can send light signals from a "center" making them reflect; the mid-time between the start and the return of the light signal is defined to be simultaneous with the reflection. Such a **nearly standard local synchronization** may have the following unpleasant properties:

(1) It exists only in neighborhood of the "center."
(2) The one-way light speed is isotropic only in the "center."
(3) Different proper times pass in different space points between two synchronization time points.
(4) Different "centers" establish different synchronizations.

In the model, the nearly standard local synchronization corresponding to a "center" is described as follows. Let r be the world line function of the "center"; a world point x is simultaneous with the occurrence $r(\boldsymbol{s})$ if $x - r(\boldsymbol{s})$ is in the hyperplane Lorentz orthogonal to $\dot{r}(\boldsymbol{s})$. In other words, the set of world points simultaneous with $r(\boldsymbol{s})$ is $r(\boldsymbol{s}) + \mathbf{S}_{\dot{r}(\boldsymbol{s})}$. It can be proved that such a synchronization exists in a neighborhood of the "center," but only in a neighborhood because the simultaneous hyperplanes can meet if the "center" is not inertial (see Figure 14.1).

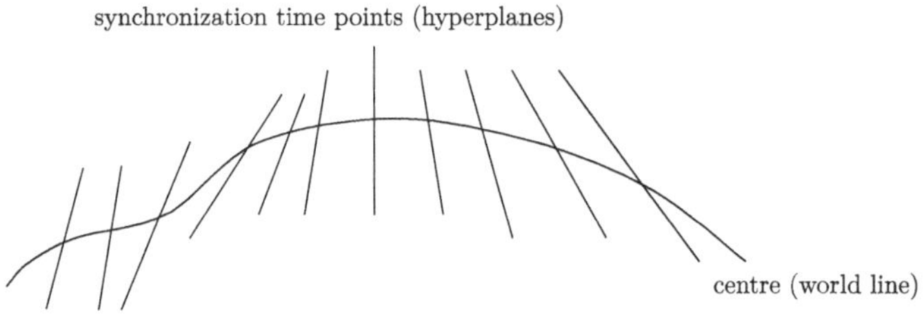

Figure 14.1: Nearly standard local synchronization.

Let us proceed in this way applying "tiny steps," from "centers" to "centers" (from Greenwich to London, from London to Dover, from Dover to Paris, etc.) and taking the

https://doi.org/10.1515/9783112219553-017

limit with infinitesimal steps we may hope to obtain a world surface whose tangent space at every point is Lorentz orthogonal to the absolute velocity of the observer. Such a globally standard synchronization, however, may not exist for a noninertial observer, as it is known by Frobenius' theorem in mathematics.

14.2 Synchronizations of a uniformly rotating observer

Recall that we mentioned two synchronizations on the Earth: the one is established by light signals (from Greenwich to Budapest, etc.), the other one is established by the position of stars (Sun) (see Subsection 10.6.1). Now we examine the relation between these synchronizations.

We may model the stars as an inertial observer. The axis of the Earth is considered to be at rest with respect to this inertial observer. The synchronization corresponding to the stars is the standard synchronization of an inertial observer.

The synchronization with a light signal is defined in such a way that one-way light speed in a chosen space point is the same in every direction. That is why this synchronization is the nearly standard local synchronization in a point of the Earth's surface.

The uniformly rotating rigid observer can be exactly defined in the spacetime model and it can be shown that the synchronizations above are theoretically different but practically are the same. Namely, the synchronization corresponding to the stars gives that the light speed on the equator towards east and west are [7]

$$c_+ = \frac{1}{1-(1.6)10^{-6}}, \quad c_- = \frac{1}{1+(1.6)10^{-6}},$$

respectively.

Lastly, an important fact: there is no globally standard synchronization for the uniformly rotating rigid observer.

15 Paradoxes

In usual treatments of relativity theory, based on intuitive notions, tacit assumptions and coordinates, some assertions contradict "common sense," which provides ample opportunities for paradoxes. Now we analyze some of them.

15.1 The tunnel paradox

One of the "classical" paradoxes is connected with Lorentz contraction.

Consider a train and a tunnel. The proper length of the train equals the proper length of the tunnel. The traveling train enters the tunnel.

The train moves with respect to the tunnel, so the (reference frame of) tunnel states that the train is contracted, and that is why the train will be entirely in the tunnel for a while.

The tunnel moves with respect to the train, so the (reference frame of) train states that the tunnel is contracted, and that is why the train never will be entirely in the tunnel.

The paradox is easily removed by making it clear that the two assertions concern two different synchronizations. Let us consider the occurrence that the front of the train meets the end of the tunnel. The end of the train (see Figure 15.1)

- Is inside the tunnel at the corresponding standard instant of the tunnel.
- Is outside the tunnel at the corresponding standard instant of the train.

Both the tunnel and the train are right with different synchronizations.

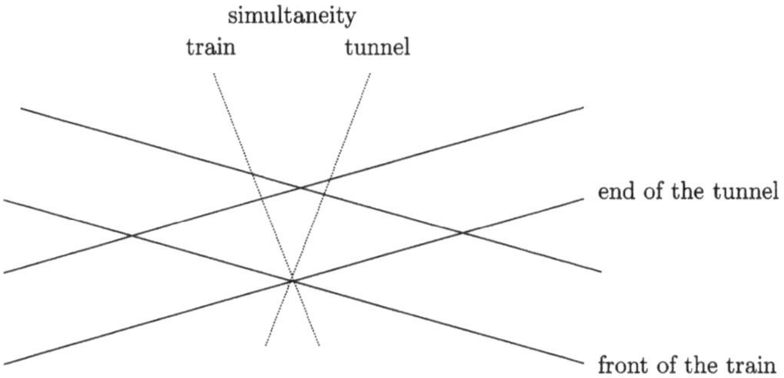

Figure 15.1: Train and tunnel.

Someone not convinced by the previous argument says: the tunnel will be right if it lowers trap doors at its both ends when the train is inside. Then the train meeting the front door, cannot break through, it stops "suddenly." How does it stop? The tunnel, taking care of the safety of the train, brakes the whole train "at an instant" according to

https://doi.org/10.1515/9783112219553-018

its standard synchronization. The whole train, however, does not stop "at an instant" according to its standard synchronization: when the locomotive stops then the first coach is still going forward, when the first coach stops then the second one is still going forward and so on: the train is contracted (see Figure 15.2).

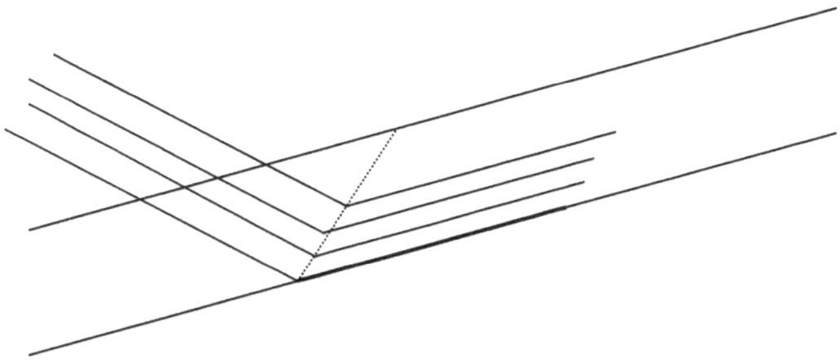

Figure 15.2: Train is braked suddenly at a standard tunnel moment.

In fact, the braked train is inside the tunnel, but braking makes the train noninertial.

15.2 The twin paradox

The other "classical" paradox is connected with time dilation.

Let us consider two twins, Alice and Bernard. Both are launched in separate spaceships. Alice says that Bernard is moving relative to her, hence Bernard's time passes more slowly and Alice finds that when she is 25, then Bernard is only 20. On the other hand, Bernard says that Alice is moving relative to him, hence he finds that when he is 25 then Alice is only 20. Which of them is right?

Both Alice and Bernard are right with different synchronizations: Alice's "when" means simultaneity with respect to her standard synchronization, Bernard's "when" means simultaneity with respect to his standard synchronization (see Figure 15.3).

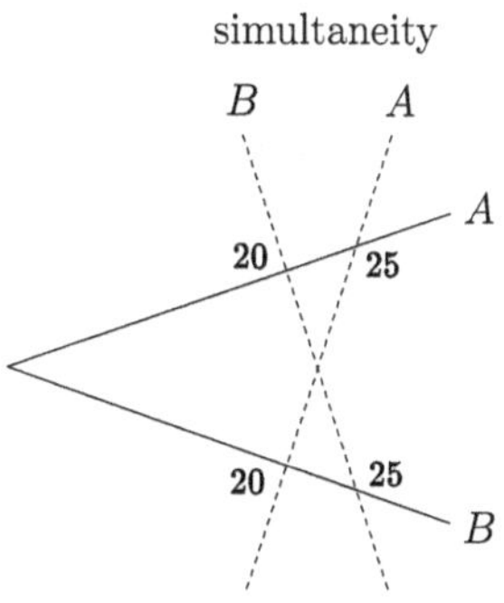

Figure 15.3: Age of twins.

Someone not convinced by this argument says: Let us see which of them is younger if they meet.

They never meet if both remain inertial. If Alice (Bernard) goes to Bernard (Alice), then she (he) will be younger because the inertial time is longer between two occurrences.

How does Alice (Bernard) find that "when she (he) is 25, then her (his) twin is only 20"? They communicate with radio messages. Alice (Bernard) on his 10th birthday sends the question: how old are you? Receiving the message, Bernard (Alice) answers: I am 20. The answer arrives to Alice (Bernard) when she (he) is 40, so she (he) calculates: the messages went there and back for 30 years, "obviously" 15 years there and 15 years back, so when my twin answered I was 25.

15.3 Velocity addition paradox

The paradox arose as follows [11]. Let us consider three standard inertial frames: Alice (A), me, and Bernard (B), for the sake of easy formulation. My velocity $\boldsymbol{v}_1$ relative to A and B's velocity $\boldsymbol{v}_2$ relative to me determine B's velocity relative to A for which the formula

$$\boldsymbol{v}_1 \oplus \boldsymbol{v}_2 := \frac{\alpha(\beta+\gamma)}{\gamma(1+\alpha)}\boldsymbol{v}_1 + \frac{\beta}{\gamma}\boldsymbol{v}_2 \tag{15.1}$$

was given where

$$\alpha := \frac{1}{\sqrt{1-|\boldsymbol{v}_1|^2}}, \quad \beta := \frac{1}{\sqrt{1-|\boldsymbol{v}_2|^2}}, \quad \gamma := \alpha\beta(1+\boldsymbol{v}_1\cdot\boldsymbol{v}_2).$$

"Evidently," A's velocity relative to me is $-\boldsymbol{v}_1$ and my velocity relative B is $-\boldsymbol{v}_2$ and A's velocity relative to B is $(-\boldsymbol{v}_2)\oplus(-\boldsymbol{v}_1)$, which, "of course," equals $-(\boldsymbol{v}_1\oplus\boldsymbol{v}_2)$; however, they are different according to formula (15.1):

$$(-\boldsymbol{v}_1)\oplus(-\boldsymbol{v}_2) \neq -(\boldsymbol{v}_2\oplus\boldsymbol{v}_1), \quad \text{or, equivalently,} \quad \boldsymbol{v}_1\oplus\boldsymbol{v}_2 \neq \boldsymbol{v}_2\oplus\boldsymbol{v}_1. \tag{15.2}$$

To explain the paradox [9], let us use notation in Subsection 10.6.4: let $\boldsymbol{u}$, $\boldsymbol{u}'$, and $\boldsymbol{u}''$ denote A, me, and B, respectively; then $\boldsymbol{v}_1 = \boldsymbol{v}_{u'u}$, $\boldsymbol{v}_2 = \boldsymbol{v}_{u''u'}$ and $\boldsymbol{v}_1\oplus\boldsymbol{v}_2 = \boldsymbol{v}_{u''u}$.

We see at once that formula (15.1) is not correct: the linear combination of $\boldsymbol{v}_{u''u'} \in \frac{\mathbf{S}_{u'}}{\mathbb{T}}$ and $\boldsymbol{v}_{u'u} \in \frac{\mathbf{S}_u}{\mathbb{T}}$, in general, cannot be $\boldsymbol{v}_{u''u}$, an element of $\frac{\mathbf{S}_u}{\mathbb{T}}$.

The right formulas are

$$\boldsymbol{v}_{u''u} = (\boldsymbol{B}_{uu'}\cdot\boldsymbol{v}_{u''u'})\oplus\boldsymbol{v}_{u'u},$$
$$\boldsymbol{v}_{uu''} = (\boldsymbol{B}_{u''u'}\cdot\boldsymbol{v}_{uu'})\oplus\boldsymbol{v}_{u'u''}.$$

The left sides of the above equalities are related by a further boost:

$$\boldsymbol{B}_{uu''} \cdot \boldsymbol{v}_{uu''} = -\boldsymbol{v}_{u''u};$$

therefore,

$$\begin{aligned}\boldsymbol{B}_{uu''} \cdot \big((\boldsymbol{B}_{u''u'} \cdot \boldsymbol{v}_{uu'}) \oplus \boldsymbol{v}_{u'u''}\big) &= (\boldsymbol{B}_{uu''} \cdot \boldsymbol{B}_{u''u'} \cdot \boldsymbol{v}_{uu'}) \oplus \boldsymbol{B}_{uu''} \cdot \boldsymbol{v}_{u'u''} \\ &= -(\boldsymbol{B}_{uu'} \cdot \boldsymbol{v}_{u''u'}) \oplus \boldsymbol{v}_{u'u};\end{aligned}$$

the first equality holds because $\oplus$ is linear in the velocities.

The "evidence" in the paradox would mean that $\boldsymbol{B}_{uu''} \cdot \boldsymbol{v}_{u'u''}$ is equal to $-\boldsymbol{B}_{uu'} \cdot \boldsymbol{v}_{u''u'}$ and $\boldsymbol{B}_{uu''} \cdot \boldsymbol{B}_{u''u'} \cdot \boldsymbol{v}_{uu'}$ is equal to $-\boldsymbol{v}_{u'u}$, which is not true because the boosts are not transitive:

$$\begin{aligned}\boldsymbol{B}_{uu''} \cdot \boldsymbol{v}_{u'u''} &= -\boldsymbol{B}_{uu''} \cdot \boldsymbol{B}_{u''u'} \cdot \boldsymbol{v}_{u''u'} \neq -\boldsymbol{B}_{uu'} \cdot \boldsymbol{v}_{u''u'}, \\ \boldsymbol{B}_{uu''} \cdot \boldsymbol{B}_{u''u'} \cdot \boldsymbol{v}_{uu'} &= -\boldsymbol{B}_{uu''} \cdot \boldsymbol{B}_{u''u'} \cdot \boldsymbol{B}_{u'u} \cdot \boldsymbol{v}_{u'u} \neq -\boldsymbol{v}_{u'u}.\end{aligned}$$

The paradox is the consequence of the fact that different observer spaces are considered to be the same ($\mathbb{R}^3$ in coordinates) and the boosts get lost.

15.4 Light propagation paradox

The paradox appeared as follows [15].

Consider a light source on the rim of a rotating disk (platform) having an inertial center. Light signals are sent around the rim (e. g., with the aid of mirrors) both forwards and backwards according to the direction of the rotation. We can measure the proper time periods of the source between the start and the arrival of the light signals. Then knowing the distance covered by the light signals (the circumference of the circle), the **round-way light speeds** c_+ (forwards) and c_- (backwards) can be determined:

$$c_+ = \frac{1}{1+\omega r}, \quad c_- = \frac{1}{1-\omega r}, \tag{15.3}$$

where ω is the magnitude of angular velocity of the rotation and r is the distance of the source from the center of rotation. These round-way speeds can be different because the platform is not an inertial observer.

It was stated that $\frac{1-\omega r}{1+\omega r}$ "does not give only the ratio of global light velocities for a full trip around the platform in the two opposite directions, but the local ratio as well: isotropy of space ensures that the velocities of light are the same in all points of the rim and, therefore, the average value coincides with the local ones."

Then it is argued: Consider uniformly rotating platforms whose angular speed ω is smaller and smaller, and take their small pieces whose distance r from the center is larger and larger, such that $\beta := \omega r$ is constant. Then the ratio of light speeds in the opposite directions is the same $\frac{1-\beta}{1+\beta} \neq 1$ for all such pieces. Those pieces, however, become more and more similar to pieces of a limit inertial observers moving with speed β with respect to the center. As a consequence, the ratio will differ from unity for that limit inertial observers, in contradiction to special relativity, which asserts that light speed is the same in all directions with respect to inertial observers. So, accepting special relativity, we have a discontinuity, which is not confirmed by experiments.

This conclusion is erroneous [7]. Namely, one speaks about light speed without specifying a synchronization and confuses round-way (in fact circle-way) speed with one-way speed. Formula (15.3) concerns the circle-way speeds of light, which does not tell anything about "local," i. e., one-way speed. The circle-way speeds are meaningful without synchronization, but one-way speed makes sense only if a synchronization is given (and depends on the synchronization); so *any assertion regarding one-way (local) light speeds—e. g., a formula for their ratio–would be meaningful if a synchronization had been specified.*

"... isotropy of space ensures ..." would mean that every point of the rim is related to the center in the same way even with respect to synchronizations, i. e., they apply the standard synchronization of the center.

Therefore, the corresponding round-way speeds are equal to the forward and backward one-way light speeds for the limit inertial observers if they apply the synchronization of the center instead of their own standard synchronization. That is why it is perfectly correct that the one-way light speed is different in different directions.

16 Discussion

16.1 Observer, reference frame, coordinate system

The following letters refer to our definitions:
A) Observer (Section 2.4)
B) Inertial frame (Subsection 3.5.5)
C) Standard inertial frame (Section 10.6)
D) Coordinate system (Section 10.11)

which are generally confused in books on relativity theory.

1. "Inertial frame" in the book [12] means
   - (B) on p. 2: "... the law of inertia, a material particle when left to itself will continue to move in a straight line with constant velocity." "All systems of reference for which the law of inertia is valid are called systems of inertia."
   - (C) on p. 30: "... the velocity of propagation of light in vacuo must be the same constant value $c = 3 \cdot 10^8$ m/s in every system of inertia."
   - (A) on p. 249: "According to the special principle of relativity, which is the base of the special theory of relativity, all systems of inertia, i. e., all rigid systems of reference moving with constant velocity to the fixed stars ..."
2. "(Inertial) frame" in the book [2] means
   - (D) on p. 3: "A *reference frame* is simply a method of assigning a position, a set of numbers, to events. Whenever you have a coordinate system, you have a reference frame, and I will use the two terms almost interchangeably."
   - (B) on p. 5: "An inertial frame is a reference frame with respect to which Newton's first law holds."
   - (C) p. 28: "There exists a constant speed $c = 299792458\ldots \mathrm{m\,s^{-1}}$ such that anything that moves at this speed in one inertial frame is measured to move at the same speed in all inertial frames."
3. "(Inertial) frame" and "observer" in the book [1] are synonyms and mean
   - (B) on p. 2: "... are defined by Newton's first law. According to this law, an *inertial frame* is one in which a free body (i. e., a body not subject to forces) does not accelerate."
   - (D) on p. 2: "Physical phenomena are normally described with respect to a given reference frame, which consists of a set of spatial axes and synchronized clocks ..."
   - (C) on p. 14: "The speed of light in vacuo has the same value $c \approx 3 \cdot 10^8$ m/s in all inertial frames, regardless of the velocity of the observer or the source."

https://doi.org/10.1515/9783112219553-019

16.2 Equivalence and others

Equivalence of inertial frames is one of the cornerstones of usual treatments. The following quotes are from the books cited above.

[1, p. 249] "... all systems of inertia ... are completely equivalent in respect of our description of nature."

[2, p. 13] "The laws of (nongravitational) physics assume the same form in all inertial frames. All inertial observers are equivalent."

[3, p. 27] "All inertial frames are equivalent for the performance of all physical experiments." "There is no physical ... experiment I can do, which will have a different result when I'm moving uniformly from when I'm stationary."

All these are intuitive notions of equivalence. We have given an exact definition (see Subsection 13.2.5) for equivalence, which fits the intuitive notion that two inertial frames are equivalent if and only if "the same experiments give the same results in both frames." In other words, two inertial frames are not equivalent if there is an experiment that gives different results in them, e. g., if the maximal and minimal one-way light speeds are different in the frames.

It is important, however, that

nonequivalence of two inertial frames does not imply that one of them is better than the other from a *theoretical point of view*,

which is well seen from our absolute treatment where everything is formulated without reference frames and everything can be translated to the language of an arbitrary reference frame. Of course, an inertial frame can be better than an other from a *practical point of view*, e. g., a standard inertial frame is better in this sense than a nonstandard one.

Further, it is worth making some other comments.

[2, p. 14]: "The speed of light in vacuo has the same value $c \approx 3 \cdot 10^8$ m/s in all inertial frames, regardless of the velocity of the observer or the source."
The velocity of anything is meaningful only relative to "something with some synchronization." What are the velocities in question related to? And a more important question: what about if the source is accelerated?

[3, p. 27]: "There is no physical ... experiment I can do, which will have a different result when I'm moving uniformly from when I'm stationary."
Nothing is moving or stationary in itself. The question is as above: with respect to what am I moving or stationary? Perhaps with respect to the rotating Earth?

[3, p. 28]: "There exists a constant speed $c = 299792458\ldots\,\mathrm{m\,s^{-1}}$ such that anything that moves at this speed in one inertial frame is measured to move at the same speed in all inertial frames."

We emphasize that speed is meaningful only with a given synchronization. Usually one speaks about light speed without synchronization, which is a consequence of the nonrelativistic point of view where one need not bother about synchronization. The usual axiom of constant light speed contradicts everyday thinking. The quote above is a tricky formulation of the contraintuitive light speed axiom. The questions arise, however: what about something that moves faster than this mystic speed? Is it possible that something moves faster in an inertial frame and slower in another one?

The seeming contradiction of the light speed axiom disappears if it is made clear that the same speed in different frames are related to different synchronizations.

[3, p. 29]: "The second axiom states that the speed of light is a universal constant, and it is this that allows us to define a procedure for synchronizing clocks."

This is an upside-down argument: first the speed, then the synchronization!

Lastly, we suggest to look again at the remarks in Sections 10.8.1, 5.9, 10.4, 10.1.2, 10.6.3, 10.7.3, 10.10, 10.3.2, 11.8.

16.3 How is it transformed?

It is a usual fundamental requirement that only **covariant** quantities and equations have a proper physical sense; covariance means that the different forms of a quantity or an equation with respect to different reference frames are **transformed** into each other by a prescribed **definition**.

Rephrased to our language, this requirement is that only **absolute** quantities and equations have a proper physical sense; reference frames **split** an absolute quantity and equation into relative ones and **transformation rules** can be **stated** for split forms with respect to different reference frames.

Relativistic thermodynamics is a very good example how one is led astray by the requirement of covariance. One puts the question [10]: how is temperature transformed?

The answer is sought based on the thermostatical relation $dE = TdS - pdV$.

Among others, three different answers are the most known: $T' = T$, $T' = T/\gamma$ and $T' = \gamma T$ where γ is the relativistic factor.

The first answer states that temperature is a scalar.

The second answer comes from the ill transformation rule $V' = V/\gamma$ based on the Lorentz contraction which, however, is not the transformation rule of the length of vectors: the Lorentz contraction, as we emphasized earlier, has no physical meaning, it depends on synchronization.

The third answer comes from the ill transformation rule $E' = \gamma E$ based on the mistaken "equivalence of mass and energy" and of the names "rest mass" m, "mass in motion" γm. Mass is a scalar, whereas energy—more precisely, relative energy—is the time-like component of absolute momentum and it is transformed together with relative momentum. Moreover, E in thermostatics is the (nonrelativistic) internal energy and not the energy.

It is not surprising that there is a controversy even nowadays because "how temperature is transformed" is a mistaken question suggested by the point of view of "covariance." The right question would be: Is temperature a scalar or (with respect to a reference frame) the time-like component of a vector or the time, time-like component of a tensor? If it is not a scalar, then it is not transformed in itself alone and the vector or the tensor in question must be specified.

16.4 Time reversal invariance

Time reversal invariance is often treated in connection with reversibility, which is declared in such a way that "if a given sequence of events is physically possible then the same sequence in opposite order is also possible."

A seemingly more concrete formulation is that "an equation or formula containing time remains valid for the reversing time, i. e., for $\boldsymbol{t}$ replaced with $-\boldsymbol{t}$." The concept of $\boldsymbol{t}$ replaced with $-\boldsymbol{t}$ concerns *only one reference frame with origin*. Invariance for a single time reversal cannot express a physical fact (e. g., reversibility). Only invariance for all time reversals can have a physical meaning.

There is an $(\boldsymbol{u}', o')$-time reversal for all $(\boldsymbol{u}', o')$ and the $(\boldsymbol{u}, o)$-split form of the $(\boldsymbol{u}', o')$-time reversal is (5.59).

The definition of reversibility by 'the' time reversal $\boldsymbol{t} \mapsto -\boldsymbol{t}$ is mistaken.

16.5 Nonrelativistic limit

It is usually stated that the nonrelativistic theory can be obtained from the relativistic one by some limit procedures. This is a consequence of always trying to explain and understand everything through the Lorentz transformation rule (1) between only two standard inertial frames in which the relativistic factor $\frac{1}{\sqrt{1-(v/c)^2}}$ is included.

16.5.1 Low speed limit

One of the limiting condition is that v should be much less than the light speed, $v \ll c$: then the relativistic factor is nearly 1. This, however, has two problems.

First: v is the relative speed between **two** standard inertial frames. The limit with "low speeds" would be acceptable only if **all** the standard inertial frame had "low speeds" relative to each other.

Let us consider standard inertial frames S_1, S_2, $S_3 \dots S_n \dots$ and let $v_{i,k}$ denote the relative speed of S_i with respect to S_k and let $v_{i,i+1} \ll c$. If n is "sufficiently large," then $v_{i,i+n}$ can be near c.

Second: $v \ll c$ as a limiting condition would mean that v tends to zero but then there would not be transformation at all.

$v \ll c$ does not send the relativistic theory to the nonrelativistic one.

16.5.2 Infinite light speed limit

Another usual condition of the limit procedure is $c \to \infty$. Then, formally, $\lim_{c\to\infty} \frac{1}{\sqrt{1-(v/c)^2}} = 1$.

A simple counterexample shows that neither this way gives a transition from the relativistic theory to the nonrelativistic one: the usual formula of relative energy is $\frac{mc^2}{\sqrt{1-(v/c)^2}}$, its limit is infinite when c tends to infinity, but the nonrelativistic energy is not infinite.

The previous formulas are valid only formally because light speed is 1 in our model, as it cannot tend to infinity.

For the exact formulation, let us rewrite our model by distinguishing the measure line of distances $\mathbb{L}$ from the measure line of time periods $\mathbb{T}$.

Then, chosen a positive element c of $\frac{\mathbb{L}}{\mathbb{T}}$, we make the relativistic spacetime model $(\mathrm{M}, \mathbb{T}, \mathbb{L}, \boldsymbol{g}_c)$ where $\boldsymbol{g}_c : \mathbf{M} \times \mathbf{M} \to \mathbb{L} \otimes \mathbb{L}$ is an arrow oriented Lorentz form and

- The set of future-like vectors $\mathrm{T}_c^{\rightarrow}$ is defined by $\boldsymbol{g}_c$ according to the sense
- The inertial time progress is $\boldsymbol{P}_c(\boldsymbol{x}_c) := \frac{\sqrt{-\boldsymbol{g}_c(\boldsymbol{x}_c, \boldsymbol{x}_c)}}{c}$ where $\boldsymbol{x}_c \in \mathrm{T}_c^{\rightarrow}$
- Then the set of absolute velocities is $\mathrm{V}(c) = \{\boldsymbol{u}_c \in \frac{\mathbf{M}}{\mathbb{T}} \mid \boldsymbol{g}_c(\boldsymbol{u}_c, \boldsymbol{u}_c) = -c^2\}$

and the semi-Euclidean forms, too, are defined according to the sense but we need not them in the sequel.

The “limit from the relativistic theory to the nonrelativistic theory” requires the existence of a linear surjection $\boldsymbol{\tau} : \mathbf{M} \to \mathbb{T}$ in such a way

$$\lim_{c\to\infty} \mathrm{T}_c^{\rightarrow} = \{\boldsymbol{x} \mid \boldsymbol{\tau} \cdot \boldsymbol{x} > 0\} =: \mathrm{T}_\infty^{\rightarrow}$$

in some sense and

$$\lim_{c\to\infty} \frac{\sqrt{-\boldsymbol{g}_c(\boldsymbol{x}_c, \boldsymbol{x}_c)}}{c} = \boldsymbol{\tau} \cdot \boldsymbol{x}, \tag{16.1}$$

if $\lim_{c\to\infty} \boldsymbol{x}_c = \boldsymbol{x} \in \mathrm{T}_\infty^{\rightarrow}$.

It is easy to say “in some sense” but it is not a simple task to define that a set is the limit of a sequence of sets. Putting aside this problem, let us examine equality (16.1). $\boldsymbol{\tau}$ is linear; thus,

$$\lim_{c\to\infty} \frac{\sqrt{-\boldsymbol{g}_c(\boldsymbol{x}_c + \boldsymbol{y}_c, \boldsymbol{x}_c + \boldsymbol{y}_c)}}{c} = \lim_{c\to\infty} \frac{\sqrt{-\boldsymbol{g}_c(\boldsymbol{x}_c, \boldsymbol{x}_c)}}{c} + \lim_{c\to\infty} \frac{\sqrt{-\boldsymbol{g}_c(\boldsymbol{y}_c, \boldsymbol{y}_c)}}{c},$$

must hold. The sum of limits is the limit of sums, so by taking the squares we get

$$\lim_{c\to\infty} \frac{-g_c(\boldsymbol{x}_c,\boldsymbol{y}_c)}{c^2} = \lim_{c\to\infty} \frac{\sqrt{-g_c(\boldsymbol{x}_c,\boldsymbol{x}_c)}\sqrt{-g_c(\boldsymbol{y}_c,\boldsymbol{y}_c)}}{c^2}. \tag{16.2}$$

According to the reversed Cauchy inequality,

$$-g_c(\boldsymbol{x}_c,\boldsymbol{y}_c) \geq \sqrt{-g_c(\boldsymbol{x}_c,\boldsymbol{x}_c)}\sqrt{-g_c(\boldsymbol{y}_c,\boldsymbol{y}_c)}$$

and equality holds if and only if $\boldsymbol{x}_c$ and $\boldsymbol{y}_c$ are parallel. For sequences in which neither $\boldsymbol{x}_c$ is parallel to $\boldsymbol{y}_c$ nor the limit value $\boldsymbol{x}$ is parallel to $\boldsymbol{y}$, we do not obtain equality (16.2); nonparallel sequences exist and if the limits of arbitrary two nonparallel sequences were parallel, then $\mathrm{T}_\infty^{\rightarrow}$ would be a half-line.

$\lim_{c\to\infty}$ does not send the relativistic theory to the nonrelativistic one.

17 Some words about curved spacetime models

17.1 Gravitation and curvature

Formulas of the nonrelativistic spacetime model are in accordance with everyday experience for

- "Slow" mechanical phenomena when the speeds of bodies with respect to each other are "not too high."
- Electromagnetic phenomena with respect to a given single observer.

Experimental and theoretical problems arise, however, for

- "Fast" mechanical phenomena.
- Electromagnetic phenomena when more observers are compared.

The special relativistic spacetime model eliminates these problems but a new problem arises, however. Namely,

- Gravitation can be described nonrelativistically by forces having absolute scalar potential.
- Gravitation cannot be described special relativistically because there are no absolute scalar potentials in that model.

It is known, however, that there is some trouble even nonrelativistically if the gravitation is "too strong": the motion of the Mercury around the Sun differs from that expected by Newton's law.

Let us return to the considerations, which led us to the base of flat spacetimes. We emphasized at the end of Section 1.1 that the notion of straight lines over human size is questionable. We see that the path of a light beam in our space is a straight line. But what is a straight line? What is the criterion for a line to be straight? There is no criterion for a single line. We have to examine the paths of a number of light beams to decide whether their relations meet the relations of straight lines in an affine space.

The common statement that "gravitation makes spacetime curved" means that gravitation has an effect on light propagation, all the paths of light beams cannot be straight lines, e. g., it can occur that two light beams meet twice. This can be roughly illustrated as follows.

Two light beams emitted from a source

- At the same proper time point in two different directions will meet later (space-like curvature).
- At different proper time points in the same direction will meet later (time-like curvature).

In "small size," however, the light paths seem to be straight lines.

https://doi.org/10.1515/9783112219553-020

17.2 Inertial mass and gravitational mass

Einstein's guiding principles for general relativity were the equivalence of inertial mass and gravitational mass as well as covariance of physical quantities.

To illustrate equivalence of inertial mass and gravitational mass, Einstein's famous example was a freely falling lift in which one does not feel gravity. This is lame, however. Instead of a lift let us consider a space cabin falling towards the Earth. Let the width of the cabin d when it is at the distance h from the center of the Earth. If the cabin is malleable and falls freely, then all of its points go towards the center of the Earth; thus, at the distance $h/2$ its width is $d/2$, so it "feels" gravity by being compressed. On the other hand, if the cabin is rigid, then it does not fall freely, and not all of its points go towards the center of the Earth.

The equivalence principle is valid only "approximately in small size and for short time," which is not too convincing.

The question of inertial mass arise in connection with noninertial reference frames, moreover, the usual names "special relativity" and "general relativity" supports the unfortunate idea that special relativity is the theory of inertial frames and general relativity is the theory of arbitrary frames, though this was refuted by an excellent author [16, 17] long ago.

The truth is that

(1) A general relativistic spacetime model is a theory describing a gravitational action whereas the special relativistic spacetime model concerns the lack of gravitation.

(2) The noninertial reference frames can be well treated in the special theory but this requires the same complicated mathematical tools as the treatment of the general theory in itself.

17.3 General relativistic spacetime models

1. The first task is to establish a new type of spacetime model based on some experience regarding gravitation and on theoretical considerations.
   "In small size," the paths of light signals seem to be straight lines.
   This is similar to that a ship on the ocean sees the surface of water to be a plane though we know that it is a part of a sphere.
   This indicates that, for describing gravitation, we have to give up the affine structure of spacetime in such a way that the new model in "small size" be "similar" to a part of the special relativistic spacetime model. Mathematically: spacetime is a four-dimensional smooth manifold and the tangent space at every world point is a vector space endowed with a Lorentz form.
   The description of gravitational actions involves the gravitational constant. Correspondingly, we choose the measure lines in such a way that this constant be the real number 1. As a consequence, the measure line $\mathbb{T}$ of the special theory is substituted

by $\mathbb{R}$ and all the time periods, space distances, and masses are measured by real numbers.

Now we are ready to put the exact definition:

A **general relativistic spacetime model** is a pair $(\mathrm{M}, \boldsymbol{g})$ where M is a four-dimensional oriented smooth manifold and $\boldsymbol{g}$ is an arrow oriented Lorentz form field on M.

In detail: Every point x of M has its own tangent space $\mathrm{T}_x(\mathrm{M})$, a four- dimensional vector space and an arrow oriented Lorentz form $\boldsymbol{g}[x] : \mathrm{T}_x(\mathrm{M}) \times \mathrm{T}_x(\mathrm{M}) \to \mathbb{R}$ is given in such a way that $x \mapsto g[x]$ is smooth (in a convenient sense).

Then all the notions of the special model can be transferred to the general model by a pointwise definition:

- For every $x \in \mathrm{M}$, $\boldsymbol{g}[x]$ determines the time-like vectors, the space-like vectors, light-like vectors, absolute velocities in the tangent space over x.
- A world line is a curve in spacetime whose every tangent vector is time-like.
- A light line is a curve in spacetime whose every tangent vector is light-like.
- An observer is a smooth absolute velocity field.
- The space points of an observer are the integral curves of the corresponding absolute velocity field.
- A synchronization is a smooth equivalence relation whose equivalence classes are three-dimensional space-like hypersurfaces.
- And so on.

The general relativistic spacetime models $(\mathrm{M},\boldsymbol{g})$ and $(\mathrm{M}',\boldsymbol{g}')$ are **isomorphic** if there is a diffeomorphism $F : \mathrm{M} \to \mathrm{M}'$, such that

$$\boldsymbol{g}'[F(x)](\mathcal{D}F[x])\boldsymbol{x}, \mathcal{D}F[x]\boldsymbol{y}) = \boldsymbol{g}[x](\boldsymbol{x},\boldsymbol{y}) \quad (x \in \mathrm{M},\ \boldsymbol{x},\boldsymbol{y} \in \mathrm{T}_x^{\rightarrow}(\mathrm{M})).$$

Two spacetime models are isomorphic both in the nonrelativistic and in the special relativistic case; this is not true in the general case, as it cannot be, because a general relativistic spacetime model is the model of a gravitational action and there are different gravitational actions, e. g., the one due to a mass point and the other due to two mass points.

2. To construct the model of a concrete gravitational action, we have to:
   (i) Specify the spacetime manifold M.
   (ii) Describe the mass distribution in M.
   (iii) Obtain $\boldsymbol{g}$ by solving Einstein's equation.

   In general, tasks (i) and (ii) are difficult from a theoretical point of view, (iii) is difficult from a technical point of view because Einstein's equation is highly complicated.

18 Mathematical tools

The fundamental notions and notation of sets theory are supposed to be known: subset, intersection, union, Cartesian product, function, injective function (injection), surjective function (surjection), bijective function (bijection), etc.

We find convenient sometimes that an element a and the set $\{a\}$ containing that element are not distinguished in notation.

18.1 Vector spaces

In this book, vector spaces are real and finite-dimensional.

18.1.1 Basic notions

The fundamental notions of linear algebra are supposed to be known: linear combination, basis, dimension, linear subspace, linear map, bilinear map, etc.

The following notation are often used: if $\boldsymbol{v}$ is a vector, α is a real number, A and B are sets of vectors, then

$$\boldsymbol{v} + \mathrm{A} := \{\boldsymbol{v} + \boldsymbol{x} \mid \boldsymbol{x} \in \mathrm{A}\}, \quad \alpha\mathrm{A} := \{\alpha\boldsymbol{x} \mid \boldsymbol{x} \in \mathrm{A}\}, \quad \mathrm{A} + \mathrm{B} := \{\boldsymbol{x} + \boldsymbol{y} \mid \boldsymbol{x} \in \mathrm{A}, \boldsymbol{y} \in \mathrm{B}\}.$$

The action of linear maps is denoted by a dot product: $\boldsymbol{L} \cdot \boldsymbol{v}$ is the value of the linear map $\boldsymbol{L}$ on the vector $\boldsymbol{v}$. Similarly, the product (composition) of linear maps is denoted by a dot product: $\boldsymbol{L} \cdot \boldsymbol{K}$.

The kernel of a linear map $\boldsymbol{L}$ is $\{\boldsymbol{v} \mid \boldsymbol{L} \cdot \boldsymbol{v} = 0\}$.

The notion of a norm and its properties on a vector space are supposed to be known. A norm admits the definition of the fundamental notions of analysis: open set, closed set, convergence, continuity, etc.

The norms on a finite-dimensional vector space are equivalent, which means that all of them give the same open sets, closed sets, convergence, continuity, etc.; thus, the notions of analysis can be used without specifying a norm. Some proofs, however, require to take a norm.

18.1.2 Identifications

Two vector spaces of equal final dimension are of the same kind, which means that there is a linear bijection between them; in fact, there are infinite many.

It occurs sometimes that we "identify" two vector spaces (suggested by their concrete definition) by a "distinguished" linear bijection between them.

https://doi.org/10.1515/9783112219553-021

Neither “distinguished” nor “identified” are mathematical notions; their meaning is the following.

We choose a linear bijection $L_{1,2}$ from the vector space $\mathbf{V}_1$ to the vector space $\mathbf{V}_2$ making a correspondence between the elements of the vector spaces: v_2 corresponds to v_1 if $v_2 = L_{1,2} \cdot v_1$; then keeping in mind that linear bijection, we omit it in notation and we write $v_1 \equiv v_2$ and $\mathbf{V}_1 \equiv \mathbf{V}_2$.

18.1.3 Complementary subspaces

Let $\mathbf{V}$ be a vector space.

The linear subspaces $\mathbf{E}$ and $\mathbf{F}$ of $\mathbf{V}$ are **transverse** if $\mathbf{E} \cap \mathbf{F} = \mathbf{0}$.

$\mathbf{E}$ and $\mathbf{F}$ are **complementary subspaces** if $\mathbf{E} \cap \mathbf{F} = \mathbf{0}$ and $\mathbf{E} + \mathbf{F} = \mathbf{V}$. Then $\dim \mathbf{E} + \dim \mathbf{F} = \dim \mathbf{V}$.

If $\mathbf{E}$ and $\mathbf{F}$ are complementary subspaces, then for every $v \in \mathbf{V}$ there are uniquely defined $v_\mathbf{E}$ in $\mathbf{E}$ and $v_\mathbf{F}$ in $\mathbf{F}$ such that $v = v_\mathbf{E} + v_\mathbf{F}$.

Then the map $v \mapsto v_\mathbf{E}$ is linear and called the **projection onto E along F**.

18.1.4 Factor spaces

Let $\mathbf{E}$ be a linear subspace of $\mathbf{V}$.

A subset of $\mathbf{V}$ is called an **affine subspace directed by** $\mathbf{E}$ if it is of the form $v + \mathbf{E}$ for some vector v. Then $v + \mathbf{E} = u + \mathbf{E}$ if and only if $v - u \in \mathbf{E}$.

An affine subspace directed by a one-dimensional linear subspace is a **straight line**; a **hyperplane** is an affine subspace directed by a linear subspace of dimension $\dim \mathbf{V} - 1$.

Affine subspaces directed by the same linear subspace are called **parallel**.

The set of affine subspaces directed by $\mathbf{E}$ is denoted by $\mathbf{V}/\mathbf{E}$ and called the **factor space** of $\mathbf{V}$ by $\mathbf{E}$.

$\mathbf{V}/\mathbf{E}$ is a vector space with the well-defined addition and multiplication by real numbers:

$$(v + \mathbf{E}) + (u + \mathbf{E}) := (v + u) + \mathbf{E}, \quad \alpha(v + \mathbf{E}) := (\alpha v) + \mathbf{E}$$

$\mathbf{V}/\mathbf{E}$ has dimension $\dim \mathbf{V} - \dim \mathbf{E}$.

If $\mathbf{F}$ and $\mathbf{E}$ are complementary subspaces, then every affine subspace directed by $\mathbf{E}$ contains a unique element of $\mathbf{F}$; more closely, the map $\mathbf{F} \to \mathbf{V}/\mathbf{E}$, $v \mapsto v + \mathbf{E}$ is a linear bijection.

18.1.5 Orientation

The ordered bases $(v_1, \dots, v_N)$ and $(v'_1, \dots, v'_N)$ of $\mathbf{V}$ are **equally oriented** if the determinant of the linear map defined by $v_i \mapsto v'_i$ $(i = 1, \dots, N)$ is positive.

An equivalence class of equally oriented bases is called an **orientation** of **V**. **V** is **oriented** if an orientation is given; the bases in the chosen equivalence class are called **positively oriented**.

In a one-dimensional vector space, two bases, i. e., two nonzero elements of are equally oriented if and only if the elements are positive multiples of each other; thus, the equally oriented bases (an equivalence class) is a "half line."

A linear bijection between oriented vector spaces is **orientation preserving** if it maps positively oriented bases into positively oriented ones.

The basis $(1, 0, 0, \dots, 0)$, $(0, 1, 0, \dots, 0)$ $\dots$ $(0, 0, 0, \dots, 1)$ gives the standard orientation of $\mathbb{R}^N$.

18.1.6 Dual spaces

The set of linear maps $\mathbf{V} \to \mathbb{R}$ is called the **dual** of **V** and is denoted by $\mathbf{V}^*$. Endowed with the pointwise operations, the dual space is a vector space for which $\dim \mathbf{V}^* = \dim \mathbf{V}$ holds.

Elements of $\mathbf{V}^*$ are called **covectors**.

The dual space separates the elements of **V**, which means that if $\boldsymbol{p} \cdot \boldsymbol{v}_1 = \boldsymbol{p} \cdot \boldsymbol{v}_2$ for all covectors $\boldsymbol{p}$, then $\boldsymbol{v}_1 = \boldsymbol{v}_2$.

For all vectors $\boldsymbol{v}$, the map $\mathbf{V}^* \to \mathbb{R}$, $\boldsymbol{p} \cdot \boldsymbol{v}$ is linear, i. e., it is an element of the **bidual** $(\mathbf{V}^*)^* =: \mathbf{V}^{**}$; since covectors separates the vectors, different vectors define different elements of the bidual. This allows us to **identify** **V** with the dual of $\mathbf{V}^*$:

$$\mathbf{V} \equiv \mathbf{V}^{**}, \quad \boldsymbol{v} \equiv (\boldsymbol{p} \mapsto \boldsymbol{p} \cdot \boldsymbol{v}).$$

According to this identification, the vectors are "co-covectors," and that is why we write $\boldsymbol{v} \cdot \boldsymbol{p} = \boldsymbol{p} \cdot \boldsymbol{v}$ for all $\boldsymbol{v} \in \mathbf{V}$ and $\boldsymbol{p} \in \mathbf{V}^*$.

The **dual basis** of an ordered basis $(\boldsymbol{v}_1, \boldsymbol{v}_2, \dots, \boldsymbol{v}_N)$ of **V** is the ordered basis $(\boldsymbol{p}^1, \boldsymbol{p}^2, \dots, \boldsymbol{p}^N)$ of $\mathbf{V}^*$ defined by $\boldsymbol{p}^k \cdot \boldsymbol{v}_i = \delta_{ki}$ (Kronecker delta). Accordingly, an orientation of **V** defines an orientation of $\mathbf{V}^*$.

The dual of $\mathbb{R}^N$ is usually identified with $\mathbb{R}^N$ in such a way that the action of a covector $(p_1, p_2, \dots, p_N)$ on a vector $(x^1, x^2, \dots, x^N)$ is

$$p_1 v^1 + p_2 v^2 + \cdots + p_N v^N = \sum_{k=1}^{N} p_k x^k.$$

Einstein's rule simplifies this formula: the symbol of summation is omitted whenever indexes up and down in a formula are equal.

18.1.7 Transposed linear maps

Let **V** and **U** be finite-dimensional vector spaces.

The **transpose** of a linear map $\boldsymbol{L} : \mathbf{V} \to \mathbf{U}$ is the linear map

$$\boldsymbol{L}^* : \mathbf{U}^* \to \mathbf{V}^*, \quad \boldsymbol{r} \mapsto \boldsymbol{r} \cdot \boldsymbol{L},$$

i. e.,

$$(\boldsymbol{L}^* \cdot \boldsymbol{r}) \cdot \boldsymbol{v} := \boldsymbol{r} \cdot (\boldsymbol{L} \cdot \boldsymbol{v}) \quad (\boldsymbol{r} \in \mathbf{U}^*,\ \boldsymbol{v} \in \mathbf{V}).$$

Briefly,

$$\boldsymbol{L}^* \cdot \boldsymbol{r} = \boldsymbol{r} \cdot \boldsymbol{L} \quad (\boldsymbol{r} \in \mathbf{U}^*), \quad \boldsymbol{L} \cdot \boldsymbol{v} = \boldsymbol{v} \cdot \boldsymbol{L}^* \quad (\boldsymbol{v} \in \mathbf{V}). \tag{18.1}$$

According to the identifications $\mathbf{V}^{**} \equiv \mathbf{V}$ and $\mathbf{U}^{**} \equiv \mathbf{U}$,

$$\boldsymbol{L}^{**} = \boldsymbol{L}.$$

Further,
- $\boldsymbol{L}$ is injective if and only if $\boldsymbol{L}^*$ is surjective.
- $\boldsymbol{L}$ is surjective if and only if $\boldsymbol{L}^*$ is injective.

If $\boldsymbol{L}$ is bijective, then

$$(\boldsymbol{L}^{-1})^* = (\boldsymbol{L}^*)^{-1}.$$

If $\boldsymbol{L}, \boldsymbol{K} : \mathbf{V} \to \mathbf{U}$ are linear maps and α is a real number, then

$$(\boldsymbol{L} + \boldsymbol{K})^* = \boldsymbol{L}^* + \mathbf{K}^*, \quad (\alpha \boldsymbol{L})^* = \alpha \boldsymbol{L}^*.$$

If $\boldsymbol{L} : \mathbf{V} \to \mathbf{U}$ and $\boldsymbol{F} : \mathbf{U} \to \mathbf{W}$ are linear maps, then

$$(\boldsymbol{F} \cdot \boldsymbol{L})^* = \boldsymbol{L}^* \cdot \boldsymbol{F}^*.$$

A linear map $\mathbb{R}^N \to \mathbb{R}^M$ is given by an $N \times M$ matrix in the well-known way. In the usual identification $(\mathbb{R}^N)^* \equiv \mathbb{R}^N$ and $(\mathbb{R}^M)^* \equiv \mathbb{R}^M$, the transpose of a linear map is just the usual transpose of the corresponding matrix.

18.1.8 Tensors

1. General formulas

For the sake of clear formulation, it is worth introducing the notation:

$$\mathrm{Lin}(\mathbf{V}, \mathbf{U}) := \{\boldsymbol{L} : \mathbf{V} \to \mathbf{U} \mid \boldsymbol{L} \text{ is linear}\},$$
$$\mathrm{Bilin}(\mathbf{V} \times \mathbf{U}, \mathbb{R}) := \{\boldsymbol{B} : \mathbf{V} \times \mathbf{U} \to \mathbb{R} \mid \boldsymbol{B} \text{ is bilinear}\}$$

for the vector spaces $\mathbf{V}, \mathbf{U}$; then the meaning of $\mathrm{Lin}(\mathbf{V}, \mathbf{U}^*)$, etc. is straightforward.

With this notation, $\mathbf{V}^* = \mathrm{Lin}(\mathbf{V}, \mathbb{R})$.

If $\boldsymbol{L} \in \mathrm{Lin}(\mathbf{V}, \mathbf{U}^*)$, then the map $(\boldsymbol{u}, \boldsymbol{v}) \mapsto \boldsymbol{u} \cdot (\boldsymbol{L} \cdot \boldsymbol{v})$ is in $\mathrm{Bilin}(\mathbf{U} \times \mathbf{V}, \mathbb{R})$.

If $\boldsymbol{B} \in \mathrm{Bilin}(\mathbf{U} \times \mathbf{V}, \mathbb{R})$, then $\boldsymbol{v} \mapsto \boldsymbol{B}(\cdot, \boldsymbol{v})$ is in $\mathrm{Lin}(\mathbf{V}, \mathbf{U}^*)$.

That is why we accept the identification

$$\mathrm{Bilin}(\mathbf{U} \times \mathbf{V}, \mathbb{R}) \equiv \mathrm{Lin}(\mathbf{V}, \mathbf{U}^*).$$

This identification can be well seen by the following: $\boldsymbol{p} \in \mathbf{V}^*$ and $\boldsymbol{r} \in \mathbf{U}^*$ determine a linear map as well as a bilinear map, called their **tensor product**:

$$\boldsymbol{r} \otimes \boldsymbol{p} : \mathbf{V} \to \mathbf{U}^*, \quad v \mapsto \boldsymbol{r}(\boldsymbol{p} \cdot v),$$
$$\boldsymbol{r} \otimes \boldsymbol{p} : \mathbf{U} \times \mathbf{V} \to \mathbb{R}, \quad (\boldsymbol{u}, v) \mapsto \boldsymbol{u} \cdot (\boldsymbol{r} \otimes \boldsymbol{p}) \cdot v = (\boldsymbol{u} \cdot \boldsymbol{r})(\boldsymbol{p} \cdot v).$$

Accordingly, we define the **tensor product space**:

$$\mathbf{U}^* \otimes \mathbf{V}^* := \mathrm{Lin}(\mathbf{V}, \mathbf{U}^*) \equiv \mathrm{Bilin}(\mathbf{U} \times \mathbf{V}, \mathbb{R}).$$

If $\boldsymbol{r} \otimes \boldsymbol{p}$ is considered a linear map $\mathbf{V} \to \mathbf{U}^*$, then its transpose is a linear map $\mathbf{U}^{**} \to \mathbf{V}^* \equiv \mathbf{U} \to \mathbf{V}^*$ and it is a simple fact that $(\boldsymbol{r} \otimes \boldsymbol{p})^* = \boldsymbol{p} \otimes \boldsymbol{r}$.

Interchanging the roles of vectors and covectors, the meaning of $\mathbf{U} \otimes \mathbf{V}^*$, $\mathbf{U}^* \otimes \mathbf{V}$, and $\mathbf{U} \otimes \mathbf{V}$ is trivial.

Arbitrary element of a tensor product space is the linear combination—in fact, the sum—of elements, which are tensor products. The dimension of a tensor product space is the product of the dimensions of the corresponding vector spaces.

It is worth noting that

- $\mathbb{R} \otimes \mathbf{V} = \mathrm{Lin}(\mathbf{V}^*, \mathbb{R}) = \mathbf{V}^{**} \equiv \mathbf{V}$
- $\mathbf{V} \otimes \mathbb{R} = \mathrm{Lin}(\mathbb{R}^*, \mathbf{V}) \equiv \mathrm{Lin}(\mathbb{R}, \mathbf{V}) \equiv \mathbf{V}$

summing up:

$$\mathbb{R} \otimes \mathbf{V} \equiv \mathbf{V} \otimes \mathbb{R} \equiv \mathbf{V}, \quad \alpha \otimes v \equiv v \otimes \alpha \equiv \alpha v. \tag{18.2}$$

It is familiar that an $M \times N$ matrix is considered a linear map $\mathbb{R}^N \to \mathbb{R}^M \equiv (\mathbb{R}^M)^*$ in the form

$$\begin{pmatrix} b_{11} & \cdots & \cdots & b_{1N} \\ \vdots & \vdots & \vdots & \vdots \\ b_{M1} & \cdots & \cdots & b_{MN} \end{pmatrix} \begin{pmatrix} v^1 \\ \vdots \\ v^N \end{pmatrix}$$

as well as a bilinear map $\mathbb{R}^M \times \mathbb{R}^N \to \mathbb{R}$ in the form

$$\begin{pmatrix} u^1 & \cdots & u^M \end{pmatrix} \begin{pmatrix} b_{11} & \cdots & \cdots & b_{1N} \\ \vdots & \vdots & \vdots & \vdots \\ b_{M1} & \cdots & \cdots & b_{MN} \end{pmatrix} \begin{pmatrix} v^1 \\ \vdots \\ v^N \end{pmatrix};$$

thus, $\mathbb{R}^M \otimes \mathbb{R}^N \equiv (\mathbb{R}^M)^* \otimes (\mathbb{R}^N)^* \equiv \mathbb{R}^{M \times N}$.

2. Tensors of a vector space

The case $\mathbf{U} = \mathbf{V}$ is particularly important; then the elements of
- $\mathbf{V} \otimes \mathbf{V}$ are called **tensors**
- $\mathbf{V}^* \otimes \mathbf{V}^*$ are called **cotensors**
- $\mathbf{V} \otimes \mathbf{V}^*$ and $\mathbf{V}^* \otimes \mathbf{V}$ are called **mixed tensors**

If $\boldsymbol{L} \in \mathbf{V}^* \otimes \mathbf{V}^*$ is considered a linear map $\mathbf{V} \to \mathbf{V}^*$, then its transpose is a linear map $\boldsymbol{L}^* : \mathbf{V}^{**} \equiv \mathbf{V} \to \mathbf{V}^*$, i. e., is also an element of $\mathbf{V}^* \otimes \mathbf{V}^*$; thus, the following definition makes sense: $\boldsymbol{L}$ is **symmetric** or **antisymmetric** if $\boldsymbol{L}^* = \boldsymbol{L}$ or $\boldsymbol{L}^* = -\boldsymbol{L}$.

The symmetric and antisymmetric elements of $\mathbf{V} \otimes \mathbf{V}$ are defined in a similar way.

Keep in mind that a mixed tensor cannot be either symmetric or antisymmetric.

The possibility of being symmetric and antisymmetric is well reflected by the possibility of interchanging the order in the tensor product. In particular, we define:
(1) The symmetric tensor product $\boldsymbol{v}_1 \vee \boldsymbol{v}_2 := \boldsymbol{v}_1 \otimes \boldsymbol{v}_2 + \boldsymbol{v}_2 \otimes \boldsymbol{v}_1 = \boldsymbol{v}_2 \vee \boldsymbol{v}_1$
(2) The antisymmetric tensor product $\boldsymbol{v}_1 \wedge \boldsymbol{v}_2 := \boldsymbol{v}_1 \otimes \boldsymbol{v}_2 - \boldsymbol{v}_2 \otimes \boldsymbol{v}_1 = -\boldsymbol{v}_2 \wedge \boldsymbol{v}_1$

of two vectors $\boldsymbol{v}_1$ and $\boldsymbol{v}_2$; correspondingly, $\mathbf{V} \vee \mathbf{V}$ and $\mathbf{V} \wedge \mathbf{V}$ denote the linear subspaces of $\mathbf{V} \otimes \mathbf{V}$ consisting of symmetric and antisymmetric elements, respectively.

For mixed tensors, we have the linear map, called **trace**,

$$\mathrm{Tr} : \mathbf{V} \otimes \mathbf{V}^* \to \mathbb{R} \quad \text{and} \quad \mathrm{Tr} : \mathbf{V}^* \otimes \mathbf{V} \to \mathbb{R},$$

defined by

$$\boldsymbol{v} \otimes \boldsymbol{p} \mapsto \boldsymbol{v} \cdot \boldsymbol{p} \quad \text{and} \quad \boldsymbol{p} \otimes \boldsymbol{v} \mapsto \boldsymbol{p} \cdot \boldsymbol{v};$$

thus, $\mathrm{Tr}(\boldsymbol{L}) = \mathrm{Tr}(\boldsymbol{L}^*)$ for $\boldsymbol{L} \in \otimes\mathbf{V}^*$.

Keep in mind that tensors and cotensors have no trace.

In general, for a natural number n, the n-**cotensors** are defined as n-linear maps $\mathbf{V}^n \to \mathbb{R}$; the 0-cotensors are the real numbers.

The n-**tensors** and **mixed** n-**tensors** are defined according to the sense.

For instance, the tensor product $\boldsymbol{q} \otimes \boldsymbol{p} \otimes \boldsymbol{r}$ of the covectors $\boldsymbol{q}, \boldsymbol{p}, \boldsymbol{r}$ is the three-linear map

$$\mathbf{V}^3 \to \mathbb{R}, \quad (\boldsymbol{u}, \boldsymbol{v}, \boldsymbol{w}) \mapsto (\boldsymbol{q} \cdot \boldsymbol{u})(\boldsymbol{p} \cdot \boldsymbol{v})(\boldsymbol{r} \cdot \boldsymbol{w}).$$

3. Tensor products of linear maps

The following notions can be defined more generally but we need only for the special case of a single vector space.

The **tensor product** of the linear maps $\boldsymbol{L} : \mathbf{V} \to \mathbf{V}$ and $\boldsymbol{K} : \mathbf{V} \to \mathbf{V}$ is the linear map defined by

$$\boldsymbol{L} \otimes \boldsymbol{K} : \mathbf{V} \otimes \mathbf{V} \to \mathbf{V} \otimes \mathbf{V}, \quad \boldsymbol{v}_1 \otimes \boldsymbol{v}_2 \mapsto (\boldsymbol{L} \cdot \boldsymbol{v}_1) \otimes (\boldsymbol{K} \cdot \boldsymbol{v}_2) = \boldsymbol{L} \cdot (\boldsymbol{v}_1 \otimes \boldsymbol{v}_2) \cdot \boldsymbol{K}^*. \tag{18.3}$$

Taking here and there $\mathbf{V}^*$ instead of $\mathbf{V}$, we get tensor products in a similar way. For instance,

$$\boldsymbol{L} \otimes \boldsymbol{K}^* : \mathbf{V} \otimes \mathbf{V}^* \to \mathbf{V} \otimes \mathbf{V}^*, \quad \boldsymbol{v} \otimes \boldsymbol{p} \mapsto (\boldsymbol{L} \cdot \boldsymbol{v}) \otimes (\boldsymbol{K}^* \cdot \boldsymbol{p}) = \boldsymbol{L} \cdot (\boldsymbol{v} \otimes \boldsymbol{p}) \cdot \boldsymbol{K}.$$

18.2 Measure lines

18.2.1 Units of measurement

Let A be the set of possible values, magnitudes of a physical quantity (length, mass, etc.). Choosing an arbitrary element a of A (meter, gram, etc.) and a positive real number α, we can define α times a, denoted by αa (3 meters, 0.3 grams, etc.). This multiplication by positive numbers has the following properties: for all $a \in A$ and $\alpha, \beta > 0$,

- $1a = a$.
- $\beta(\alpha a) = (\beta\alpha)a$.
- If $\alpha \neq \beta$ then $\alpha a \neq \beta a$.
- For every element a' of A, there is a number $\alpha' > 0$ such that $a' = \alpha' a$.

In general, multiplication by negative numbers is not defined because a physical quantity has no negative values: there is no negative length, etc. For a convenient mathematical application, however, we introduce negative values, too.

Let us define $-A$ as the set of pairs $(-1, a)$ where $a \in A$ and let us use the notation $-a := (-1, a)$. Then let $\mathbb{A}$ be the union of A and $-A$, rounded off with an element zero. Then multiplication by negative numbers can be defined as an extension of the multiplication by positive numbers in A: for $\alpha \in \mathbb{R}$ and $a \in A$,

- $0a := 0, \quad \alpha 0 := 0$.
- $\alpha a := -|\alpha| a$ if $\alpha < 0$.
- $\alpha(-a) := -\alpha a$ if $\alpha > 0$.
- $\alpha(-a) := |\alpha| a$ if $\alpha < 0$.

As a consequence, every element of $\mathbb{A}$ can be given as a multiple of a nonzero $\boldsymbol{a} \in \mathbb{A}$. This allows us to introduce an **addition** in $\mathbb{A}$: choosing a nonzero $\boldsymbol{a}$. we put $\alpha\boldsymbol{a}+\beta\boldsymbol{a} := (\alpha+\beta)\boldsymbol{a}$. It is simple that this definition does not depends on $\boldsymbol{a}$.

Then $\mathbb{A}$ becomes a one-dimensional vector space whose two “halves” have different physical meanings: $\mathbb{A}$ is oriented by A. Elements of A are **units of measurement** and $\mathbb{A}$ is called a **measure line**.

If the values of a physical quantity form a one-dimensional vector as a matter of course, e. g., in case of electric charge, then nature chooses an orientation (by distinguishing between a proton and an electron).

A nonzero element $\boldsymbol{a}$ of $\mathbb{A}$ is called **positive**, in notation $0 < \boldsymbol{a}$, if as a basis is positively oriented. Then the meaning of $\boldsymbol{a} \le 0$ is straightforward.

Further, let $\boldsymbol{b} \le \boldsymbol{a}$ mean that $0 \le \boldsymbol{b} - \boldsymbol{a}$. This defines a total ordering in $\mathbb{A}$ for which

- If $\boldsymbol{a} \le \boldsymbol{b}$ and $\boldsymbol{b} \le \boldsymbol{c}$ then $\boldsymbol{a} \le \boldsymbol{c}$.
- If $\boldsymbol{a} \le \boldsymbol{b}$ and α is a positive number, then $\alpha\boldsymbol{a} \le \alpha\boldsymbol{b}$.

We introduce the notation

$$\mathbb{A}^+ := \{\boldsymbol{a} \in \mathbb{A} \mid 0 < \boldsymbol{a}\}$$

and for $\boldsymbol{a} \in \mathbb{A}$

$$|\boldsymbol{a}| := \begin{cases} \boldsymbol{a} & \text{if } \boldsymbol{a} \in \mathbb{A}^+, \\ -\boldsymbol{a} & \text{if } \boldsymbol{a} \notin \mathbb{A}^+. \end{cases}$$

18.2.2 Quotients and products of measure lines

In practice, some units of measurement are obtained from other units of measurement by multiplication and division, e. g., $\frac{\mathrm{kg\,m}}{\mathrm{s}^2}$ is a unit of force. Such multiplications and divisions are defined by tensorial operations as follows.

Let $\mathbb{A}$ and $\mathbb{B}$ be measure lines. Then $\mathbb{A} \otimes \mathbb{B}$ and $\frac{\mathbb{B}}{\mathbb{A}}$ are measure lines, oriented by the tensor product, and the quotient of positive elements, respectively.

Every element of $\mathbb{A} \otimes \mathbb{B}$ is of the form $\boldsymbol{a} \otimes \boldsymbol{b}$.

The tensor product of measure lines is considered commutative by the identification $\mathbb{B} \otimes \mathbb{A} \equiv \mathbb{A} \otimes \mathbb{B}$, $\boldsymbol{b} \otimes \boldsymbol{a} \equiv \boldsymbol{a} \otimes \boldsymbol{b}$.

We retain the notation $\mathbb{A} \otimes \mathbb{B}$ but we write $\boldsymbol{ab}$ instead of $\boldsymbol{a} \otimes \boldsymbol{b}$.

If $0 \neq \boldsymbol{a} \in \mathbb{A}$, then all $\boldsymbol{a}' \in \mathbb{A}$ is a unique multiple of $\boldsymbol{a}$: the corresponding real number is denoted by $\frac{\boldsymbol{a}'}{\boldsymbol{a}}$; thus, $\frac{\boldsymbol{a}'}{\boldsymbol{a}}\boldsymbol{a} = \boldsymbol{a}'$.

As usual for numbers, we write $\frac{1}{\boldsymbol{a}}\boldsymbol{a}' = \frac{\boldsymbol{a}'}{\boldsymbol{a}}$; therefore, $\frac{1}{\boldsymbol{a}} : \mathbb{A} \to \mathbb{R}$ is a linear map, i. e., is an element of $\mathbb{A}^*$.

That is why we accept the definition $\frac{\mathbb{R}}{\mathbb{A}} \cup \{0\} := \mathbb{A}^*$.

Even $\frac{\mathbb{B}}{\mathbb{A}} := \mathbb{B} \otimes \mathbb{A}^* = \mathrm{Lin}(\mathbb{A}, \mathbb{B})$, $\frac{\boldsymbol{b}}{\boldsymbol{a}}\boldsymbol{a}' := \frac{\boldsymbol{a}'}{\boldsymbol{a}}\boldsymbol{b}$.

In case of $\mathbb{A}\otimes\mathbb{A}$, $\mathbb{A}\otimes\mathbb{A}\otimes\mathbb{A}$, etc. we use the notation $\mathbb{A}^{(2)}$, $\mathbb{A}^{(3)}$, and $\boldsymbol{a}^2 := \boldsymbol{aa}$, $\boldsymbol{a}^3 := \boldsymbol{aaa}$, etc.

$$\mathbb{A}^+ \to (\mathbb{A} \otimes \mathbb{A})^+, \quad \boldsymbol{a} \mapsto \boldsymbol{a}^2$$

is a bijection having the inverse $\sqrt{} : (\mathbb{A} \otimes \mathbb{A})^+ \to \mathbb{A}^+$.

Then it is obvious what $\boldsymbol{a}^n$ and $\sqrt[n]{}$ mean.

We can go further.

If $\mathbf{V}$ is a vector space, then every element of $\mathbb{A} \otimes \mathbf{V}$ is of the form $\boldsymbol{a} \otimes \boldsymbol{v}$, and we can deal with the elements of $\mathbb{A}$ as if they were real numbers.

Then we consider the tensor product to be commutative,

$$\mathbb{A} \otimes \mathbf{V} \equiv \mathbf{V} \otimes \mathbb{A}, \quad \boldsymbol{a} \otimes \boldsymbol{v} \equiv \boldsymbol{v} \otimes \boldsymbol{a},$$

moreover, we write

$$\boldsymbol{a}\boldsymbol{v} := \boldsymbol{a} \otimes \boldsymbol{v}.$$

If $\boldsymbol{v}$ is in $\mathbf{V}$ and $0 \neq \boldsymbol{a}$ is in $\mathbb{A}$, then we define the linear map

$$\frac{\boldsymbol{v}}{\boldsymbol{a}} : \mathbb{A} \to \mathbf{V}, \quad \boldsymbol{b} \mapsto \frac{\boldsymbol{b}}{\boldsymbol{a}}\boldsymbol{v},$$

called the **tensorial quotient** of $\boldsymbol{v}$ by $\boldsymbol{a}$.

We say that an element of $\mathbb{A} \otimes \mathbf{V}$ or $\frac{\mathbf{V}}{\mathbb{A}}$ is **parallel** to a vector $\boldsymbol{v}$ of $\mathbf{V}$ if it is of the form $\boldsymbol{a}\boldsymbol{v}$ or $\frac{\boldsymbol{v}}{\boldsymbol{a}}$, respectively.

Every nonzero linear map $\mathbb{A} \to \mathbf{V}$ is of the form $\frac{\boldsymbol{v}}{\boldsymbol{v}}$, and that is why the set of such linear maps is denoted by $\frac{\mathbf{V}}{\mathbb{A}}$.

Products and quotients by elements of measure lines follow the rules of products and quotients by real numbers.

This is true for the action of linear maps, too: the linear map $\boldsymbol{L} : \mathbf{V} \to \mathbf{U}$ can be considered a linear map:

– $\mathbb{A} \otimes \mathbf{V} \to \mathbb{A} \otimes \mathbf{U}$ in such a way that

$$\boldsymbol{L} \cdot (\boldsymbol{a}\boldsymbol{v}) := \boldsymbol{a}\boldsymbol{L} \cdot \boldsymbol{v}.$$

– $\frac{\mathbf{V}}{\mathbb{A}} \to \frac{\mathbf{U}}{\mathbb{A}}$ in such a way that

$$\boldsymbol{L} \cdot \frac{\boldsymbol{v}}{\boldsymbol{a}} := \frac{\boldsymbol{L} \cdot \boldsymbol{v}}{\boldsymbol{a}}.$$

Remark. The formulas in this subsection are valid for arbitrary oriented one-dimensional vector spaces instead of measure lines.

18.3 Euclidean vector spaces

18.3.1 Fundamental properties

A Euclidean vector space is a triplet $(\mathbf{S}, \mathbb{L}, \boldsymbol{h})$ where:

(1) $\mathbf{S}$ is a finite-dimensional real vector space.

(2) $\mathbb{L}$ is a measure line.

(3) $\boldsymbol{h} : \mathbf{S}\times\mathbf{S} \to \mathbb{L}\otimes\mathbb{L}$ is a symmetric, positive definite bilinear map, called the **Euclidean form**.

Since $\mathbb{L} \otimes \mathbb{L}$ is also a measure line, positive definiteness is meaningful: $\boldsymbol{h}(\boldsymbol{q}, \boldsymbol{q}) > 0$ for all nonzero $\boldsymbol{q}$.

The Euclidean form $\boldsymbol{h}$ admits us to make the identification

$$\frac{\mathbf{S}}{\mathbb{L}\otimes\mathbb{L}} \equiv \mathbf{S}^*, \quad \frac{\boldsymbol{p}}{d^2} \equiv \boldsymbol{q} \mapsto \frac{\boldsymbol{h}(\boldsymbol{p},\boldsymbol{q})}{d^2}. \tag{18.4}$$

In other words, the elements of $\mathbf{S}$ can be identified with linear maps $\mathbf{S} \to \mathbb{L}\otimes\mathbb{L}$, $\boldsymbol{p} \equiv \boldsymbol{h}(\boldsymbol{p},\,\cdot\,)$. Accordingly, we find convenient to write a dot product instead of $\boldsymbol{h}$:

$$\boldsymbol{p}\cdot\boldsymbol{q} := \boldsymbol{h}(\boldsymbol{p},\boldsymbol{q}) \in \mathbb{L}\otimes\mathbb{L} \quad (\boldsymbol{p},\boldsymbol{q} \in \mathbf{S}).$$

The **length** of a vector $\boldsymbol{q}$ is

$$|\boldsymbol{q}| := \sqrt{\boldsymbol{q}\cdot\boldsymbol{q}} \in \mathbb{L}.$$

The length has the fundamental properties:
(i) $|\boldsymbol{q}| \geq 0$ and is zero if and only if $\boldsymbol{q} = \mathbf{0}$
(ii) $|\alpha\boldsymbol{q}| = |\alpha||\boldsymbol{q}|$
(iii) $|\boldsymbol{q}+\boldsymbol{p}| \leq |\boldsymbol{q}| + |\boldsymbol{p}|$

for all vectors $\boldsymbol{q},\boldsymbol{p}$ and real numbers α. The last relation is called the **triangle inequality** in which equality occurs if and only if $\boldsymbol{q}$ and $\boldsymbol{p}$ are parallel; it is a consequence of the **Cauchy–Schwartz inequality**:

$$|\boldsymbol{q}\cdot\boldsymbol{p}| \leq |\boldsymbol{q}|\,|\boldsymbol{p}|,$$

where equality occurs if and only if $\boldsymbol{q}$ and $\boldsymbol{p}$ are parallel.

Then the **angle** between the vectors $\boldsymbol{q} \neq 0$ and $\boldsymbol{p} \neq 0$ is

$$\arg(\boldsymbol{q},\boldsymbol{p}) := \arccos\frac{\boldsymbol{q}\cdot\boldsymbol{p}}{|\boldsymbol{q}|\,|\boldsymbol{p}|}.$$

$\boldsymbol{q}$ and $\boldsymbol{p}$ are **perpendicular** or **orthogonal** to each other, $\boldsymbol{q} \perp \boldsymbol{p}$, if $\boldsymbol{q}\cdot\boldsymbol{p} = 0$; for nonzero vectors this means that $\pi/2$ is the angle between them.

For arbitrarily given $\mathrm{m} \in \mathbb{L}^+$, a basis $\boldsymbol{e}_1,\dots,\boldsymbol{e}_N$ in $\mathbf{S}$ is called **orthogonal and normalized to** m if $\boldsymbol{e}_i\cdot\boldsymbol{e}_i = \mathrm{m}^2$ and $\boldsymbol{e}_i\cdot\boldsymbol{e}_k = 0$ ($i \neq k = 1,\dots,N := \dim(\mathbf{S})$).

For an arbitrary measure line $\mathbb{A}$, the Euclidean form can be transported to $\mathbb{A}\otimes\mathbf{S}$ and $\frac{\mathbf{S}}{\mathbb{A}}$:

$$(\boldsymbol{a}\boldsymbol{q})\cdot(\boldsymbol{b}\boldsymbol{p}) := \boldsymbol{ab}(\boldsymbol{q}\cdot\boldsymbol{p}),$$

and

$$\left(\frac{\boldsymbol{q}}{\boldsymbol{a}}\right)\cdot\left(\frac{\boldsymbol{p}}{\boldsymbol{b}}\right) := \frac{\boldsymbol{q}\cdot\boldsymbol{p}}{\boldsymbol{ab}}.$$

Correspondingly,

$$|\boldsymbol{a}\boldsymbol{q}| = |\boldsymbol{a}||\boldsymbol{q}| \in \mathbb{A} \otimes \mathbb{L}, \qquad \left|\frac{\boldsymbol{q}}{\boldsymbol{a}}\right| = \frac{|\boldsymbol{q}|}{|\boldsymbol{a}|} \in \frac{\mathbb{L}}{\mathbb{A}}.$$

In particular, the Euclidean form on $\frac{\mathbf{S}}{\mathbb{L}}$ has real values. According to the identification (18.4),

$$\frac{\mathbf{S}}{\mathbb{L}} \equiv \left(\frac{\mathbf{S}}{\mathbb{L}}\right)^{*}.$$

18.3.2 Adjoint maps

The ***h*-adjoint** of a linear map $\boldsymbol{L} : \mathbf{S} \to \mathbf{S}$ is the linear map $\boldsymbol{L}^{\star} : \mathbf{S} \to \mathbf{S}$ defined by $\boldsymbol{q} \cdot (\boldsymbol{L} \cdot \boldsymbol{p}) = (\boldsymbol{L}^{\star} \cdot \boldsymbol{q}) \cdot \boldsymbol{p}$ $(\boldsymbol{q}, \boldsymbol{p} \in \mathbf{S})$.

$\boldsymbol{L}$ is ***h*-symmetric** or ***h*-antisymmetric** if $\boldsymbol{L}^{\star} = \boldsymbol{L}$ or $\boldsymbol{L}^{\star} = -\boldsymbol{L}$, respectively.

$\boldsymbol{L}$ is ***h*-orthogonal** if $\boldsymbol{L}^{\star} \cdot \boldsymbol{L}$ is the identity of $\mathbf{S}$; in other words, $\boldsymbol{L}^{\star} = \boldsymbol{L}^{-1}$ and $(\boldsymbol{L} \cdot \boldsymbol{q}) \cdot (\boldsymbol{L} \cdot \boldsymbol{p}) = \boldsymbol{q} \cdot \boldsymbol{p}$ for all $\boldsymbol{q}, \boldsymbol{p} \in \mathbf{S}$, i. e., it preserves the Euclidean form.

The adjoint corresponds to the transpose, due to the identification of vectors and covectors. In general, for the sake of simplicity, we omit referring to the Euclidean form and we say symmetric and antisymmetric and even we write $\boldsymbol{L}^{*}$ instead of $\boldsymbol{L}^{\star}$. This is supported by the fact that the adjoint of a tensor product is just its transpose: for $\boldsymbol{v}, \boldsymbol{v}' \in \frac{\mathbf{S}}{\mathbb{L}}$,

$$\begin{aligned}((\boldsymbol{v} \otimes \boldsymbol{v}')^{\star} \cdot \boldsymbol{q}) \cdot \boldsymbol{p} &= \boldsymbol{q} \cdot ((\boldsymbol{v} \otimes \boldsymbol{v}') \cdot \boldsymbol{p}) = (\boldsymbol{q} \cdot \boldsymbol{v})(\boldsymbol{v}' \cdot \boldsymbol{p}) \\ &= ((\boldsymbol{v}' \otimes \boldsymbol{v}) \cdot \boldsymbol{q}) \cdot \boldsymbol{p}.\end{aligned}$$

It is remarkable that—contrary to the general case—here it makes sense that a linear map $\mathbf{S} \to \mathbf{S}$ is symmetric or antisymmetric.

For arbitrary measure lines $\mathbb{A}$ and $\mathbb{B}$, the adjoint of linear maps $\mathbb{A} \otimes \mathbf{S} \to \mathbb{B} \otimes \mathbf{S}$ and $\frac{\mathbf{S}}{\mathbb{A}} \to \frac{\mathbf{S}}{\mathbb{B}}$ are defined according to the sense.

18.3.3 Axial vectors

In what follows, $\mathbf{S}$ is a three-dimensional and oriented Euclidean vector space.

It is convenient to introduce the notation $\mathbf{N} := \frac{\mathbf{S}}{\mathbb{L}}$ because the Euclidean form on it is real valued, the length of vectors is a real number. Moreover, $\mathbf{N}^{*} \equiv \mathbf{N}$; thus, cotensors and mixed tensors, too, are identified with tensors.

Since $\mathbf{N}$ is three-dimensional, $\mathbf{N} \wedge \mathbf{N}$ is three-dimensional $\mathbf{N} \wedge \mathbf{N} \wedge \mathbf{N}$ is one-dimensional.

For every positively oriented, orthogonal basis $(\boldsymbol{n}_1, \boldsymbol{n}_2, \boldsymbol{n}_3)$ of N, normalized to 1,

$$\boldsymbol{\epsilon} := \boldsymbol{n}_1 \wedge \boldsymbol{n}_2 \wedge \boldsymbol{n}_3$$

is the same (is independent of the basis); it is called the **Levi–Civita tensor.**

Further, it can be shown that for every such a basis the linear bijection defined by

$$\boldsymbol{j} : \mathrm{N} \wedge \mathrm{N} \to \mathrm{N}, \quad \boldsymbol{n}_1 \wedge \boldsymbol{n}_2 \mapsto \boldsymbol{n}_3,\ \boldsymbol{n}_2 \wedge \boldsymbol{n}_3 \mapsto \boldsymbol{n}_1,\ \boldsymbol{n}_3 \wedge \boldsymbol{n}_1 \mapsto \boldsymbol{n}_2 \tag{18.5}$$

is the same (is independent of the basis).

This allows us to define the **vectorial product**

$$\mathrm{N} \times \mathrm{N} \to \mathrm{N}, \quad (\boldsymbol{k}, \boldsymbol{n}) \mapsto \boldsymbol{k} \times \boldsymbol{n} := \boldsymbol{j}(\boldsymbol{k} \wedge \boldsymbol{n})$$

for which the following important relations hold for $\boldsymbol{A}, \boldsymbol{B} \in \mathrm{N} \wedge \mathrm{N}$, and $\boldsymbol{n} \in \mathrm{N}$, with the notation

$$[\boldsymbol{A}, \boldsymbol{B}] := \boldsymbol{A} \cdot \boldsymbol{B} - \boldsymbol{B} \cdot \boldsymbol{A} :$$

(i) $|\boldsymbol{j}(\boldsymbol{A})| = |\boldsymbol{A}| := \sqrt{-\frac{1}{2}\,\mathrm{Tr}(\boldsymbol{A}^2)}$
(ii) $\boldsymbol{A} \cdot \boldsymbol{n} = -\boldsymbol{j}(\boldsymbol{A}) \times \boldsymbol{n}$
(iii) $\boldsymbol{A} \cdot \boldsymbol{j}(\boldsymbol{B}) = -\boldsymbol{j}([\boldsymbol{A}, \boldsymbol{B}])$
(iv) $\boldsymbol{j}([\boldsymbol{A}, \boldsymbol{B}]) = \boldsymbol{j}(\boldsymbol{A}) \times \boldsymbol{j}(\boldsymbol{B})$
(v) $\boldsymbol{j}(\boldsymbol{A}) \wedge \boldsymbol{j}(\boldsymbol{B}) = [\boldsymbol{A}, \boldsymbol{B}]$
(vi) $\boldsymbol{A} \wedge \boldsymbol{n} = (\boldsymbol{j}(\boldsymbol{A}) \cdot \boldsymbol{n})\boldsymbol{\epsilon}$

Of course, for arbitrary measure lines $\mathbb{A}$ and $\mathbb{B}$ we have the linear bijection

$$\boldsymbol{j} : (\mathbb{A} \otimes \mathrm{N}) \wedge (\mathbb{B} \otimes \mathrm{N}) = (\mathbb{A} \otimes \mathbb{B}) \otimes \mathrm{N}$$

and the vectorial product

$$(\mathbb{A} \otimes \mathrm{N}) \times (\mathbb{B} \otimes \mathrm{N}) \to (\mathbb{A} \otimes \mathbb{B}) \otimes \mathrm{N};$$

for instance, $\mathbf{S} = \mathbb{L} \otimes \mathrm{N}$, so the vectorial product in $\mathbf{S}$ is

$$\mathbf{S} \times \mathbf{S} \to \mathbb{L} \otimes \mathbf{S}.$$

In physical applications, elements of N corresponding to antisymmetric tensors are called **axial vectors.** It is often said that an axial vector is a vector, which does not change for space inversion. This, however, makes no sense because space inversion is the linear map $-\mathrm{id}_{\mathrm{N}}$, the multiplication by -1. According to Subsection 18.3, tensors do not change for space inversion: $(-\mathrm{id}_{\mathrm{N}}) \otimes (-\mathrm{id}_{\mathrm{N}})(\boldsymbol{n} \otimes \boldsymbol{k}) = (-\boldsymbol{n}) \otimes (-\boldsymbol{k}) = \boldsymbol{n} \otimes \boldsymbol{k}$, briefly,

$$(-\mathrm{id}_{\mathrm{N}}) \otimes (-\mathrm{id}_{\mathrm{N}}) = \mathrm{id}_{\mathrm{N}\otimes\mathrm{N}}.$$

18.4 Minkowski vector spaces

18.4.1 Fundamental properties

A Minkowski vector space is a triplet $(\mathbf{M}, \mathbb{T}, \boldsymbol{g})$ where:

(1) $\mathbf{M}$ is a finite dimensional real vectors space having at least two dimensions.

(2) $\mathbb{T}$ is a measure line.

(3) $\boldsymbol{g} : \mathbf{M} \times \mathbf{M} \to \mathbb{T} \otimes \mathbb{T}$ is a Lorentz form, a symmetric, bilinear map of type $(1, 3)$.

In the sequel, $\mathbf{M}$ will be four-dimensional.

Type $(1, 3)$ means that if $\boldsymbol{e}_0$, $\boldsymbol{e}_1$, $\boldsymbol{e}_2$, $\boldsymbol{e}_3$ are vectors such that $\boldsymbol{g}(\boldsymbol{e}_i, \boldsymbol{e}_k) = 0$ for $i \neq k$, then $\boldsymbol{g}(\boldsymbol{e}_0, \boldsymbol{e}_0) < 0$ and $\boldsymbol{g}(\boldsymbol{e}_i, \boldsymbol{e}_i) > 0$ if $i = 1, 2, 3$.

The Lorentz form is nondegenerate, i. e., if $\boldsymbol{g}(\boldsymbol{y}, \boldsymbol{x}) = 0$ for all $\boldsymbol{x} \in \mathbf{M}$, then $\boldsymbol{y} = 0$.

The Lorentz form admits us to make the identification

$$\frac{\mathbf{M}}{\mathbb{T} \otimes \mathbb{T}} \equiv \mathbf{M}^*, \quad \frac{\boldsymbol{y}}{\mathrm{s}^2} \equiv \boldsymbol{x} \mapsto \frac{\boldsymbol{g}(\boldsymbol{y}, \boldsymbol{x})}{\mathrm{s}^2}. \tag{18.6}$$

Then, similar to Euclidean vector spaces, we write a dot product instead of $\boldsymbol{g}$:

$$\boldsymbol{x} \cdot \boldsymbol{y} := \boldsymbol{g}(\boldsymbol{x}, \boldsymbol{y}) \in \mathbb{T} \otimes \mathbb{T} \quad (\boldsymbol{x}, \boldsymbol{y} \in \mathbf{M}).$$

The **pseudolength** of a vector $\boldsymbol{x}$ is

$$|\boldsymbol{x}| := \sqrt{|\boldsymbol{x} \cdot \boldsymbol{x}|}.$$

The pseudolength has the properties:

(i) $|\boldsymbol{x}| = 0$ if $\boldsymbol{x} = \mathbf{0}$ but $|\boldsymbol{x}| = 0$ does not imply $\boldsymbol{x} = \mathbf{0}$.

(ii) $|\alpha \boldsymbol{x}| = |\alpha||\boldsymbol{x}|$ for all $\alpha \in \mathbb{R}$.

(iii) There is no general relation between $|\boldsymbol{x} + \boldsymbol{y}|$ and $|\boldsymbol{x}| + |\boldsymbol{y}|$.

The vectors $\boldsymbol{x}$ and $\boldsymbol{y}$ are called **$\boldsymbol{g}$-orthogonal** if $\boldsymbol{x} \cdot \boldsymbol{y} = 0$. Contrary to the Euclidean case, there are nonzero vectors $\boldsymbol{x}$, $\boldsymbol{g}$-orthogonal to themselves: $\boldsymbol{x} \cdot \boldsymbol{x} = 0$.

Let us consider the set

$$\mathrm{T} := \{\boldsymbol{x} \in \mathbf{M} \mid \boldsymbol{x} \cdot \boldsymbol{x} < 0\}.$$

The elements $\boldsymbol{x}$ and $\boldsymbol{y}$ of T are called having **the same arrow** if $\boldsymbol{x} \cdot \boldsymbol{y} < 0$.

Having the same arrow is an *equivalence relation*: it is evidently reflexive and symmetric. As concerns transitivity, let $\boldsymbol{x}, \boldsymbol{y}, \boldsymbol{z}$ be elements of T having the same arrow. Then $\boldsymbol{q} := (\boldsymbol{y} \cdot \boldsymbol{z})\boldsymbol{x} - \boldsymbol{y}(\boldsymbol{x} \cdot \boldsymbol{z})$ is $\boldsymbol{g}$-orthogonal to $\boldsymbol{z}$; since the Lorentz form is of $(1, 3)$ type, $\boldsymbol{q} \cdot \boldsymbol{q} \geq 0$ and is zero if and only if $\boldsymbol{q} = 0$. Then we infer

$$2(\boldsymbol{x} \cdot \boldsymbol{y})(\boldsymbol{y} \cdot \boldsymbol{z})(\boldsymbol{z} \cdot \boldsymbol{x}) \leq (\boldsymbol{x} \cdot \boldsymbol{z})^2(\boldsymbol{y} \cdot \boldsymbol{x}) + (\boldsymbol{y} \cdot \boldsymbol{z})^2(\boldsymbol{x} \cdot \boldsymbol{x}) \leq 0 \tag{18.7}$$

and equality holds if and only if $\boldsymbol{x}$ and $\boldsymbol{y}$ are parallel. So, if $\boldsymbol{x} \cdot \boldsymbol{y} < 0$ and $\boldsymbol{y} \cdot \boldsymbol{z} < 0$, then $\boldsymbol{x} \cdot \boldsymbol{z} < 0$.

Evidently, there are two equivalence classes. An equivalence class is called an **arrow orientation** of $\boldsymbol{g}$. $\boldsymbol{g}$ is **arrow oriented** if there is given an arrow orientation. We say the vectors in the chosen equivalence class have **positive arrow** and we denote their set by $\mathrm{T}^{\rightarrow}$.

$\mathrm{T}^{\rightarrow}$ is open, being the preimage of an open subset by a continuous function $\boldsymbol{y} \mapsto \boldsymbol{x}\cdot\boldsymbol{y}$.

$\mathrm{T}^{\rightarrow}$ is a convex cone: if $\boldsymbol{x}$ and $\boldsymbol{y}$ have the same arrow, α, β are nonnegative numbers, $\alpha + \beta \neq 0$, then

$$(\alpha\boldsymbol{x} + \beta\boldsymbol{y}) \cdot (\alpha\boldsymbol{x} + \beta\boldsymbol{y}) = \alpha^2(\boldsymbol{x} \cdot \boldsymbol{x}) + 1\alpha\beta(\boldsymbol{x} \cdot \boldsymbol{y}) + \beta^2 = \boldsymbol{y} \cdot \boldsymbol{y}) < 0, \tag{18.8}$$
$$\boldsymbol{x} \cdot (\alpha\boldsymbol{x} + \beta\boldsymbol{y}) = \alpha(\boldsymbol{x} \cdot \boldsymbol{x}) + \beta(\boldsymbol{x} \cdot \boldsymbol{y}) < 0.$$

The two most important properties are the **reversed Cauchy inequality** and the **reversed triangle inequality**: if $\boldsymbol{x}, \boldsymbol{y} \in \mathrm{T}^{\rightarrow}$, then

$$|\boldsymbol{x} \cdot \boldsymbol{y}| \geq |\boldsymbol{x}||\boldsymbol{y}| > 0, \tag{18.9}$$
$$|\boldsymbol{x} + \boldsymbol{y}| \geq |\boldsymbol{x}| + |\boldsymbol{y}|; \tag{18.10}$$

in both cases, equality occurs if and only if $\boldsymbol{x}$ and $\boldsymbol{y}$ are parallel.

Inequality (18.9) is obtained from (18.7) with $\boldsymbol{z} = \boldsymbol{x}$.

Inequality (18.10) follows from the previous one and (18.8):

$$|\boldsymbol{x} + \boldsymbol{y}|^2 = -(\boldsymbol{x} + \boldsymbol{y}) \cdot (\boldsymbol{x} + \boldsymbol{y}) = |\boldsymbol{x}|^2 - 2\boldsymbol{x} \cdot \boldsymbol{y} + |\boldsymbol{y}|^2 \geq |\boldsymbol{x}|^2 + 2|\boldsymbol{x}||\boldsymbol{y}| + |\boldsymbol{y}|^2.$$

Let us consider the set

$$\mathrm{L} := \{\boldsymbol{x} \in \mathbf{M} \mid \boldsymbol{x} \cdot \boldsymbol{x} = 0,\ \boldsymbol{x} \neq 0\}.$$

If $\boldsymbol{y} \in \mathrm{T}$ and $\boldsymbol{x} \in \mathrm{L}$, then $\boldsymbol{y} \cdot \boldsymbol{x} \neq 0$ (if it were zero then $\boldsymbol{y} \cdot \boldsymbol{y}$ would be nonnegative). The elements of

$$\mathrm{L}^{\rightarrow} := \{\boldsymbol{x} \in \mathrm{L} \mid \boldsymbol{y} \cdot \boldsymbol{x} < 0 \text{ for } \boldsymbol{y} \in \mathrm{T}^{\rightarrow}\}$$

are said to have **positive arrow**.

If $\boldsymbol{y} \in \mathrm{T}^{\rightarrow}$ and $\boldsymbol{x} \in \mathrm{L}^{\rightarrow}$, then $\boldsymbol{x} + \boldsymbol{y} \in \mathrm{T}^{\rightarrow}$ because

$$(\boldsymbol{x} + \boldsymbol{y}) \cdot (\boldsymbol{x} + \boldsymbol{y}) = 2\boldsymbol{y} \cdot \boldsymbol{x} + \boldsymbol{x} \cdot \boldsymbol{x} < 0,$$
$$\boldsymbol{y} \cdot (\boldsymbol{x} + \boldsymbol{y}) = \boldsymbol{y} \cdot \boldsymbol{x} + \boldsymbol{x} \cdot \boldsymbol{x} < 0.$$

So, if $\boldsymbol{x} \in \mathrm{L}^{\rightarrow}, \boldsymbol{y} \in \mathrm{T}^{\rightarrow}$ and $0 < \alpha_n$ $(n = 1, 2, \dots)$ is sequence for which $\lim_n \alpha_n = 0$, then $x + \alpha_n\boldsymbol{y} \in \mathrm{T}^{\rightarrow}$ and $\lim_n(x + \alpha_n\boldsymbol{y}) = \boldsymbol{x}$. This means that $\mathrm{L}^{\rightarrow}$ is the boundary of $\mathrm{T}^{\rightarrow}$, except zero:

$$\mathrm{L}^{\rightarrow} = \partial\mathrm{T}^{\rightarrow} \setminus \{0\}.$$

For an arbitrary measure line $\mathbb{A}$, the Lorentz form can be transferred to $\mathbb{A} \otimes \mathbf{M}$ and $\frac{\mathbf{M}}{\mathbb{A}}$:

$$(a\boldsymbol{x}) \cdot (b\boldsymbol{y}) := ab(\boldsymbol{x} \cdot \boldsymbol{y}),$$

and

$$\left(\frac{\boldsymbol{x}}{a}\right) \cdot \left(\frac{\boldsymbol{y}}{b}\right) := \frac{\boldsymbol{x} \cdot \boldsymbol{y}}{ab}.$$

Correspondingly,

$$|a\boldsymbol{x}| = |a||\boldsymbol{x}| \in \mathbb{A} \otimes \mathbb{T}, \qquad \left|\frac{\boldsymbol{x}}{a}\right| = \frac{|\boldsymbol{x}|}{|a|} \in \frac{\mathbb{T}}{\mathbb{A}}.$$

The Lorentz form on $\frac{\mathbf{M}}{\mathbb{T}}$ has real values, thus according to the identification (18.6),

$$\frac{\mathbf{M}}{\mathbb{T}} \equiv \left(\frac{\mathbf{M}}{\mathbb{T}}\right)^{*}.$$

18.4.2 Adjoint maps

The g**-adjoint** or **Lorentz adjoint** of a linear map $\boldsymbol{L} : \mathbf{M} \to \mathbf{M}$ is the linear map $\boldsymbol{L}^{\star} : \mathbf{M} \to \mathbf{M}$ defined by $\boldsymbol{x} \cdot (\boldsymbol{L} \cdot \boldsymbol{y}) = (\boldsymbol{L}^{\star} \cdot \boldsymbol{x}) \cdot \boldsymbol{y}$ $(\boldsymbol{x}, \boldsymbol{y} \in \mathbf{M})$.

$\boldsymbol{L}$ is g**-symmetric** or g**-antisymmetric** if $\boldsymbol{L}^{\star} = \boldsymbol{L}$ or $\boldsymbol{L}^{\star} = -\boldsymbol{L}$, respectively.

$\boldsymbol{L}$ is g**-orthogonal** if and only if $\boldsymbol{L}^{\star} \cdot \boldsymbol{L}$ is the identity of $\mathbf{M}$; in other words, $\boldsymbol{L}^{\star} = \boldsymbol{L}^{-1}$ and $(\boldsymbol{L} \cdot \boldsymbol{x}) \cdot (\boldsymbol{L} \cdot \boldsymbol{y}) = \boldsymbol{x} \cdot \boldsymbol{y}$ for all $\boldsymbol{x}, \boldsymbol{y} \in \mathbf{M}$, i. e., it preserves the Lorentz form.

The adjoint corresponds to the transpose, due to the identification of vectors and covectors. In general, for the sake of simplicity, we omit referring to the Lorentz form and we say symmetric and antisymmetric and even we write $\boldsymbol{L}^{*}$ instead of $\boldsymbol{L}^{\star}$. This is supported by the fact that the adjoint of a tensor product is just its transpose: for $\boldsymbol{u}, \boldsymbol{u}' \in \frac{\mathbf{M}}{\mathbb{T}}$,

$$\begin{aligned}((\boldsymbol{u} \otimes \boldsymbol{u}')^{\star} \cdot \boldsymbol{x}) \cdot \boldsymbol{y} = \boldsymbol{x} \cdot ((\boldsymbol{u} \otimes \boldsymbol{u}') \cdot \boldsymbol{y}) &= (\boldsymbol{x} \cdot \boldsymbol{u})(\boldsymbol{u}' \cdot \boldsymbol{p}) \\ &= ((\boldsymbol{u}' \otimes \boldsymbol{u}) \cdot \boldsymbol{x}) \cdot \boldsymbol{y}.\end{aligned}$$

It is remarkable that—contrary to the general case—here it makes sense that a linear map $\mathbf{M} \to \mathbf{M}$ is symmetric or antisymmetric.

For arbitrary measure lines $\mathbb{A}$ and $\mathbb{B}$, the adjoint of linear maps $\mathbb{A} \otimes \mathbf{M} \to \mathbb{B} \otimes \mathbf{M}$ and $\frac{\mathbf{M}}{\mathbb{A}} \to \frac{\mathbf{S}}{\mathbb{B}}$ are defined according to the sense.

18.5 Affine spaces

18.5.1 Fundamental properties

An **affine space** is a triplet (V, **V**, −) where:

(1) V is a nonvoid set.

(2) **V** is a vector space.

(3) − is a map from V × V onto **V**, which is written in the form

$$(x, y) \mapsto x - y \tag{18.11}$$

and has the properties

(i) For all $y \in \mathrm{V}$, the map $\mathrm{V} \to \mathbf{V}$, $x \mapsto x - y$ is a bijection.

(ii) $(x - y) + (y - z) + (z - x) = \mathbf{0}$ for all $x, y, z \in \mathrm{V}$.

As usual, the affine space is denoted by a single letter, saying that V is an affine space over the vector space **V** (**V** is the underlying vector space of V) and the map (18.11) is called **subtraction**.

The **dimension** of V is the dimension of **V** and V is **oriented** if **V** is oriented.

Properties (i) and (ii) imply that for all $x, y \in \mathrm{V}$:

(1) $x - y = \mathbf{0}$ if and only if $x = y$

(2) $x - y = -(y - x)$

furthermore, for all natural numbers $n \geq 3$ and for elements $x_1, x_2, \ldots, x_n$ of V

$$(x_1 - x_2) + (x_2 - x_3) + \cdots + (x_n - x_1) = \mathbf{0}$$

holds.

For an arbitrary element y of V, the map $x \mapsto x - y$ is called the **vectorization of** V **with origin** y.

The inverse of the vectorization by y is denoted by

$$\mathbf{V} \to \mathrm{V}, \quad \boldsymbol{x} \mapsto y + \boldsymbol{x}. \tag{18.12}$$

According to the definition,

$$y + (x - y) = x$$

for all $y, x \in \mathrm{V}$.

Note that

- Addition, subtraction, and multiplication by real numbers make sense for vectors, the results are vectors.
- Subtraction makes sense for affine space elements, the result is a vector (addition and multiplication by real numbers make no sense).

- The sum of an affine space element and a vector makes sense, the result is an element of the affine space.

The notation are determined in such a way that the usual rules of operations are valid if the operations make sense.

For instance, $(x+\mathbf{x})+\mathbf{y} = x+(\mathbf{x}+\mathbf{y})$ (the signs + on the left side denote the operation (18.12), whereas on the right-hand side, the first one denotes the operation (18.12) and the second one denotes the addition of vectors).

Another example: $(x-y)+(u-v) = (x-v)-(y-u)$ is all right but $(x-y)+(u-v)$ is not equal to $(x+u)-(y+v)$ because the sum of affine space elements is not meaningful.

A subset E of the affine space V is an **affine subspace** if there is linear subspace **E** in **V** in such a way that E is an affine space over **E**.

A vector space is an affine space over itself with the vectorial subtraction. Affine spaces are—without being vector spaces—the affine subspaces of a vector spaces (see Subsection 18.1.4).

18.5.2 Factor spaces

Let **E** be a linear subspace of **V**.

A set of the affine space V is **an affine subspace directed by E** if it is of the form $x + \mathbf{E}$ for some element x of V. Then $x + \mathbf{E} = y + \mathbf{E}$ if and only if $x - y \in \mathbf{E}$.

Affine subspaces directed by a one-dimensional linear subspace are called **straight lines; hyperplanes** are the affine subspaces directed by linear subspaces of dimension $\dim \mathbf{V} - 1$.

Affine subspaces directed by the same linear subspace are called **parallel.**

The set of affine subspaces directed by **E** is denoted by $\mathrm{V}/\mathbf{E}$ and called the **factor space** of V by **E**.

$\mathrm{V}/\mathbf{E}$ is an affine space over $\mathbf{V}/\mathbf{E}$ with the well-defined subtraction

$$(y + \mathbf{E}) - (x + \mathbf{E}) := (y - x) + \mathbf{E}.$$

$\mathrm{V}/\mathbf{E}$ has dimension $\dim \mathbf{V} - \dim \mathbf{E}$.

18.5.3 Affine maps

Let V and U be affine spaces over the vector spaces **V** and **U**, respectively. A map $L : \mathrm{V} \to \mathrm{U}$ is an **affine map** if there is linear map $\mathbf{L} : \mathbf{V} \to \mathbf{U}$ such that

$$L(y) - L(x) = \mathbf{L} \cdot (y - x) \quad (x, y \in \mathrm{V}).$$

The linear map $\boldsymbol{L}$ is unique. We say that L is an affine map over the linear map $\boldsymbol{L}$ ($\boldsymbol{L}$ is the linear map under L). The formula above is equivalent to

$$L(x + \mathbf{x}) = L(x) + \boldsymbol{L} \cdot \mathbf{x} \quad (x \in \mathrm{V}, \mathbf{x} \in \mathbf{V}).$$

It is trivial that a linear map between vector spaces is an affine map over itself.

An affine map L is injective or surjective if and only if the underlying linear map $\boldsymbol{L}$ is injective or surjective. If L is bijective, L^{-1} is an affine map over $\boldsymbol{L}^{-1}$.

If V and U are oriented and L is a bijection, then L **preserves orientation** if $\boldsymbol{L}$ preserves orientation.

The image of an affine subspace by an affine map is an affine subspace and the preimage of an affine subspace by an affine map is an affine subspace.

In particular, for all u in the range of the affine map L,

$$\{x \in \mathrm{V} \mid L(x) = u\}$$

is an affine subspace directed by the kernel of $\boldsymbol{L}$.

18.6 Differentiation

18.6.1 Differentiation with real variables

Let U be an affine space over (the finite-dimensional) vector space $\mathbf{U}$. A function $f : \mathbb{R} \to \mathrm{U}$ is differentiable at an interior point t of the domain of f if the limit

$$f'(t) := \lim_{s \to t} \frac{f(t) - f(s)}{t - s} \in \mathbf{U}$$

exists. Then the function $x \mapsto f'(x)$ is the derivative of f. In some cases, the notation $\dot{f}$ is also used instead of f'.

The reader is supposed to be familiar with the properties of such differentiation. One of the most important rules is that $(\alpha f + \beta g)' = \alpha f' + \beta g'$ for functions f, g and real numbers α, β.

18.6.2 Differentiation in general

First of all, we introduce some fundamental notions.

If V is affine space over (the finite-dimensional) vector space $\mathbf{V}$ and $\| \ \|$ is a norm on $\mathbf{V}$, then $\mathrm{V} \times \mathrm{V} \to \mathbb{R}$, $(x, y) \mapsto \|x - y\|$ is a metric on V.

Arbitrary two norms on a finite-dimensional vector space are equivalent, i. e., they define the same open sets, closed sets, convergent sequences, continuous functions, etc.; therefore, we can take these notions—supposed to be known—even in affine spaces without specifying a norm.

Linear, bilinear, multilinear maps as well as affine maps are automatically continuous (in finite dimension).

If $\mathbf{V}$ and $\mathbf{U}$ are finite-dimensional vector spaces,

$\mathrm{Ordo} : \mathbf{V} \to \mathbf{U}$ denotes a function such that

- It is defined in a neighborhood of $\mathbf{0} \in \mathbf{V}$.
- $\lim_{\boldsymbol{x}\to\mathbf{0}} \mathrm{Ordo}(\boldsymbol{x}) = \mathbf{0}$.

$\mathrm{ordo} : \mathbf{V} \to \mathbf{U}$ denotes a function such that

- It is defined in a neighborhood of $\mathbf{0} \in \mathbf{V}$.
- $\lim_{\boldsymbol{x}\to\mathbf{0}} \frac{\mathrm{ordo}(\boldsymbol{x})}{\|\boldsymbol{x}\|} = \mathbf{0}$ for some (hence for every) norm $\| \ \|$ on $\mathbf{V}$.

Let V and U be affine spaces. A function $F : \mathrm{V} \to \mathrm{U}$ is **differentiable** at an interior point x of its domain if there is a linear map $\mathcal{D}F[x] : \mathbf{V} \to \mathbf{U}$—in other words, an element of $\mathbf{U} \otimes \mathbf{V}^*$—in such a way that

$$F(y) - F(x) = \mathcal{D}F[x] \cdot (y - x) + \mathrm{ordo}(y - x) \quad (y \in \mathrm{V});$$

in another form,

$$F(x + \boldsymbol{x}) - F(x) = \mathcal{D}F[x] \cdot \boldsymbol{x} + \mathrm{ordo}(\boldsymbol{x}) \quad (\boldsymbol{x} \in \mathbf{V}).$$

$\mathcal{D}F[x]$ is the **derivative** of F at x.

The function F is:

(i) **Differentiable** if it is differentiable at every point of its domain.

(ii) **Continuously differentiable** if it is differentiable and the function $\mathrm{V} \to \mathbf{U} \otimes \mathbf{V}^*$, $x \mapsto \mathcal{D}F[x]$ is continuous (which makes sense because $\mathbf{U}\otimes\mathbf{V}^*$ is a finite-dimensional vector space).

(iii) **Twice differentiable** if it is differentiable and the function $\mathrm{V} \to \mathbf{U}\otimes\mathbf{V}^*, x \mapsto \mathcal{D}F[x]$ is differentiable. The second derivative of F at x, denoted by $\mathcal{D}^2F[x]$, is an element of $(\mathbf{U} \otimes \mathbf{V}^*) \otimes \mathbf{V}^* = \mathbf{U} \otimes (\mathbf{V}^* \otimes \mathbf{V}^*)$.

(iv) n **times differentiable**, where n is a natural number n, if it is $n - 1$ times and the function $\mathrm{V} \to \mathbf{U} \otimes (\mathbf{V}^*)^{\otimes(n-1)n}$, $x \mapsto \mathcal{D}^{n-1}[x]$ is differentiable.

(v) **Smooth** (or infinitely differentiable) if it is differentiable for all n.

An affine map is differentiable, and its derivative at every point is the underlying linear map; as a consequence, it is smooth, its derivatives of higher rank are zero.

Since a vector space is an affine space over itself, everything remains valid if we take $\mathbf{U}$ and/or $\mathbf{V}$ instead of U and/or V.

The derivative of a function f defined in a one-dimensional affine space A can be obtained similar to real variable functions:

$$(\dot{f}(a) =) f'(a) := \lim_{b\to a} \frac{f(b) - f(a)}{b - a} \in \frac{\mathbf{V}}{\mathbf{A}}.$$

18.6.3 Differentiation of tensor fields

In view of the notation used in spacetime models, now we consider functions defined in an affine space M and having values in $\mathbb{R}$, $\mathbf{M}$, $\mathbf{M}^*$, $\mathbf{M} \otimes \mathbf{M}$, $\mathbf{M}^* \otimes \mathbf{M}^*$, etc. (scalar fields, vector fields, covector fields, tensor fields, cotensor fields, etc.).

The derivative of the function $f : \mathrm{M} \to \mathbb{R}$ is the covector field $\mathcal{D}f : \mathrm{M} \to \mathbf{M}^*$.

The derivative of the vector field $\boldsymbol{J} : \mathrm{M} \to \mathbf{M}$ is the mixed tensor field $\mathcal{D}\boldsymbol{J} : \mathrm{M} \to \mathbf{M} \otimes \mathbf{M}^*$.

The derivative of the covector field $\boldsymbol{K} : \mathrm{M} \to \mathbf{M}^*$ is the cotensor field $\mathcal{D}\boldsymbol{K} : \mathrm{M} \to \mathbf{M}^* \otimes \mathbf{M}^*$.

The derivative of the tensor field $\boldsymbol{G} : \mathrm{M} \to \mathbf{M} \otimes \mathbf{M}$ is the multiple mixed tensor field $\mathcal{D}\boldsymbol{G} : \mathrm{M} \to \mathbf{M} \otimes \mathbf{M} \otimes \mathbf{M}^*$.

The derivative of the cotensor field $\boldsymbol{F} : \mathrm{M} \to \mathbf{M}^* \otimes \mathbf{M}^*$ is the cotensor field $\mathcal{D}\boldsymbol{F} : \mathrm{M} \to \mathbf{M}^* \otimes \mathbf{M}^* \otimes \mathbf{M}^*$.

And so on: The differentiation can be conceived as a formal covector, which acts quasi as a tensorial multiplication. Unfortunately, the notation of the derivatives are not in accordance with the tensorial notation: $\mathcal{D}$ is written from the left but it acts in a tensorial multiplication from the right. A simple trick remedies this fault: We define

$$\mathcal{D} \otimes \boldsymbol{J} := (\mathcal{D}\boldsymbol{J})^* : \mathrm{M} \to \mathbf{M}^* \otimes \mathbf{M},$$
$$\mathcal{D} \otimes \boldsymbol{K} := (\mathcal{D}\boldsymbol{K})^* : \mathrm{M} \to \mathbf{M}^* \otimes \mathbf{M}^*,$$
$$\mathcal{D} \otimes \boldsymbol{G} : \mathrm{M} \to \mathbf{M}^* \otimes \mathbf{M} \otimes \mathbf{M}^*,$$
$$\mathcal{D} \otimes \boldsymbol{F} : \mathrm{M} \to \mathbf{M}^* \otimes \mathbf{M}^* \otimes \mathbf{M}^*,$$

then $\boldsymbol{x} \cdot (\mathcal{D} \otimes \boldsymbol{J}[x]) = (\boldsymbol{x} \cdot \mathcal{D})\boldsymbol{J}[x]$, etc. appears instead of $(\mathcal{D}\boldsymbol{J}[x]) \cdot \boldsymbol{x}$.

The values of the derivative of the vector field $\boldsymbol{J}$ are mixed tensors, so we can take their trace; in this way, we obtain the **divergence** of $\boldsymbol{J}$:

$$\operatorname{div}\boldsymbol{J} := \mathcal{D} \cdot \boldsymbol{J} := \operatorname{Tr}(\mathcal{D} \otimes \boldsymbol{J}).$$

Similarly, according to the sense, we define the divergence $\operatorname{div}\boldsymbol{G} := \mathcal{D} \cdot \boldsymbol{G}$ of a tensor field $\boldsymbol{G}$.

The **exterior derivative** of a cotensor field $\boldsymbol{K}$ is the antisymmetrization of the derivative of $\boldsymbol{K}$:

$$\mathcal{D} \wedge \boldsymbol{K} := \mathcal{D} \otimes \boldsymbol{K} - \mathcal{D}\boldsymbol{K}.$$

Similarly, according to the sense, we define the exterior derivative $\mathcal{D} \wedge \boldsymbol{F}$ of an antisymmetric cotensor field $\boldsymbol{F}$.

Similar formulas, according to the sense, hold for vector fields, etc. multiplied or divided by a measure line, $\mathbf{M} \otimes \mathbb{A}$ or $\frac{\mathbf{M}}{\mathbb{A}}$, etc.

18.6.4 Differentiation in three-dimensional Euclidean space

In the case of an affine space over a three-dimensional oriented Euclidean vector space, the differentiation is usually denoted by the symbol ∇.

Then the derivative of a vector field $\boldsymbol{D}$ is $\nabla\boldsymbol{D}$, its divergence is $\operatorname{div}\boldsymbol{D} = \nabla\cdot\boldsymbol{D}$.

The exterior derivative of a covector field $\boldsymbol{E}$ is $\nabla\wedge\boldsymbol{E}$; applying the map (18.5), we get the **curl** of $\boldsymbol{E}$: $\boldsymbol{j}(\nabla\wedge\boldsymbol{E}) = \nabla\times\boldsymbol{E} =: \operatorname{curl}\boldsymbol{E}$.

The derivative of an antisymmetric tensor field $\boldsymbol{H}$ is $\nabla\boldsymbol{H}$, its divergence is the vector field $\nabla\cdot\boldsymbol{H}$. According to the formula (ii) in Subsection 18.3.3, $\nabla\cdot\boldsymbol{H} = -\nabla\times\boldsymbol{j}(\boldsymbol{H}) =: -\operatorname{curl}\boldsymbol{j}(\boldsymbol{H})$.

The exterior derivative of an antisymmetric cotensor field $\boldsymbol{B}$ is the antisymmetric 3-cotensor field $\nabla\wedge\boldsymbol{B}$. According to the formula (vi) in Subsection 18.3.3, this is the Levi–Civita tensor multiplied by $-\nabla\cdot\boldsymbol{j}(\boldsymbol{B}) =: -\operatorname{div}\boldsymbol{j}(\boldsymbol{B})$.

18.7 Submanifolds

18.7.1 Curves

Let V be an affine space with dimension greater than 1.

A subset C of V is a **curve** or **line** if there is a function $p : \mathbb{R} \to \mathrm{V}$, a **parametrization** of C, such that:

(1) Its domain is an open interval, its range is C.

(2) It is continuously differentiable and its derivative $\dot{p}$ is nowhere zero.

(3) It is injective and its inverse is continuous.

Let x be an element of C. The vectors parallel to $\dot{p}(p^{-1}(x))$, i. e., the real multiples of $\dot{p}(p^{-1}(x))$ are called the **tangent vectors** of C at x; their set is the **tangent space** of C over x.

Though the tangent spaces are defined by a parametrization, they are independent of the parametrization: it can be shown that if p as well as q are parametrizations of C, then $p^{-1}\circ q : \mathbb{R} \to \mathbb{R}$ is continuously differentiable and $\dot{p}(p^{-1}(x))$ and $\dot{q}(q^{-1}(x))$ are parallel for all $x \in \mathrm{C}$.

p and q are **equally oriented** parametrizations if $\dot{p}(p^{-1}(x))$ and $\dot{q}(q^{-1}(x))$ are positive multiples of each other for all $x \in \mathrm{C}$. An equivalence class of equally oriented parametrizations is called an **orientation** of C and C is **oriented** if an orientation is given.

A special curve is a straight line, i. e., a one-dimensional affine subspace in V: $\mathrm{C} = x + \mathbf{A}$ where x an arbitrary element of C and $\mathbf{A}$ is one-dimensional linear subspace in $\mathbf{V}$. Taking an arbitrary element $\boldsymbol{c}$ of $\mathbf{A}$, the function $\alpha \mapsto x + \alpha\boldsymbol{c}$ is a parametrization of C. The tangent space of the straight line is $\mathbf{A}$ at every point. An orientation of C is given by an orientation of $\mathbf{A}$.

Lastly, we mention that parametrizations can also be given by one-dimensional oriented affine spaces, as it often occurs in practice,

18.7.2 Higher-dimensional submanifolds

Let V be an affine space with dimension N greater than 1 and let M be a natural number $1 \le M \le N$.

A subset H of V is an M-**dimensional simple submanifold** if there is a function $p : \mathbb{R}^M \to \mathrm{V}$, a **parametrization** of H, such that:
(1) The domain of p is open and connected, the range of p is H.
(2) p is continuously differentiable and $\mathcal{D}p[\xi]$ is injective for all ξ in the domain of p.
(3) p is injective and p^{-1} is continuous.

According to this definition, the curves are one-dimensional simple submanifolds; a simple submanifold of dimension $N-1$ is called a **hypersurface**.

A subset H of V is an M-**dimensional submanifold** if every $x \in \mathrm{H}$ has a neighborhood $G(x)$ in V in such a way that $G(x) \cap \mathrm{H}$ is an M-dimensional simple submanifold; a parametrization of $G(x) \cap \mathrm{H}$ is called a **local parametrization** of H.

The inverse of a local parametrization is called a **local coordination** of H.

Let x be in the range of a local parametrization p. Recall that $\mathcal{D}p[p^{-1}(x)]$ is a linear injection $\mathbb{R}^M \to \mathbf{V}$; the elements in its range are called the **tangent vectors** of H at x; their set is the **tangent space** of H over x:

$$\mathrm{T}_x(\mathrm{H}) := \mathrm{Ran}(\mathcal{D}p[p^{-1}(x)]).$$

Though the tangent vectors are defined by a parametrization, they are independent of the parametrization: it can be shown that if p as well as q are local parametrizations of C and their ranges are not disjoint, then $p^{-1} \circ q : \mathbb{R}^M \to \mathbb{R}^M$ is continuously differentiable and

$$\mathrm{Ran}(\mathcal{D}p[p^{-1}(x)]) = \mathrm{Ran}(\mathcal{D}q[q^{-1}(x)])$$

for all $x \in \mathrm{Ran}(p) \cap \mathrm{Ran}(q)$. The tangent spaces of the submanifold are M-dimensional linear subspaces of **V**.

Let U be an affine space, $\dim \mathrm{U} = N - M$ and let $S : \mathrm{V} \to \mathrm{U}$ be a continuously differentiable function such that $\mathcal{D}S[x]$ is surjective for all x in the domain of S. If $u \in \mathrm{Ran}(S)$, then

$$\mathrm{H} := \{x \in \mathrm{V} \mid S(x) = u\}$$

(a level of S) is an M-dimensional submanifold in V and

$$\mathrm{T}_x(\mathrm{H}) = \mathrm{Ker}\, \mathcal{D}S[x].$$

Even, it can be shown that every point x of a submanifold H has a neighborhood $G(x)$ in such a way that $G(x) \cap \mathrm{H}$ is a level of some function S.

A Solutions of exercises in common sense relativity and spacetime

Exercises 3.7

1. Equal absolute velocities are parallel.
   Let the absolute velocities $\boldsymbol{u}_1$ and $\boldsymbol{u}_2$ be parallel. Then one of them is a nonzero multiple of the other, say $\boldsymbol{u}_2 = \alpha\boldsymbol{u}_1$. α must be positive, because on the contrary, a future-like vector would be equal to a past-like. Since the time progress is positive homogeneous, $1 = \boldsymbol{P}(\boldsymbol{u}_2) = \boldsymbol{P}(\alpha\boldsymbol{u}_1) = \alpha$, i. e., $\boldsymbol{u}_2 = \boldsymbol{u}_1$.
2. Let $\boldsymbol{u}_1 + \mathbb{R}\boldsymbol{u}$ be parallel to $\boldsymbol{u}_2 + \mathbb{R}\boldsymbol{u}$. Then there is a real number α_1 such that $\boldsymbol{u}_2 + \mathbb{R}\boldsymbol{u} = \alpha_1(\boldsymbol{u}_1 + \mathbb{R}\boldsymbol{u}) = \alpha_1\boldsymbol{u}_1 + \mathbb{R}\boldsymbol{u}$, which means that there is a real number α such that $\boldsymbol{u}_2 - \alpha_1\boldsymbol{u}_1 = \alpha\boldsymbol{u}$: the three absolute velocities are coplanar.
   Let $\boldsymbol{u}_1$, $\boldsymbol{u}_2$, and $\boldsymbol{u}$ be coplanar, i. e., $\lambda_1\boldsymbol{u}_1 + \lambda_2\boldsymbol{u}_2 + \lambda\boldsymbol{u} = 0$ for some nonzero real numbers λ_1, λ_2. Dividing by λ_2 and taking $\alpha_1 := -\lambda_1/\lambda_2$. $\alpha := -\lambda/\lambda_2$, we get $\boldsymbol{u}_2 = \alpha_1\boldsymbol{u}_1 + \alpha\boldsymbol{u}$, which results in $\boldsymbol{u}_2 + \mathbb{R}\boldsymbol{u} = \alpha_1(\boldsymbol{u}_1 + \mathbb{R}\boldsymbol{u})$: $\boldsymbol{u}_1 + \mathbb{R}\boldsymbol{u}$ and $\boldsymbol{u}_2 + \mathbb{R}\boldsymbol{u}$ are parallel.
3. Figures A.1 and A.2 show possible solutions.

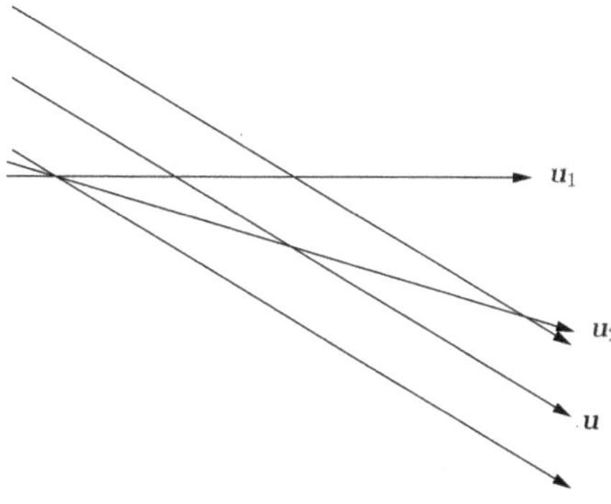

Figure A.1: According to $\boldsymbol{u}$, $\boldsymbol{u}_1$ is faster than $\boldsymbol{u}_2$.

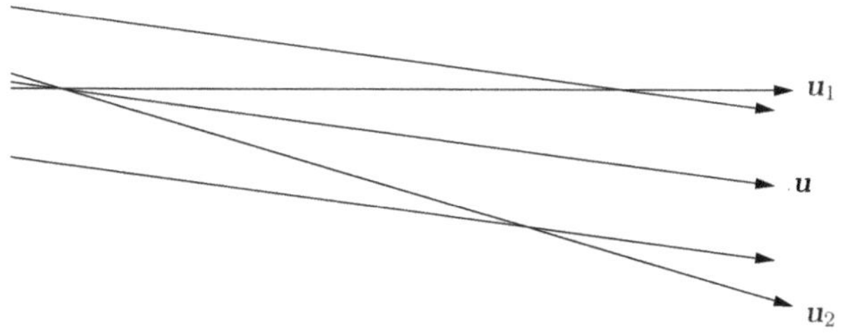

Figure A.2: $\boldsymbol{u}$ cannot decide which of $\boldsymbol{u}_1$ and $\boldsymbol{u}_2$ is faster.

https://doi.org/10.1515/9783112219553-022

4. After the meeting of the material points, at every $\boldsymbol{u}$-space point, the proper time period between the arrivals of $\boldsymbol{u}_2$ and $\boldsymbol{u}_1$ is twice the proper time period between the arrivals of $\boldsymbol{u}_3$ and $\boldsymbol{u}_2$ (see Figure A.3).

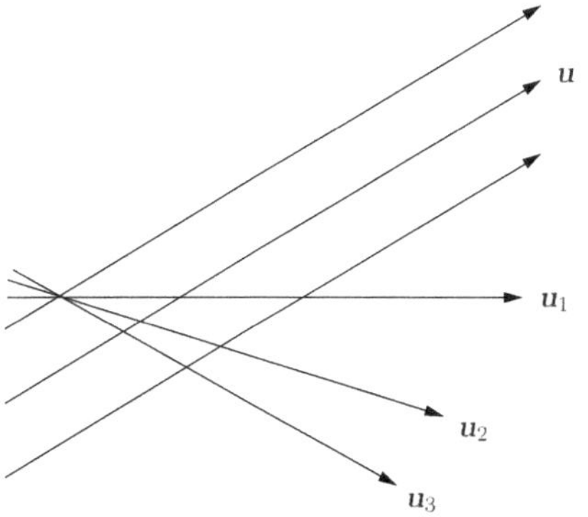

Figure A.3: According to $\boldsymbol{u}$, $\boldsymbol{u}_2$ is three times faster than $\boldsymbol{u}_1$ as $\boldsymbol{u}_3$ is faster than $\boldsymbol{u}_2$.

5. Every spacetime symmetry $\boldsymbol{L}$ for which $\boldsymbol{L} \cdot \boldsymbol{u} = \boldsymbol{u}$ is a symmetry of $\boldsymbol{u}$.
6. $\mathbf{S}_s$ is transverse to all absolute velocities, i. e., to all time-like vectors, too. Therefore, $y - x$ is time-like for all x in t and y in s. To achieve a contradiction, let us suppose that for some $x_1, x_2 \in t$ and $y_1, y_2 \in s$, $y_1 - x_1$ is future-like and $y_2 - x_2$ is past-like. Then $-(y_2 - x_2) = x_2 - y_2$ is future-like, thus, $(y_1 - x_1) + (x_2 - y_2) = (y_1 - y_2) + (x_2 - x_1)$ would also be future-like but it is in $\mathbf{S}_s$ by the definition of the synchronization.
7. (i) $\mathrm{T}^{\rightarrow}$ is he preimage of an open subset by a continuous function, so it is open. A direct proof is the following. Let $\boldsymbol{x} = (x^1, x^2, x^3, x^4)$ be in $\mathrm{T}^{\rightarrow}$: $x^1 > x^2 > 0$. Then there is a positive real number λ such that $x^1 - \lambda > x^2 - \lambda > 0$. As a consequence, $x^1 \pm \delta > x^2 \pm \delta$ for all $0 < \delta < \lambda$ and $\boldsymbol{x} \pm (\delta, \delta, \delta, \delta) = (x^1 \pm \delta, x^2 \pm \delta, x^3 \pm \delta, x^4 \pm \delta)$ is future-like, i. e., a neighborhood of $\boldsymbol{x}$ is in the given set; thus, it is open.
   Let $\boldsymbol{x} = (x^1, x^2, x^3, x^4)$ and $\boldsymbol{y} = (y^1, y^2, y^3, y^4)$ be in $\mathrm{T}^{\rightarrow}$. If α and β are positive real numbers then $\alpha x^1 > \alpha x^2$ and $\beta y^1 > \beta y^2$, so $\alpha x^1 + \beta y^1 > \alpha x^2 + \beta y^2$. Therefore, $\alpha \boldsymbol{x} + \beta \boldsymbol{y} = (\alpha x^1 + \beta y^1, \alpha x^2 + \beta y^2, \alpha x^3 + \beta y^3, \alpha x^4 + \beta y^4)$ is in $\mathrm{T}^{\rightarrow}$; thus, it is a convex cone with apex at zero.
   $\boldsymbol{P}(\alpha \boldsymbol{x}) = \alpha x^1 + \alpha x^2 = \alpha(x^1 + x^2) = \alpha \boldsymbol{P}(\boldsymbol{x})$ for all positive α.
   (ii) $\boldsymbol{u} = (u^1, u^2, u^3, u^4)$ is an absolute velocity if and only if $u^1 > u^2 > 0$ and $u^1 + u^2 = 1$.
   (iii) $(\boldsymbol{u} \cdot \boldsymbol{u})\boldsymbol{x} - \boldsymbol{u}(\boldsymbol{u} \cdot \boldsymbol{x})$ is zero if and only if $\boldsymbol{x}$ is parallel to $\boldsymbol{u}$. The scalar product is positive definite, therefore the given form is bilinear, semidefinite and having $\mathbb{T}\boldsymbol{u}$ as its kernel; lastly, $\boldsymbol{b}_{\boldsymbol{u}}$ depends continuously on $\boldsymbol{u}$.
   (iv) $\boldsymbol{b}_{\boldsymbol{u}'}(\boldsymbol{u}, \boldsymbol{u}) = \boldsymbol{b}_{\boldsymbol{u}}(\boldsymbol{u}', \boldsymbol{u}')$ if and only if $(\boldsymbol{u}' \cdot \boldsymbol{u}')^2(\boldsymbol{u} \cdot \boldsymbol{u}) - (\boldsymbol{u} \cdot \boldsymbol{u}')^2(\boldsymbol{u}' \cdot \boldsymbol{u}') = (\boldsymbol{u} \cdot \boldsymbol{u})^2(\boldsymbol{u}' \cdot \boldsymbol{u}') - (\boldsymbol{u} \cdot \boldsymbol{u}')^2(\boldsymbol{u} \cdot \boldsymbol{u})$. This does not hold for $\boldsymbol{u} = (2/3, 1/3, 0, 0)$ and $\boldsymbol{u}' = (3/4, 1/4, 0, 0)$, so the model is not fair (see Chapter 15).
   (v) $(r^1, r^2, r^3, r^4) : \mathbb{R} \to \mathbb{R}^4$ is a world line function if and only if it is continuously differentiable and $\dot{r}^1(s) > \dot{r}^2(s) > 0$, $\dot{r}^1(s) + \dot{r}^2(s) = 1$ for all proper time values s.

Then $\dot{r}_1 = 1 - \dot{r}_2 > \dot{r}_2$, so if r_2 is a function such that $\dot{r}_2 < 1/2$; then $r_1(s) := s - r_2(s) + \text{const}$; the other components r_3 and r_4 can be arbitrary, Simple examples are $r_2(s) = (1/2)\log(s+1)$ and $r_2(s) = -(1/2)\exp(-s)$ for $s > 0$.

(vi) The simplest three-dimensional linear subspace transverse to all time-like vectors is $\{\boldsymbol{x} \mid x^1 = 0\}$.

Exercises 5.1.7

1. Let the bases $(\boldsymbol{e}_0, \boldsymbol{e}_1, \boldsymbol{e}_2, \boldsymbol{e}_3)$ and $(\boldsymbol{e}'_0, \boldsymbol{e}'_1, \boldsymbol{e}'_2, \boldsymbol{e}'_3)$ of $\mathbf{M}$ be equally oriented where $\boldsymbol{e}_0$ and $\boldsymbol{e}'_0$ are future-like and the other vectors are in $\mathbf{S}$. Applying $\boldsymbol{\tau}$ to $\boldsymbol{e}'_i = \sum_{ik} A_{ki}\boldsymbol{e}_k$, we get that $A_{00} > 0$ and $A_{k0} = 0$ for $k = 1, 2, 3$. Then

$$0 < \det\begin{pmatrix} A_{00} & A_{01} & A_{01} & A_{03} \\ 0 & A_{11} & A_{12} & A_{13} \\ 0 & A_{21} & A_{22} & A_{23} \\ 0 & A_{31} & A_{32} & A_{33} \end{pmatrix} = A_{00}\det\begin{pmatrix} A_{11} & A_{12} & A_{13} \\ A_{21} & A_{22} & A_{23} \\ A_{31} & A_{32} & A_{33} \end{pmatrix};$$

thus, the determinant of the 3×3 matrix is positive.

2. Transitivity of boosts: using $\boldsymbol{\tau} \cdot (\boldsymbol{u}' - \boldsymbol{u}) = 0m$, we have

$$\begin{aligned} \boldsymbol{B}_{u''u'} \cdot \boldsymbol{B}_{u'u} &= (\mathbf{1} + (\boldsymbol{u}'' - \boldsymbol{u}') \otimes \boldsymbol{\tau}) \cdot (\mathbf{1} + (\boldsymbol{u}' - \boldsymbol{u}) \otimes \boldsymbol{\tau}) \\ &= \mathbf{1} + (\boldsymbol{u}'' - \boldsymbol{u}') \otimes \boldsymbol{\tau} + (\boldsymbol{u}' - \boldsymbol{u}) \otimes \boldsymbol{\tau} + (\boldsymbol{u}'' - \boldsymbol{u}')(\boldsymbol{\tau} \cdot (\boldsymbol{u}' - \boldsymbol{u})) \otimes \boldsymbol{\tau} \\ &= \mathbf{1} + ((\boldsymbol{u}'' - \boldsymbol{u}') + (\boldsymbol{u}' - \boldsymbol{u})) \otimes \boldsymbol{\tau} = \mathbf{1} + (\boldsymbol{u}'' - \boldsymbol{u}) \otimes \boldsymbol{\tau} = \boldsymbol{B}_{u''u}. \end{aligned}$$

3. (i) $\mathrm{T} = \{\boldsymbol{x} \mid x^1 + x^2 > 0\}$.
   (ii) $\mathrm{V}(1) = \{\boldsymbol{u} \mid u^1 + u^2 = 1\} = \{(1 - u^2, u^2, u^3, u^4) \mid (u^2, u^3, u^4) \in \mathbb{R}^3\}$.
   (iii) $\mathbf{S} = \{(q^1, q^2, q^3, q^4) \mid q^1 + q^2 = 0\} = \{(-q^2, q^2, q^3, q^4) \mid (q^2, q^3, q^4) \in \mathbb{R}^3\}$.
   (iv) $\boldsymbol{\tau} = (1, 1, 0, 0)$, so $e\boldsymbol{\tau} = \{(e, e, 0, 0) \mid e \in \mathbb{R}\}$ is the set of absolute time-like covectors.
   (v) Let us write the vectors as columns and the covectors as rows. Then with the usual rules of matrix multiplication

$$\boldsymbol{u} \otimes \boldsymbol{\tau} = \begin{pmatrix} 1 - u_2 & 1 - u_2 & 0 & 0 \\ u^2 & u^2 & 0 & 0 \\ u^3 & u^3 & 0 & 0 \\ u^4 & u^4 & 0 & 0 \end{pmatrix};$$

and the same holds for $\boldsymbol{u}'$, so

$$\boldsymbol{B}_{u'u} = \mathbf{1} + (\boldsymbol{u}' - \boldsymbol{u}) \otimes \boldsymbol{\tau} = \begin{pmatrix} 1 & 0 & 0 & 0 \\ 0 & 1 & 0 & 0 \\ 0 & 0 & 1 & 0 \\ 0 & 0 & 0 & 1 \end{pmatrix} + \begin{pmatrix} u^2 - (u')^2 & u^2 - (u')^2 & 0 & 0 \\ (u')^2 - u^2 & (u')^2 - u^2 & 0 & 0 \\ (u')^3 - u^3 & (u')^3 - u^3 & 0 & 0 \\ (u')^4 - u^4 & (u')^4 - u^4 & 0 & 0 \end{pmatrix}.$$

(vi) The absolute time point corresponding to the world point $\{\xi^1, \xi^2, \xi^3, \xi^4\}$ is

$$\{(\xi^1, \xi^2, \xi^3, \xi^4)\} + \mathbf{S} = \{(\xi^1 - q^2, \xi^2 + q^2, \xi^3 + q^3, \xi^4 + q^4) \mid q^1, q^2, q^3 \in \mathbb{R}\}.$$

The simplest absolute Euclidean form is $\boldsymbol{h}(\boldsymbol{q}, \boldsymbol{p}) := q^2p^2 + q^3p^3 + q^4p^4$.

4. (i) The future-like vectors are matrices for which $x^4 > 0$, so the absolute velocities are of the form $\begin{pmatrix} u^1 & u^2 \\ u^3 & 1 \end{pmatrix}$.
   (ii) $\tau = \begin{pmatrix} 0 & 0 \\ 0 & 1 \end{pmatrix}$ acting by matrix multiplication, so the absolute time-like covectors are of the form $e\tau = \begin{pmatrix} 0 & 0 \\ 0 & e \end{pmatrix}$.
   (iii) The boost from $\boldsymbol{u}$ to $\boldsymbol{u}'$ is the multiplication by the matrix

$$\begin{pmatrix} 1 & 0 \\ 0 & 1 \end{pmatrix} + \begin{pmatrix} (u')^1 - u^1 & (u')^2 - u^2 \\ (u')^3 - u^3 & 0 \end{pmatrix} \begin{pmatrix} 0 & 0 \\ 0 & 1 \end{pmatrix}.$$

   (iv) The absolute time point corresponding to the world point $\begin{pmatrix} \xi^1 & \xi^2 \\ \xi^3 & \xi^4 \end{pmatrix}$ is

$$\left\{ \begin{pmatrix} \xi^1 + q^1 & \xi^2 + q^2 \\ \xi^3 + q^3 & \xi^4 \end{pmatrix} \,\middle|\, q^1, q^2, q^3 \in \mathbb{R} \right\}.$$

   (v) The inertial world line with absolute velocity $\boldsymbol{u} = \begin{pmatrix} u^1 & u^2 \\ u^3 & 1 \end{pmatrix}$ and passing through the world point $x = \begin{pmatrix} \xi^1 & \xi^2 \\ \xi^3 & \xi^4 \end{pmatrix}$ is $\left\{ \begin{pmatrix} \xi^1 & \xi^2 \\ \xi^3 & \xi^4 \end{pmatrix} + \alpha \begin{pmatrix} u^1 & u^2 \\ u^3 & 1 \end{pmatrix} \mid \alpha \in \mathbb{R} \right\}$.

Exercise in Section 5.2

The boost in question is

$$\begin{pmatrix} 1 & 0 & 0 & 0 \\ (v')^1 - v^1 & 1 & 0 & 0 \\ (v')^2 - v^2 & 0 & 1 & 0 \\ (v')^5 - v^3 & 0 & 0 & 1 \end{pmatrix}.$$

Exercises 5.3.3

1. Let $\boldsymbol{F}$ and $\boldsymbol{Z}$ in the isomorphism $(L, \boldsymbol{F}, \boldsymbol{Z})$ be the identities of $\mathbb{R}$ and let $L(\xi^1, \xi^2, \xi^3, \xi^4) := (\xi^1 + \xi^2, \xi^2, \xi^2, \xi^4)$ be the affine bijection, so the underlying linear bijection is formally the same: $\boldsymbol{L} \cdot (x^1, x^2, x^3, x^4) := (x^1 + x^2, x^2, x^3, x^4)$.
2. Let $\boldsymbol{F}$ and $\boldsymbol{Z}$ be the identities of $\mathbb{R}$ and let $L\begin{pmatrix} \xi^1 & \xi^2 \\ \xi^3 & \xi^4 \end{pmatrix} := (\xi^4, \xi^1, \xi^2, \xi^3)$.

Exercises 5.4.3

1. A symmetry of the observer is a proper Noether transformation L over the Galilei transformation $\boldsymbol{L}$ such that $\boldsymbol{U}(L(x)) = \boldsymbol{L}\cdot\boldsymbol{U}(x)$.
   $L(x+\boldsymbol{q}) = L(x) + \boldsymbol{L}\cdot\boldsymbol{q}$, so $\boldsymbol{U}(L(x)+\boldsymbol{L}\cdot\boldsymbol{q}) = \boldsymbol{L}\cdot\boldsymbol{U}(x+\boldsymbol{q})$ and
   $$\boldsymbol{U}\big(L(x)+\boldsymbol{L}\cdot\boldsymbol{q}\big) - \boldsymbol{U}\big(L(x)\big) = \boldsymbol{L}\cdot\boldsymbol{U}(x+\boldsymbol{q}) - \boldsymbol{L}\cdot\boldsymbol{U}(x) = \boldsymbol{L}\cdot\boldsymbol{\Omega}\cdot\boldsymbol{q}.$$
   On the other hand, by the very definition of the observer,
   $$\boldsymbol{U}\big(L(x)+\boldsymbol{L}\cdot\boldsymbol{q}\big) - \boldsymbol{U}\big(L(x)\big) = \boldsymbol{\Omega}\cdot(\boldsymbol{L}\cdot\boldsymbol{q}),$$
   thus, $\boldsymbol{L}\cdot\cdot\boldsymbol{\Omega} = \boldsymbol{\Omega}\cdot(\boldsymbol{L}\cdot\boldsymbol{q})$ for all space-like $\boldsymbol{q}$, which yields
   $$\boldsymbol{R}^{-1}\cdot\boldsymbol{\Omega}\cdot\boldsymbol{R} = \boldsymbol{\Omega} \tag{A.1}$$
   where $\boldsymbol{R}$ is the restriction of $\boldsymbol{L}$ to the absolute space-like vectors (it is a rotation). We can reformulate our result in terms of ω: since $\boldsymbol{R}$ is a rotation,
   $$\boldsymbol{U}\big(L(x)+\boldsymbol{L}\cdot\boldsymbol{q}\big) - \boldsymbol{U}\big(L(x)\big) = -\boldsymbol{R}\cdot(\boldsymbol{\omega}\times\boldsymbol{q}) = -(\boldsymbol{R}\cdot\boldsymbol{\omega})\times(\boldsymbol{R}\cdot\boldsymbol{q})$$
   and
   $$\boldsymbol{U}\big(L(x)+\boldsymbol{L}\cdot\boldsymbol{q}\big) - \boldsymbol{U}\big(L(x)\big) = -\boldsymbol{\omega}\times(\boldsymbol{R}\cdot\boldsymbol{q});$$
   thus,
   $$\boldsymbol{R}\cdot\boldsymbol{\omega} = \boldsymbol{\omega}. \tag{A.2}$$
   The symmetries of the uniformly rotating observers are the proper Noether transformation L over the Galilei transformation $\boldsymbol{L}$ such that (A.1) or (A.2) holds where $\boldsymbol{R} := \boldsymbol{L}|_S$ is a rotation around the kernel of Ω spanned by ω.
2. $\boldsymbol{\tau}\cdot(\boldsymbol{T_u}\cdot\boldsymbol{x}) = \boldsymbol{\tau}\cdot(\boldsymbol{x} - 2(\boldsymbol{\tau}\cdot\boldsymbol{x})\boldsymbol{u}) = \boldsymbol{\tau}\cdot\boldsymbol{x} - 2\boldsymbol{\tau}\cdot\boldsymbol{x} = -\boldsymbol{\tau}\cdot\boldsymbol{x}$; thus, $\boldsymbol{T_u}[\mathrm{T}] = -\mathrm{T}$.
   A basis $(\boldsymbol{e}_0, \boldsymbol{e}_1, \boldsymbol{e}_2, \boldsymbol{e}_3)$ where $\boldsymbol{e}_0$ is future-like and the other vectors are absolute space-like is mapped by $\boldsymbol{T_u}$ into the oppositely oriented $(-\boldsymbol{e}_0, \boldsymbol{e}_1, \boldsymbol{e}_2, \boldsymbol{e}_3)$ basis.
3. $\boldsymbol{T_u}\cdot\boldsymbol{u}' = \boldsymbol{u}' - 2\boldsymbol{u} = -(\boldsymbol{u} - (\boldsymbol{u}' - \boldsymbol{u})) = -(\boldsymbol{u} - \boldsymbol{v}_{u'u})$ where $\boldsymbol{v}_{u'u}$ is the relative velocity of $\boldsymbol{u}'$ with respect to $\boldsymbol{u}$.
4. $\boldsymbol{P_u}$ maps the basis $(\boldsymbol{e}_0, \boldsymbol{e}_1, \boldsymbol{e}_2, \boldsymbol{e}_3)$ above into $(\boldsymbol{e}_0, -\boldsymbol{e}_1, -\boldsymbol{e}_2, -\boldsymbol{e}_3)$.
5. $\boldsymbol{P}_{u,o}(x) := o + \boldsymbol{P_u}\cdot(x-o) = o - (x-o) + 2\boldsymbol{\tau}\cdot(x-o)\boldsymbol{u} = x + 2\boldsymbol{\tau}\cdot(x-o)\boldsymbol{u}$.
6. The "little group" corresponding to $(1, v, 0, 0)$ consists of Galilei transformations (5.21) with +1 and $v^1 + R^1{}_1 v = v$, $v^2 + R^2{}_1 v = 0$, $v^3 + R^3{}_1 v = 0$.

Exercise in Section 5.6

The vectorial transformation rule is

$$\left(\begin{pmatrix} t \\ q^1 \\ q^2 \\ q^3 \end{pmatrix}\right) \mapsto \left(\begin{pmatrix} t \\ q^1 - t((v')^1 - v^1) \\ q^2 - t((v')^2 - v^2) \\ q^3 - t((v')^3 - v^3) \end{pmatrix}\right).$$

The covectorial transformation rule is

$$(e\,(p_1\,p_2\,p_3)) \mapsto (e + p_1((v')^1 - v^1) + p_2((v')^2 - v^2) + p_3((v')^3 - v^3), (p_1, p_2, p_3)).$$

Exercises 5.7.3

1. According to the splitting formula (5.47), the $\boldsymbol{u}$-split components of the $\boldsymbol{u}'$-time reversal are

$$\begin{aligned} \boldsymbol{\tau} \cdot (\mathbf{1} - 2\boldsymbol{u}' \otimes \boldsymbol{\tau}) \cdot \boldsymbol{u} &= 1, \quad \boldsymbol{\tau} \cdot (\mathbf{1} - 2\boldsymbol{u}' \otimes \boldsymbol{\tau}) \cdot \boldsymbol{i} = 0, \\ \boldsymbol{\pi}_u \cdot (\mathbf{1} - 2\boldsymbol{u}' \otimes \boldsymbol{\tau}) \cdot \boldsymbol{u} &= -2\boldsymbol{v}_{u'u}, \quad \boldsymbol{\pi}_u \cdot (\mathbf{1} - 2\boldsymbol{u}' \otimes \boldsymbol{\tau}) \cdot \boldsymbol{i} = \mathbf{1}_S. \end{aligned}$$

2. Equality (5.42) $\boldsymbol{D} = \boldsymbol{\tau} \cdot \boldsymbol{G}$ and $\boldsymbol{H}_u = \boldsymbol{G} - \boldsymbol{u} \wedge (\boldsymbol{\tau} \cdot \boldsymbol{G})$; thus, $\boldsymbol{H}_u + \boldsymbol{u} \wedge \boldsymbol{D} = \boldsymbol{G}$.
   Equality (5.45) can be proved by the splitting $\boldsymbol{E}_u = \boldsymbol{i}^* \cdot \boldsymbol{F} \cdot \boldsymbol{u}$ and $\boldsymbol{B} = \boldsymbol{i}^* \cdot \boldsymbol{F} \cdot \boldsymbol{i}$:

$$\boldsymbol{i}^* \cdot (\boldsymbol{\pi}_u^* \cdot \boldsymbol{B} \cdot \boldsymbol{\pi}_u + \boldsymbol{\pi}_u^* \cdot \boldsymbol{E}_u \otimes \boldsymbol{\tau} - \boldsymbol{\tau}^* \otimes (\boldsymbol{E}_u \cdot \boldsymbol{\pi}_u)) \cdot \boldsymbol{u} = 0 + \boldsymbol{E}_u - 0$$

   and

$$\boldsymbol{i}^* \cdot (\boldsymbol{\pi}_u^* \cdot \boldsymbol{B} \cdot \boldsymbol{\pi}_u + \boldsymbol{\pi}_u^* \cdot \boldsymbol{E}_u \otimes \boldsymbol{\tau} - \boldsymbol{\tau}^* \otimes (\boldsymbol{E}_u \cdot \boldsymbol{\pi}_u)) \cdot \boldsymbol{i} = \boldsymbol{B} + 0 - 0.$$

Exercises 5.8.3

3

1. $\tau(x) - t_o = \tau(x) - \tau(o) = \boldsymbol{\tau} \cdot (x - o), \pi_u(x) - q_o = \pi_u(x) - \pi_u(o) = \boldsymbol{\pi}_u \cdot (x - o).$
2.

$$\begin{aligned} \xi_{u,o}(x) - \xi_{u,o}(y) &= (\boldsymbol{\tau} \cdot (x - o), \boldsymbol{\pi}_u \cdot (x - o)) - (\boldsymbol{\tau} \cdot (y - o), \boldsymbol{\pi}_u \cdot (y - o)) \\ &= (\boldsymbol{\tau} \cdot (x - o) - \boldsymbol{\tau} \cdot (y - o), \boldsymbol{\pi}_u \cdot (x - o) - \boldsymbol{\pi}_u \cdot (y - o)) \\ &= (\boldsymbol{\tau} \cdot (x - y), \boldsymbol{\pi}_u \cdot (x - y)) = \boldsymbol{\xi}_u \cdot (x - y). \end{aligned}$$

3. The $\boldsymbol{u}$-split form of the time reversal in question is $\xi_{\boldsymbol{u},o} \cdot T_{\boldsymbol{u}',o'}\xi_{\boldsymbol{u},o}^{-1}$.
First step:

$$T_{\boldsymbol{u}',o'}\xi_{\boldsymbol{u},o}^{-1}(t,\boldsymbol{q}) = T_{\boldsymbol{u}',o'}(o + t\boldsymbol{u} + \boldsymbol{q}) = o' + \boldsymbol{T}_{\boldsymbol{u}'} \cdot (o - o') + \boldsymbol{T}_{\boldsymbol{u}'} \cdot (t\boldsymbol{u} + \boldsymbol{q}).$$

Second step:

$$\boldsymbol{T}_{\boldsymbol{u}'} \cdot (o - o') = o - o' - 2\boldsymbol{u}'\boldsymbol{\tau} \cdot (o - o') = o - o' - 2\boldsymbol{u}'\boldsymbol{t}_0.$$

Third step:

$$T_{\boldsymbol{u}'} \cdot (t\boldsymbol{u} + \boldsymbol{q}) = t\boldsymbol{u} + \boldsymbol{q} - 2\boldsymbol{u}'\boldsymbol{t}.$$

Summing up,

$$T_{\boldsymbol{u}',o'}\xi_{\boldsymbol{u},o}^{-1}(\boldsymbol{t},\boldsymbol{q}) = o - 2\boldsymbol{u}'(\boldsymbol{t} - \boldsymbol{t}_0) + \boldsymbol{u}\boldsymbol{t} + \boldsymbol{q}.$$

Fourth step:

$$\begin{aligned}\xi_{\boldsymbol{u},o}(o - 2\boldsymbol{u}'(\boldsymbol{t} - \boldsymbol{t}_0) + \boldsymbol{u}\boldsymbol{t} + \boldsymbol{q}) &= -\boldsymbol{\xi}_{\boldsymbol{u}} \cdot (2\boldsymbol{u}'(\boldsymbol{t} - \boldsymbol{t}_0) + \boldsymbol{u}\boldsymbol{t} + \boldsymbol{q}) \\ &= -\begin{pmatrix}\boldsymbol{\tau} \\ \boldsymbol{\pi}_{\boldsymbol{u}}\end{pmatrix} \cdot (2\boldsymbol{u}'(\boldsymbol{t} - \boldsymbol{t}_0) + \boldsymbol{u}\boldsymbol{t} + \boldsymbol{q}) = \begin{pmatrix}-\boldsymbol{t} + 2\boldsymbol{t}_0 \\ \boldsymbol{q} - 2\boldsymbol{v}_{\boldsymbol{u}'\boldsymbol{u}}(\boldsymbol{t} - \boldsymbol{t}_0)\end{pmatrix}.\end{aligned}$$

Exercises in Section 5.10

1. If the coordinate systems are equivalent, then there is a proper Noether transformation L over the Galilei transformation $\boldsymbol{L}$ in such a way that $L(o) = o'$ and
   - $\boldsymbol{L} \cdot \boldsymbol{e}_0 = \boldsymbol{e}_0'$ from which $\mathrm{s} = \boldsymbol{\tau} \cdot \boldsymbol{e}_0 = \boldsymbol{\tau} \cdot \boldsymbol{L} \cdot \boldsymbol{e}_0 = \boldsymbol{\tau} \cdot \boldsymbol{e}_0' = \mathrm{s}'$.
   - There is an even permutation Π of $\{1,2,3\}$ such that $\boldsymbol{L} \cdot \boldsymbol{e}_k = \boldsymbol{e}_{\Pi(k)}'$ from which $\mathrm{m}^2 = \boldsymbol{e}_i \cdot \boldsymbol{e}_i = (\boldsymbol{L} \cdot \boldsymbol{e}_i) \cdot (\boldsymbol{L} \cdot \boldsymbol{e}_i) = \boldsymbol{e}_i' \cdot \boldsymbol{e}_i' = (\mathrm{m}')^2$ for $i = 1,2,3$.

   If $\mathrm{s} = \mathrm{s}'$ and $\mathrm{m} = \mathrm{m}'$ then $\boldsymbol{L} \cdot \boldsymbol{e}_0 := \boldsymbol{e}_0', \boldsymbol{L} \cdot \boldsymbol{e}_i := \boldsymbol{e}_i'$ $(i = 1,2,3)$ defines a proper Galilei transformation and $L(o + \boldsymbol{q}) := o' + \boldsymbol{L} \cdot \boldsymbol{q}$ $(\boldsymbol{q} \in \mathbf{S})$ is the Noether transformation establishing the equivalence.
2. At least one of the basis vectors must be time-like because there are no four linearly independent space-like vectors. The general case will be reduced to the case with three space-like basis vectors and a time-like one as follows. Let $\boldsymbol{n}_4$ be time-like; then

$$\hat{\boldsymbol{n}}_k := \boldsymbol{n}_k - \frac{\boldsymbol{\tau} \cdot \boldsymbol{n}_k}{\boldsymbol{\tau} \cdot \boldsymbol{n}_4} \quad (k = 1,2,3) \tag{A.3}$$

are space-like and $\hat{\boldsymbol{n}}_1, \hat{\boldsymbol{n}}_2, \hat{\boldsymbol{n}}_2, \boldsymbol{n}_4$ are linearly independent because if

$$0 = \alpha^1\hat{\boldsymbol{n}}_1 + \alpha^2\hat{\boldsymbol{n}}_2 + \alpha^3\hat{\boldsymbol{n}}_3 + \alpha^4\boldsymbol{n}_4 = \alpha^1\boldsymbol{n}_1 + \alpha^2\boldsymbol{n}_2 + \alpha^3\boldsymbol{n}_3 + \left(\alpha^4 - \frac{\tau\cdot\boldsymbol{n}_1 + \tau\cdot\boldsymbol{n}_2 + \tau\cdot\boldsymbol{n}_3}{\tau\cdot\boldsymbol{n}_4}\right)\boldsymbol{n}_4$$

then $0 = \alpha^1 = \alpha^2 = \alpha^3 = \alpha^4 - \frac{\tau\cdot\boldsymbol{n}_1+\tau\cdot\boldsymbol{n}_2+\tau\cdot\boldsymbol{n}_3}{\tau\cdot\boldsymbol{n}_4}$ implying that $\alpha_4 = 0$ as well.
With the notation $\boldsymbol{u} := \frac{\boldsymbol{n}_4}{\tau\cdot\boldsymbol{n}_4}$, the coordinates of x in the basis $\hat{\boldsymbol{n}}_1, \hat{\boldsymbol{n}}_2, \hat{\boldsymbol{n}}_3, \boldsymbol{n}_4$ are

$$\frac{\tau\cdot(x-o)}{\tau\cdot\boldsymbol{n}_4}, \quad \frac{\hat{\boldsymbol{n}}_k\cdot\pi_{\boldsymbol{u}}\cdot(x-o)}{\hat{\boldsymbol{n}}_k\cdot\hat{\boldsymbol{n}}_k} \quad (k=1,2,3),$$

from which the coordinates in the original basis can be calculated by (A.3).

3. Take an observer $\boldsymbol{u}$ and a world point o. Let $(\xi^0, \xi^1, \xi^2, \xi^3)$ be the coordinates of x in the standard coordinate system $(o, \mathrm{s}, \mathrm{m}, \boldsymbol{e}_0, \boldsymbol{e}_1, \boldsymbol{e}_2, \boldsymbol{e}_3)$ where $\boldsymbol{e}_0 = \mathrm{s}\boldsymbol{u}$. Then $o + \mathbb{T}\boldsymbol{u}$ is the $\boldsymbol{u}$-space origin and the spherical coordinates of $x \notin o + \mathbb{T}\boldsymbol{u}$ can be defined as follows:
   - $t := \xi^0$.
   - $r := \sqrt{(\xi^1)^2 + (\xi^2)^2 + (\xi^3)^2}$.
   - $\vartheta := \arccos\frac{\xi^3}{r}$ $(-1 < \frac{\xi^3}{r} \le 1,\ r \ne 0)$.
   - $\varphi := \arcsin\frac{\xi^2}{\sqrt{(\xi^1)^2+(\xi^2)^2}}$ $((\xi^1)^2 + (\xi^2)^2 \ne 0)$.

   Then $x = o + t\boldsymbol{e}_0 + (r\sin\vartheta\cos\varphi)\boldsymbol{e}_1 + (r\sin\vartheta\sin\varphi)\boldsymbol{e}_2 + (r\cos\vartheta)\boldsymbol{e}_3$.

Exercises in Section 6.1

1. An affine map is smooth, so r_L is continuously differentiable and with equality (5.27) of the book, $\tau(L(r(t - \boldsymbol{t}_L))) = \tau(r(t - \boldsymbol{t}_L)) + \boldsymbol{t}_L = (t - \boldsymbol{t}_L) + \boldsymbol{t}_L = t$.
2. $T_{t_o}^{-1} = T_{t_o}$, since $T_{t_o}(t_o + (t_o - t)) = t_o + (t_o - (t_o + (t_o - t))) = t_o - (t_o - t) = t$.
   Then $r_{\boldsymbol{u},o}(t) = r(\tau(o) - (t - \tau(o)) + 2(t - \tau(o)\boldsymbol{u}$, consequently, $\tau(r_{\boldsymbol{u},o}(t)) = \tau(o) - (t - \tau(o)) + 2(t - \tau(o)) = t$.

Exercises 6.5.3

1. The elastic forces are equivalent if and only if they have the same elasticity factor and their centers “live in the same way somewhere in spacetime.” This is proved as follows: there is proper Noether transformation such that

$$k\boldsymbol{L}\cdot(x - r_c(\tau(x))) = k'(L(x) - r_{c'}(\tau(L(x))) = k'\boldsymbol{L}\cdot(x - L^{-1}r_{c'}c(\tau(x) + \boldsymbol{t}_L)).$$

   $\boldsymbol{L}$ is a bijection; thus,

$$k(x - r_c(\tau(x)) = k'(x - L^{-1}r_{c'}(\tau(x) + \boldsymbol{t}_L)) \qquad \text{(A.4)}$$

   holds as well where equality (5.24) is applied.
   The left-hand side is zero for $x = r_c(\tau(x))$; thus, the right-hand side is zero, so $r_c(t) = L^{-1}r_{c'}(t + \boldsymbol{t}_L)$ for all t, which means that

(i) $r_{c'}$ is a proper Noether transformed of r_c.
Further, taking an arbitrary instant t (a hyperplane), (A.4) holds for all x in that hyperplane: $k(x - r_c(t)) = k'(x - L^{-1}r_{c'}(t + \boldsymbol{t}_L)) = k'(x - r_c(t))$ from which
(ii) $k' = k$.

3. The $(\boldsymbol{u}, o)$-time reversal invariance of a force reads as follows:

$$\boldsymbol{f}(x - 2\boldsymbol{u}(\boldsymbol{\tau} \cdot (x - o)), 2\boldsymbol{u} - \dot{x}) = \boldsymbol{f}(x, \dot{x}).$$

(i) Suppose this is true for a given $\boldsymbol{u}$ and for all o. Since for every x, $\boldsymbol{\tau} \cdot (x - o)$ runs over $\mathbb{T}$ as o runs over M, we have

$$\boldsymbol{f}(x + \boldsymbol{t}\boldsymbol{u}, 2\boldsymbol{u} - \dot{x}) = \boldsymbol{f}(x, \dot{x}) \tag{A.5}$$

for all $\boldsymbol{t}$: the force is constant on the straight lines directed by $\boldsymbol{u}$. Moreover, $2\boldsymbol{u} - \dot{x} = \boldsymbol{u} - \boldsymbol{v}_{\dot{x}\boldsymbol{u}}$, so the force depends on $\dot{x}$ through the relative velocities with respect to $\boldsymbol{u}$.

(ii) If (A.5) holds for all $\boldsymbol{u}$, then $x + \boldsymbol{t}\boldsymbol{u}$ runs over M as $\boldsymbol{t}\boldsymbol{u}$ runs over all time-like vectors, so $\boldsymbol{f}$ does not depend on spacetime points. Further, for a given $\dot{x}$, $2\boldsymbol{u} - \dot{x} =: \boldsymbol{u}'$ runs over V(1) as $\boldsymbol{u}$ runs over all absolute velocities. As a consequence, the force does not depend on absolute velocities.

Exercise 10.1.7

(i) The given function is evidently bilinear and symmetric, and

$$\boldsymbol{e}_0 := (1, 0, 1, 0), \quad \boldsymbol{e}_1 := (1, 0, -1, 0), \quad \boldsymbol{e}_2 := (0, 1, 0, 1), \quad \boldsymbol{e}_3 := (0, 1, 0, -1)$$

is a Lorentz orthogonal basis.

(ii) The set of time-like vectors is given by $(x^2)^2 + (x^4)^2 < 2x^1x^3$; the left side is nonnegative, so $x^1x^3 > 0$ must hold. Let the arrow orientation be defined by $x^1 > 0, x^3 > 0$. This corresponds to the choice that $\boldsymbol{e}_0$ be future-like.

(iii) The set of absolute velocities is

$$\mathrm{V}(1) = \{(u^1, u^2, u^3, u^4) \mid u^1 > 0,\ u^3 > 0,\ 2u^1u^3 = 1 + (u^2)^2 + (u^4)^2\}.$$

(iv) (q^0, q^1, q^2, q^3) is Lorentz-orthogonal to (u^1, u^2, u^3, u^4) if

$$u^1q^3 + u^3q^1 = u^2q^2 + u^4q^4.$$

(v) The covector $(\hat{x}_1, \hat{x}_2, \hat{x}_3, \hat{x}_4)$ corresponding to the vector (x^1, x^2, x^3, x^4) is determined by

$$\hat{x}_1y^1 + \hat{x}_2y^2 + \hat{x}_3y^3 + \hat{x}_4y^4 = -x^1y^3 - x^3y^1 + x^2y^2 + x^4y^4,$$

which results in $\hat{x}_1 = -x^3$, $\hat{x}_2 = x^2$, $\hat{x}_3 = -x^1$ and $\hat{x}_4 = x^4$.

Exercise in Subsection 10.3.2

The simplest isomorphism is the one consisting of:
(i) The identity of $\mathbb{R}$ (the time unit is 1).
(ii) The origin $(0,0,0,0)$.
(iii) The Lorentz orthogonal basis given in Exercise 10.1.7.

Then the corresponding linear bijection $\mathbb{R}^4 \to \mathbb{R}^4$ maps the given basis to the standard basis $(1,0,0,0)$, $(0,1,0,0)$, $(0,0,1,0)$, $(0,0,0,1)$ of the arithmetic model. This linear bijection is given by the matrix

$$\frac{1}{2}\begin{pmatrix} 1 & 0 & 1 & 0 \\ 1 & 0 & -1 & 0 \\ 0 & 1 & 0 & 1 \\ 0 & 1 & 0 & -1 \end{pmatrix}.$$

Exercises 10.4.3

1. If $\boldsymbol{x}$ is a future-like vector, then $\boldsymbol{u}\cdot\boldsymbol{x} < 0$. For the $\boldsymbol{u}$-time reversal of $\boldsymbol{x}$, we have $\boldsymbol{u}\cdot(\boldsymbol{x}+2\boldsymbol{u}(\boldsymbol{u}\cdot\boldsymbol{x})) = -\boldsymbol{u}\cdot\boldsymbol{x} > 0$, so $\boldsymbol{T}_{\boldsymbol{u}}\cdot\boldsymbol{x}$ is past-like.
   A basis $(t\boldsymbol{u}, \boldsymbol{e}_1, \boldsymbol{e}_2, \boldsymbol{e}_3)$ where $\boldsymbol{t} \neq 0$ and $\boldsymbol{e}_1$, $\boldsymbol{e}_2$, $\boldsymbol{e}_3$ are in $\mathbf{S}_{\boldsymbol{u}}$ is mapped by $\boldsymbol{T}_{\boldsymbol{u}}$ into the oppositely oriented basis $(-t\boldsymbol{u}, \boldsymbol{e}_1, \boldsymbol{e}_2, \boldsymbol{e}_3)$.
2. $\boldsymbol{T}_{\boldsymbol{u}}\cdot\boldsymbol{u}' = \boldsymbol{u}' + 2\boldsymbol{u}(\boldsymbol{u}\cdot\boldsymbol{u}') = -(\boldsymbol{u}\cdot\boldsymbol{u}')(\frac{\boldsymbol{u}'}{-\boldsymbol{u}\cdot\boldsymbol{u}'} - \boldsymbol{u} - \boldsymbol{u}) = (-\boldsymbol{u}\cdot\boldsymbol{u}')(\boldsymbol{u} - \boldsymbol{v}_{\boldsymbol{u}'\boldsymbol{u}})$ where $\boldsymbol{v}_{\boldsymbol{u}'\boldsymbol{u}}$ is the standard relative velocity of $\boldsymbol{u}'$ with respect to $\boldsymbol{u}$.
3. $\boldsymbol{P}_{\boldsymbol{u}}$ maps the basis $(t\boldsymbol{u}, \boldsymbol{e}_1, \boldsymbol{e}_2, \boldsymbol{e}_3)$ above into $(t\boldsymbol{u}, -\boldsymbol{e}_1, -\boldsymbol{e}_2, -\boldsymbol{e}_3)$ basis.
4. $P_{\boldsymbol{u},o}(x) := o + \boldsymbol{P}_{\boldsymbol{u}}\cdot(x-o) = x - 2\boldsymbol{u}\cdot(x-o)\boldsymbol{u}$ $(x \in \mathrm{M})$.
5. The matrix of the $\gamma'(1,0,v',0)$-time reversal is

$$\begin{pmatrix} 1-2(\gamma')^2 & 0 & 2(\gamma')^2 v' & 0 \\ 0 & 1 & 0 & 0 \\ -2(\gamma')^2 v' & 0 & 1+2(\gamma')^2(v')^2 & 0 \\ 0 & 0 & 0 & 1 \end{pmatrix}.$$

   The matrix of the $\gamma(1,v,0,0)$-time reversal is given at the end of Subsection 10.4.1. The product of the matrices in question shows how complicated arithmetic formulas can be.
6. The matrix entries of a Lorentz group are denoted by $L^k{}_i$, $k, i = 0,1,2,3$. Then the "little group" corresponding to $\gamma(1,v,0,0)$ consists of Lorentz transformations for which

$$\begin{pmatrix} L^0{}_0 & L^0{}_1 & L^0{}_2 & L^0{}_2 \\ L^1{}_0 & L^1{}_1 & L^1{}_2 & L^1{}_3 \\ L^2{}_0 & L^2{}_1 & L^2{}_2 & L^2{}_3 \\ L^3{}_0 & L^3{}_1 & L^3{}_2 & L^3{}_3 \end{pmatrix}\begin{pmatrix} 1 \\ v \\ 0 \\ 0 \end{pmatrix} = \begin{pmatrix} 1 \\ v \\ 0 \\ 0 \end{pmatrix},$$

which yields ${L^0}_0+{L^0}_1v = 1, {L^1}_0+{L^1}_1v = v, {L^2}_0+{L^2}_1v = 0, {L^3}_0+{L^3}_1v = 0$ and the other matrix entries are arbitrary (of course in such a way that the matrix be a Lorentz transformation, i. e., preserve the Lorentz form).

Exercises 10.6.6

1.

$$\mathbf{B}_{\boldsymbol{uu}'} \cdot \mathbf{v}_{\boldsymbol{u}''\boldsymbol{u}'} = \left(1 + \frac{(\boldsymbol{u}+\boldsymbol{u}') \otimes (\boldsymbol{u}+\boldsymbol{u}')}{1+\alpha} - 2\boldsymbol{u} \otimes \boldsymbol{u}'\right) \cdot \left(\frac{\boldsymbol{u}''}{\beta} - \boldsymbol{u}'\right)$$
$$= \left(\frac{\boldsymbol{u}''}{\beta} - \boldsymbol{u}'\right) + \left(-\frac{(\boldsymbol{u}+\boldsymbol{u}')(\gamma+\beta)}{\beta(1+\alpha)} + \boldsymbol{u} + \boldsymbol{u}'\right) + 0,$$

which results in (10.38).
Then

$$\mathbf{v}_{\boldsymbol{u}''\boldsymbol{u}} = \lambda\left(\boldsymbol{u} + \frac{\boldsymbol{u}''}{\beta} - \frac{(\boldsymbol{u}+\boldsymbol{u}')(\gamma+\beta)}{\beta(1+\alpha)}\right) + \mu\left(\frac{\boldsymbol{u}'}{\alpha} - \boldsymbol{u}\right)$$

gives

$$\frac{1}{\gamma} = \frac{\lambda}{\beta}, \quad -1 = \lambda - \frac{\gamma+\beta}{\beta(1+\alpha)} - \mu, \quad -\frac{\gamma+\beta}{\beta(1+\alpha)} = \frac{\mu}{\alpha}$$

(10.39) is obtained by a straightforward calculation.

2. Use the previous notations and exclude the trivial cases when $\boldsymbol{u}' = \boldsymbol{u}$ or $\boldsymbol{u}'' = \boldsymbol{u}$ or $\boldsymbol{u}'' = \boldsymbol{u}'$.
   (⇐) If $\frac{\boldsymbol{u}''}{\gamma} - \boldsymbol{u}$ is parallel to $\frac{\boldsymbol{u}'}{\alpha} - \boldsymbol{u}$, then there is number λ such that $\frac{\boldsymbol{u}''}{\gamma} - \boldsymbol{u} = \lambda(\frac{\boldsymbol{u}'}{\alpha} - \boldsymbol{u})$. Consequently, the three absolute velocities are coplanar.
   (⇒) If the absolute velocities are coplanar, then there are numbers λ and μ such that $\boldsymbol{u}'' = \lambda\boldsymbol{u}' + \mu\boldsymbol{u}$. Then Lorentz multiplying by $\boldsymbol{u}$ we get $\gamma = \lambda\alpha + \mu$ and

$$\frac{\lambda\boldsymbol{u}' + \mu\boldsymbol{u}}{\lambda\alpha+\mu} - \boldsymbol{u} = \frac{\lambda\alpha}{\lambda\alpha+\mu}\left(\frac{\boldsymbol{u}'}{\alpha} - \boldsymbol{u}\right).$$

   We can argue similarly for the other parallelisms.

3. $-(1,1,0,0)\cdot\gamma(1,v,0,0) = \gamma(1-v), -(1,1,0,0)\cdot\gamma'(1,0,v',0) = \gamma'$. Therefore, the relative velocities are

$$\frac{(1,1,0,0)}{\gamma(1-v)} - \gamma(1,v,0,0) = \left(\frac{v}{\sqrt{1-v^2}}, \frac{1}{\sqrt{1-v^2}}, 0.0\right),$$
$$\frac{(1,1,0,0)}{\gamma'} - \gamma'(1,0,v',0)) = \frac{1}{\sqrt{1-(v')^2}}((v')^2, 1, 0, 0).$$

Exercises 10.8.3

1.

$$-\boldsymbol{u}\cdot\boldsymbol{G}\cdot(\mathbf{1}+\boldsymbol{u}\otimes\boldsymbol{u}) = -\boldsymbol{u}\cdot\boldsymbol{G}-(\boldsymbol{u}\cdot\boldsymbol{G}\cdot\boldsymbol{u})\boldsymbol{u} \quad \text{and}$$
$$(\mathbf{1}+\boldsymbol{u}\otimes\boldsymbol{u})\cdot\boldsymbol{G}\cdot(\mathbf{1}+\boldsymbol{u}\otimes\boldsymbol{u}) = \boldsymbol{G}+\boldsymbol{u}\otimes(\boldsymbol{u}\cdot\boldsymbol{G})+(\boldsymbol{G}\cdot\boldsymbol{u})\otimes\boldsymbol{u}+\boldsymbol{u}\otimes\boldsymbol{u}(\boldsymbol{u}\cdot\boldsymbol{G}\cdot\boldsymbol{u}).$$

2. If $\boldsymbol{D}_u = -\boldsymbol{G}\cdot\boldsymbol{u}$ and $\boldsymbol{H}_u = \boldsymbol{G}-\boldsymbol{u}\wedge(\boldsymbol{G}\cdot\boldsymbol{u}) = \boldsymbol{G}+\boldsymbol{u}\wedge(\boldsymbol{u}\cdot\boldsymbol{G}) = \boldsymbol{G}+\boldsymbol{u}\wedge\boldsymbol{D}_u$, then equality (10.60) holds.
   Similarly, if $\boldsymbol{E}_u = \boldsymbol{F}\cdot\boldsymbol{u}$ and $\boldsymbol{B}_u = \boldsymbol{F}+\boldsymbol{u}\wedge(\boldsymbol{F}\cdot\boldsymbol{u}) = \boldsymbol{F}+\boldsymbol{u}\wedge(\boldsymbol{E}_u)$, then equality (10.61) holds.
3.

$$-\boldsymbol{u}\cdot(\mathbf{1}+2\boldsymbol{u}'\otimes\boldsymbol{u}')\cdot\boldsymbol{u} = 1-2(\boldsymbol{u}\cdot\boldsymbol{u}')^2 = 1-\frac{2}{1-v^2} = -\frac{1+v^2}{1-v^2},$$
$$-\boldsymbol{u}\cdot(\mathbf{1}+2\boldsymbol{u}'\otimes\boldsymbol{u}')\cdot\boldsymbol{\pi}_u = -2(\boldsymbol{u}\cdot\boldsymbol{u}')\boldsymbol{\pi}_u\cdot\boldsymbol{u}' = 2(\boldsymbol{u}\cdot\boldsymbol{u}')^2\frac{\boldsymbol{\pi}_u\cdot\boldsymbol{u}'}{-\boldsymbol{u}\cdot\boldsymbol{u}'} = \frac{2\boldsymbol{v}_{u'u}}{1-v^2},$$
$$\boldsymbol{\pi}_u\cdot(\mathbf{1}+2\boldsymbol{u}'\otimes\boldsymbol{u}')\cdot\boldsymbol{u} = -\frac{2\boldsymbol{v}_{u'u}}{1-v^2},$$
$$\boldsymbol{\pi}_u\cdot(\mathbf{1}+2\boldsymbol{u}'\otimes\boldsymbol{u}')\cdot\boldsymbol{\pi}_u = (\boldsymbol{\pi}_u)+2(\boldsymbol{\pi}_u\cdot\boldsymbol{u}')\otimes(\boldsymbol{\pi}_u\cdot\boldsymbol{u}') = \mathbf{1}+(1-v^2)\boldsymbol{v}_{u'u}\otimes\boldsymbol{v}_{u'u},$$

   where $(\boldsymbol{\pi}_u)$ is considered the identity of $\mathbf{S}_u$, denoted simply by $\mathbf{1}$.

Exercises 10.9.3

1.

$$\tau_u(x)-t_o = \tau_u(x)-\tau(o) = \boldsymbol{\tau}_u\cdot(x-o) = -\boldsymbol{u}\cdot(x-o),$$
$$\pi_u(x)-q_o = \pi_u(x)-\pi_u(o) = \boldsymbol{\pi}_u\cdot(x-o).$$

2.

$$\begin{aligned}\xi_{u,o}(x)-\xi_{u,o}(y) &= (-\boldsymbol{u}\cdot(x-o),\boldsymbol{\pi}_u\cdot(x-o))-(-\boldsymbol{u}\cdot(y-o),\boldsymbol{\pi}_u\cdot(y-o))\\ &= (-\boldsymbol{u}\cdot(x-o)-\boldsymbol{u}\cdot(y-o),\boldsymbol{\pi}_u\cdot(x-o)-\boldsymbol{\pi}_u\cdot(y-o))\\ &= (-\boldsymbol{u}\cdot(x-y),\boldsymbol{\pi}_u\cdot(x-y)) = \boldsymbol{\xi}_u\cdot(x-y).\end{aligned}$$

Exercises in Section 10.11

1. The coordinate systems are equivalent if and only if there is a proper Poincaré transformation L over the Lorentz transformation $\boldsymbol{L}$ in such a way that $L(o) = o'$ and there is an even permutation Π of $(0,1,2,3)$ such that $\boldsymbol{L}\cdot\boldsymbol{e}_k = \boldsymbol{e}'_{\Pi(k)}$. Since the Lorentz transformation preserves Lorentz product, $\mathrm{s}' = \mathrm{s}$ must hold.
   Conversely, if $\mathrm{s}' = \mathrm{s}$ then the Poincaré transformation L defined by $L(o) := o'$ and $\boldsymbol{L}\cdot\boldsymbol{e}'_k := \boldsymbol{e}_k$ $(k = 0,1,2,3)$ gives the equivalence of the standard coordinate systems.
2. The basis can be arbitrary, e. g., it is possible that all the basis elements are time-like or space-like or light-like. If $x - o = \alpha^1\boldsymbol{n}_1 + \alpha^2\boldsymbol{n}_2 + \alpha^3\boldsymbol{n}_3 + \alpha^4\boldsymbol{n}_4$, the coefficients α^1, $\alpha^2, \alpha^3, \alpha^4$ are uniquely determined and can be computed from the system of linear equations $\boldsymbol{n}_k\cdot(x - o) = \alpha^k$ $(k = 1,2,3,4)$.
3. Take an observer $\boldsymbol{u}$ and a world point o. Let $(\xi^0,\xi^1,\xi^2,\xi^3)$ be the coordinates of x in the standard coordinate system $(o,\mathrm{s},\boldsymbol{e}_0,\boldsymbol{e}_1,\boldsymbol{e}_2,\boldsymbol{e}_3)$ where $\boldsymbol{e}_0 := \mathrm{s}\boldsymbol{u}$. Then $o + \mathbb{T}\boldsymbol{u}$ is the $\boldsymbol{u}$-space origin and the spherical coordinates of $x \notin o + \mathbb{T}\boldsymbol{u}$ can be defined as follows:
   - $t := \xi^0$.
   - $r := \sqrt{(\xi^1)^2 + (\xi^2)^2 + (\xi^3)^2}$.
   - $\vartheta := \arccos\frac{\xi^3}{r}$ $(-1 < \frac{\xi^3}{r} \le 1,\ r \neq 0)$.
   - $\varphi := \arcsin\frac{\xi^2}{\sqrt{(\xi^1)^2+(\xi^2)^2}}$ $((\xi^1)^2 + (\xi^2)^2 \neq 0)$.

   Then $x = o + (t\boldsymbol{e}_0) + (r\sin\vartheta\cos\varphi)\boldsymbol{e}_1 + (r\sin\vartheta\sin\varphi)\boldsymbol{e}_2 + (r\cos\vartheta)\boldsymbol{e}_3$.

Exercises in Section 11.1

1. $\dot{r}_L = \boldsymbol{L}\cdot\dot{r}$; $\boldsymbol{L}$ is arrow orientation preserving, hence it maps absolute velocities into absolute velocities.
2. $\dot{r}_{\boldsymbol{u},o}(\boldsymbol{s}) = -\boldsymbol{T}_{\boldsymbol{u}}\cdot\dot{r}(\boldsymbol{s}_0 - (\boldsymbol{s} - \boldsymbol{s}_0))$: the $\boldsymbol{u}$-time reversal of a future-like vector is past-like and the negative of a past-like vector is future-like.

Exercises 11.2.3

1. (i) According to the definition of the boost, $\boldsymbol{B}_{\dot{r}(\boldsymbol{t})\dot{r}(\boldsymbol{s})}\cdot\dot{r}(\boldsymbol{s}) = \dot{r}(\boldsymbol{t})$.
   (ii) $\boldsymbol{x}(\boldsymbol{t}) = \boldsymbol{x}(\boldsymbol{s}) + \frac{\dot{r}(\boldsymbol{t})+\dot{r}(\boldsymbol{s})}{1-\dot{r}(\boldsymbol{t})\cdot\dot{r}(\boldsymbol{s})}(\dot{r}(\boldsymbol{t}) + \dot{r}(\boldsymbol{s}))\cdot\boldsymbol{x}(\boldsymbol{s})$ and then $\boldsymbol{x}(\boldsymbol{t}) - \boldsymbol{x}(\boldsymbol{s}) + \frac{\dot{r}(\boldsymbol{t})+\dot{r}(\boldsymbol{s})}{1-\dot{r}(\boldsymbol{t})\cdot\dot{r}(\boldsymbol{s})}(\dot{r}(\boldsymbol{t}) - \dot{r}(\boldsymbol{s}))\cdot\boldsymbol{x}(\boldsymbol{s})$; dividing by $\boldsymbol{t} - \boldsymbol{s}$ and then taking the limit $\boldsymbol{t} \to \boldsymbol{s}$, we get the desired result.
   (iii) Since the Lorentz square of a vector running parallel along a world line is constant, with the notation $\alpha := |\ddot{r}|$, we have by the previous result that $\dddot{r} = \alpha^2\dot{r}$; the solution of this second-order differential equation for $\dot{r}$ with the initial conditions $\dot{r}(0) =: \boldsymbol{u}_0$, $\ddot{r}(0) =: \boldsymbol{a}_0$ is $\dot{r}(\boldsymbol{s}) = \boldsymbol{u}_0\cosh(\alpha\boldsymbol{s}) + \boldsymbol{a}_0\sinh(\alpha\boldsymbol{s})$, which gives the desired result by integration.

2. The relative velocity is $(0, -v\sin\frac{vt}{\rho}, \cos\frac{vt}{\rho}, 0)$ having the magnitude v. Then $t(s) = \frac{s}{\sqrt{1-v^2}}$ by (11.1), so according to (10.36), the absolute velocity is

$$\frac{(1, -v\sin\frac{vs}{\rho\sqrt{1-v^2}}, \cos\frac{vs}{\rho\sqrt{1-v^2}}, 0)}{\sqrt{1-v^2}}$$

from which the world line function is

$$s \mapsto \left(\frac{s}{\sqrt{1-v^2}}, \rho\cos\frac{vs}{\rho\sqrt{1-v^2}}, \rho\sin\frac{vs}{\rho\sqrt{1-v^2}}, 2\right).$$

3. The relative velocity is $(0, -\sin\frac{t}{\rho}, \cos\frac{t}{\rho}, 0)$ having the magnitude 1 as it must be. Then the absolute light direction is $(1, -\sin\frac{t}{\rho}, \cos\frac{t}{\rho}, 0)$; thus, the parametrization of the light line is $t \mapsto (t, \rho\cos\frac{t}{\rho}, \rho\sin\frac{t}{\rho}, 2)$.

Exercises 11.5.3

1. For the sake of simpler formulas, let us write $\boldsymbol{u}$ and $\boldsymbol{u}'$ instead of $\boldsymbol{u}_{\mathcal{C}}$ and $\boldsymbol{u}_{\mathcal{C}'}$, respectively.
   The forces are equivalent if there is a proper Poincaré transformation such that

$$k'[\boldsymbol{u}'(L(x)-o')\cdot(\boldsymbol{L}\dot{x})-(L(x)-o')(\boldsymbol{u}'\cdot(\boldsymbol{L}\dot{x}))] = k[(\boldsymbol{L}\cdot\boldsymbol{u})(x-o)\cdot\dot{x}-(\boldsymbol{L}\cdot(x-o))\boldsymbol{u}\cdot\dot{x}] \quad \text{(A.6)}$$

   for all x and $\dot{x}$.
   For $x = o$ and $\boldsymbol{L}\cdot\dot{x} = \boldsymbol{u}'$, we obtain $\boldsymbol{u}'((L(o)-o')\cdot\boldsymbol{u}') + (L(o)-o') = 0$. This is true if $L(o)-o' = 0$ but o' is not unique, so we can choose it such a way that $L(o)-o' \neq 0$. Then there is a $\boldsymbol{t}'$ such that $L(o)-o' = \boldsymbol{t}'\boldsymbol{u}'$.
   Similarly, for $L(x) = o'$ and $\dot{x} = \boldsymbol{u}$, we have $\boldsymbol{L}\cdot\boldsymbol{u}(L^{-1}(o)-o')\cdot\boldsymbol{u} + (o'-L(o)) = 0$; thus, there is a $\boldsymbol{t}$ such that $o'-L(o) = \boldsymbol{t}\boldsymbol{L}\cdot\boldsymbol{u}$.
   Then we conclude that $\boldsymbol{u}' = \boldsymbol{L}\cdot\boldsymbol{u}$.
   Further, put $L(x)-o' = \boldsymbol{L}\cdot(x-o)+L(o)-o' = \boldsymbol{L}\cdot(x-o)+\boldsymbol{t}'\boldsymbol{L}\cdot\boldsymbol{u}$ and $(\boldsymbol{L}\cdot\boldsymbol{u})\cdot(\boldsymbol{L}\cdot\dot{x}) = \boldsymbol{u}\cdot\dot{x}$ into (A.6) to have $k' = k$.
   Since for arbitrary $\boldsymbol{u}$ and $\boldsymbol{u}'$, there is a proper Lorentz transformation such that $\boldsymbol{u}' = \boldsymbol{L}\cdot\boldsymbol{u}$, the forces are equivalent if ad only if $k' = k$.
2. Since the inverse of the $\boldsymbol{u}$-time reversal equals the $\boldsymbol{u}$-time reversal, we find convenient to write the $(\boldsymbol{u}, o)$-time reversal invariance of a force as follows:

$$(\mathbf{1}+2\boldsymbol{u}\otimes\boldsymbol{u})\cdot\boldsymbol{f}\Big(x+2\boldsymbol{u}(\boldsymbol{u}\cdot(x-o)), -(\mathbf{1}+2\boldsymbol{u}\otimes\boldsymbol{u})\cdot\dot{x}\Big) = \boldsymbol{f}(x,\dot{x}).$$

   (i) Suppose this is true for a given $\boldsymbol{u}$ and for all o. Since for every x, $\boldsymbol{u}\cdot(x-o)$ runs over $\mathbb{T}$ as o runs over M, we have

$$(\mathbf{1} + 2\boldsymbol{u} \otimes \boldsymbol{u}) \cdot \boldsymbol{f}(x + t\boldsymbol{u}, -(\mathbf{1} + 2\boldsymbol{u} \otimes \boldsymbol{u}) \cdot \dot{x}) = \boldsymbol{f}(x, \dot{x}) \tag{A.7}$$

for all $\boldsymbol{t}$: the force is constant on the straight lines directed by $\boldsymbol{u}$. Moreover, $(\mathbf{1} + 2\boldsymbol{u} \otimes \boldsymbol{u} \cdot \dot{x}) = -(\boldsymbol{u} \cdot \dot{x})(\boldsymbol{u} - \boldsymbol{v}_{\dot{x}\boldsymbol{u}})$ and $-\boldsymbol{u} \cdot \dot{x} = \frac{1}{\sqrt{1-|\boldsymbol{v}_{\dot{x}\boldsymbol{u}}|^2}}$, so the force depends on $\dot{x}$ through the $\boldsymbol{u}$-relative velocities.

(ii) If (A.7) holds for all $\boldsymbol{u}$, then $x + t\boldsymbol{u}$ runs over M as $t\boldsymbol{u}$ runs over all time-like vectors, so $\boldsymbol{f}$ does not depend on spacetime points. Further, for a given $\dot{x}$, $-(\dot{x} + 2\boldsymbol{u}(\boldsymbol{u} \cdot \dot{x})) =: \boldsymbol{u}'$ runs over V(1) as $\boldsymbol{u}$ runs over all absolute velocities. Indeed, Lorentz multiplying by $\dot{x}$ we get $1-2(\boldsymbol{u}\cdot\dot{x})+2 = \boldsymbol{u}'\cdot\dot{x}$ from which $-\boldsymbol{u}\cdot\dot{x} = \sqrt{\frac{1-\boldsymbol{u}'\cdot\dot{x}}{2}}$, and finally $\boldsymbol{u} = \frac{\dot{x}+\boldsymbol{u}'}{\sqrt{2}\sqrt{1-\boldsymbol{u}'\cdot\dot{x}}}$. As a consequence, the force does not depend on $\dot{x}$, which is possible only if $\boldsymbol{f}$ is zero.

Exercises 13.2.6

1. $\xi^{-1}_{\boldsymbol{u},\boldsymbol{u}_s}(\boldsymbol{t}, \boldsymbol{q}_S) = \boldsymbol{t}\boldsymbol{u} + \boldsymbol{q}_S$ because $\tau_{\boldsymbol{u},\boldsymbol{u}_s} \cdot (\boldsymbol{t} + \boldsymbol{q}_S) = \boldsymbol{t}$. Note that this formula is similar to the standard one.
2. The covectorial splitting is $(\xi^{-1}_{\boldsymbol{u},\boldsymbol{u}_s})^*$. Then for a covector $\boldsymbol{k}$, $((\xi^{-1}_{\boldsymbol{u},\boldsymbol{u}_s})^* \cdot \boldsymbol{k})(\boldsymbol{t}, \boldsymbol{q}_S) = \boldsymbol{k}\cdot(\xi\boldsymbol{u}, \boldsymbol{u}_s^{-1})(\boldsymbol{t}, \boldsymbol{q}_S) = \boldsymbol{k}\cdot(\boldsymbol{t}\boldsymbol{u}+\boldsymbol{q}_S) = \boldsymbol{k}\cdot\boldsymbol{u}\boldsymbol{t}+\boldsymbol{k}\cdot\boldsymbol{q}_S = (\boldsymbol{k}\cdot\boldsymbol{u})\boldsymbol{t}+(\boldsymbol{k}\cdot\boldsymbol{i}_{\boldsymbol{u}_s})\cdot\boldsymbol{q}$ where $\boldsymbol{i}_{\boldsymbol{u}_s} : \mathbf{S}_S \to \mathbf{M}$ is the canonical embedding. This formula, too, is similar to the standard one.
3. The covectorial time-like component is
$$(\boldsymbol{t}\boldsymbol{u} + \boldsymbol{q}_S) \cdot \boldsymbol{u} = -\boldsymbol{t} - \boldsymbol{q}_S \cdot \left(\frac{\boldsymbol{u}_S}{-\boldsymbol{u} \cdot \boldsymbol{u}_S} - \boldsymbol{u}\right) = -(\boldsymbol{t} + \boldsymbol{q}_S \cdot \boldsymbol{v}_{\boldsymbol{u}_s\boldsymbol{u}})$$
where $\boldsymbol{u}_S \cdot \boldsymbol{q}_S = 0$ has been exploited.
$\boldsymbol{u}_S \cdot \boldsymbol{i}_{\boldsymbol{u}_s} = 0$; thus, the covectorial space-like component is
$$(\boldsymbol{t}\boldsymbol{u} + \boldsymbol{q}_S) \cdot \boldsymbol{i}_{\boldsymbol{u}_s} = \boldsymbol{t}\left(\frac{\boldsymbol{u}_S}{-\boldsymbol{u} \cdot \boldsymbol{u}_S} + \boldsymbol{q}_S\right) = \boldsymbol{t}\boldsymbol{v}_{\boldsymbol{u}_s\boldsymbol{u}} + \boldsymbol{q}_S.$$
4. $$\begin{aligned}\xi_{\boldsymbol{u}',\boldsymbol{u}}\xi_{\boldsymbol{u}}^{-1}(\boldsymbol{t}, \boldsymbol{q}) &= \xi_{\boldsymbol{u}',\boldsymbol{u}}(\boldsymbol{t}\boldsymbol{u} + \boldsymbol{q}) = (\tau_{\boldsymbol{u}'\boldsymbol{u}}, \pi_{\boldsymbol{u}'\boldsymbol{u}})(\boldsymbol{t}\boldsymbol{u} + \boldsymbol{q}) \\ &= \left(\frac{\boldsymbol{t}}{-\boldsymbol{u}' \cdot \boldsymbol{u}}, \boldsymbol{t}\boldsymbol{u} + \boldsymbol{q} - \frac{\boldsymbol{t}\boldsymbol{u}'}{-\boldsymbol{u}' \cdot \boldsymbol{u}}\right) = (\sqrt{1 - |\boldsymbol{v}_{\boldsymbol{u}'\boldsymbol{u}}|^2}\boldsymbol{t}, \boldsymbol{q} - \boldsymbol{v}_{\boldsymbol{u}'\boldsymbol{u}}\boldsymbol{t}).\end{aligned}$$
5. $-\boldsymbol{u}_1 \cdot \boldsymbol{u}_2 = \gamma$, $-\boldsymbol{u}_1 \cdot \boldsymbol{u}_3 = \gamma'$, $-\boldsymbol{u}_2 \cdot \boldsymbol{u}_3 = \gamma\gamma'$, so
$$\begin{aligned}\boldsymbol{v}_{12,3} &= \frac{\gamma\gamma'(1, 0, 0, 0)}{\gamma'} - \gamma(1, v, 0, 0) = (0, \gamma v, 0, 0), \\ \boldsymbol{v}_{31,2} &= (0, 0, v, 0), \quad \boldsymbol{v}_{23,1} = \gamma'(0, v, -v', 0) \quad \text{etc.}\end{aligned}$$

6. $\tau_{u,w_s} = \frac{-w_s}{-u\cdot w_s}$, $\pi_{u,w_s} = \mathbf{1} + \frac{u\otimes w_s}{-u\cdot w_s}$, so the formulas for $\boldsymbol{u}$ and $\boldsymbol{u}'$ are the same as for space-like synchronizations. There is a difference, however, if one of $\boldsymbol{u}$ and $\boldsymbol{u}'$ is replaced by $\boldsymbol{w}_s$: the nonstandard relative velocity corresponding to a light line with light direction $\boldsymbol{w}_s$ makes no sense (the relative speed is "infinite").
7. The tangent space of H at $\boldsymbol{x}$ is the space-like linear subspace $\{\boldsymbol{y} \mid \boldsymbol{x}\cdot\boldsymbol{y} = 0\}$, so the tangent spaces are transverse to all time-like vectors. The simultaneity defined by H is evidently an equivalence relation.
   The proper time interval $\boldsymbol{s}_0$ elapsed on $o + \mathbb{T}\boldsymbol{u}$ from o to the synchronization instant dt_0 is determined by $o+\boldsymbol{s}_0\boldsymbol{u} \in o+\boldsymbol{t}_0\boldsymbol{u}_0+H$, so $\boldsymbol{s}_0\boldsymbol{u}-\boldsymbol{t}_0\boldsymbol{u}_0 \in H$, yielding $\boldsymbol{s}_0 = \boldsymbol{t}_0(-2\boldsymbol{u}\cdot\boldsymbol{u}_0)$.
   The proper time interval $\boldsymbol{s}$ elapsed on $o+\mathbb{T}\boldsymbol{u}$ from o to the synchronization instant dt is determined by $o+\boldsymbol{s}\boldsymbol{u} \in o+\boldsymbol{t}\boldsymbol{u}_0+H$, so $\boldsymbol{s}\boldsymbol{u}-\boldsymbol{t}\boldsymbol{u}_0 \in H$, yielding $\boldsymbol{s}^2+2\boldsymbol{s}\boldsymbol{t}(\boldsymbol{u}\cdot\boldsymbol{u}_0)+\boldsymbol{t}^2 = \boldsymbol{t}_0^2$.
   The proper time interval $\bar{\boldsymbol{s}}_0$ elapsed on $o+\boldsymbol{t}_0+\mathbb{T}\boldsymbol{u}$ from $o+\boldsymbol{t}_0\boldsymbol{u}_0$ to the synchronization instant $\boldsymbol{t}_0$ is determined by $o+\boldsymbol{t}_0\boldsymbol{u}_0+\bar{\boldsymbol{s}}_0\boldsymbol{u} \in o+\boldsymbol{t}_0\boldsymbol{u}_0+H$, so $\bar{\boldsymbol{s}}_0\boldsymbol{u} \in H$, yielding $\bar{\boldsymbol{s}}_0 = \boldsymbol{t}_0$.
   The proper time interval $\bar{\boldsymbol{s}}$ elapsed on $o+\boldsymbol{t}_0+\mathbb{T}\boldsymbol{u}$ from $o+\boldsymbol{t}_0\boldsymbol{u}_0$ to the synchronization instant $\boldsymbol{t}$ is determined by $o+\boldsymbol{t}_0\boldsymbol{u}_0+\bar{\boldsymbol{s}}\boldsymbol{u} \in o+\boldsymbol{t}\boldsymbol{u}_0+H$, so $\bar{\boldsymbol{s}}\boldsymbol{u}+(\boldsymbol{t}_0-\boldsymbol{t})\boldsymbol{u}_0 \in H$, yielding $\bar{\boldsymbol{s}}^2 + 2\bar{\boldsymbol{s}}(\boldsymbol{t}_0-\boldsymbol{t})(\boldsymbol{u}\cdot\boldsymbol{u}_0) + (\boldsymbol{t}_0-\boldsymbol{t})^2 = \boldsymbol{t}_0^2$.
   To see that $\boldsymbol{s} - \boldsymbol{s}_0 \neq \bar{\boldsymbol{s}} - \bar{\boldsymbol{s}}_0$, put $\boldsymbol{t} = a\boldsymbol{t}_0$ for $a > 1$, solve the quadratic equations, and do the calculations.

Bibliography

[1] Faraoni, V.: *Special Relativity*. Springer, Cham (2013)
[2] Grey, N.: *A Student's Guide to Special Relativity*. Cambridge University Press (2022)
[3] Hecht, E.: Am. J. Phys. **77**, 779 (2009)
[4] Landau, L. D., Lifshitz, E. M.: *Mechanics*. Addison-Wesley (1960)
[5] Marinov, S.: Found. Phys. **9**, 445 (1979)
[6] Marinov, S.: Prog. Phys. **1**, 31 (2007)
[7] Matolcsi, T.: Found. Phys. **27**, 1685 (1998)
[8] Matolcsi, T.: Atti Accad. Pelorit. Pericol. **101**(1), AI (2023)
[9] Matolcsi, T., Goher, A.: Stud. Hist. Philos. Mod. Phys. **32**, 83 (2001)
[10] McDonalds, K. T.: *Temperature and Special Relativity*. J. H. Laboratories, Princeton University (2020)
[11] Mocanu, C. I.: Found. Phys. Lett. **5**, 73 (1992)
[12] Møller, M. C.: *The Theory of Relativity*. Clarendon Press, Oxford (1960)
[13] Reichenbach, H.: *The Philosophy of Space and Time*. Dover (1957)
[14] Selleri, F.: Found. Phys. **26**, 628 (1996)
[15] Selleri, F.: Found. Phys. Lett. **10**, 73 (1997)
[16] Synge, J. L.: *Relativity: The Special Theory*. North Holland (1955)
[17] Synge, J. L.: *Relativity: The General Theory*. North-Holland (1964)
[18] Wald, R. M.: *General Relativity*. The University of Chicago Press (1984)
[19] Weinberg, S.: *The Quantum Theory of Fields*. Cambridge University Press (1995)
[20] Woodhouse, N. M. J.: *Special Relativity*. Springer, Berlin–Heidelberg (1992)

https://doi.org/10.1515/9783112219553-023

Index

https://doi.org/10.1515/9783112219553-024

List of symbols

$\boldsymbol{b}_u$	$\boldsymbol{u}$-Euclidean structures, Subsection 2.4.4
$\boldsymbol{B}_{u'u}$	boost in general Subsection 3.4, nonrelativistic (5.10), relativistic (10.17)
$\mathcal{D}$	symbol of differentiation
$\mathcal{D}_u$	$\boldsymbol{u}$-time-like differentiation, Subsections 7.2 and 12.2
$\boldsymbol{h}$	absolute Euclidean structure (4.6)
$\boldsymbol{i} : \mathbf{S} \to \mathbf{M}$	nonrelativistic imbedding (5.12)
$\boldsymbol{i}_u : \mathbf{S}_u \to \mathbf{M}$	relativistic imbedding, Subsection 10.7.1
$\mathbb{L}$	measure line of distances
T_s	synchronization time, Subsection 3.5.4
T_a	absolute time, Subsection 5.1.6
T_u	standard $\boldsymbol{u}$-time, Subsection 10.6.2
$\mathrm{L}^{\to}, \mathrm{L}^{\leftarrow}, \mathrm{L}$	future-light-like, past-light-like, light-like vectors (10.6) and (10.7)
M	spacetime
$\mathbf{M}$	spacetime vectors
$\boldsymbol{P}$	inertial time progress, Subsection 2.3.1
$\boldsymbol{\pi}_u : \mathbf{M} \to \mathbf{S}$	nonrelativistic $\boldsymbol{u}$-projection (5.6)
$\boldsymbol{\pi}_u : \mathbf{M} \to \mathbf{S}_u$	relativistic $\boldsymbol{u}$-projection (10.10)
$\pi_u : \mathrm{M} \to \mathrm{S}_u$	nonrelativistic (5.8), relativistic (10.12)
$\mathbf{S}$	nonrelativistic space-like vectors (5.1)
$\mathbf{S}_u$	relativistic $\boldsymbol{u}$-space-like vectors (10.9)
S_u	inertial space, Subsection 2.4.3
$\mathbb{T}$	measure line of time periods
$\mathrm{T}^{\to}, \mathrm{T}^{\leftarrow}, \mathrm{T}^{\leftarrow} \cup \mathrm{T}^{\to}$	future-like, past-like, time-like vectors, Subsection 2.2.2
$\boldsymbol{\tau} : \mathbf{M} \to \mathbb{T}$	nonrelativistic time progress, Subsection 5.1.1
τ	nonrelativistic time evaluation (5.13)
τ_u	standard relativistic time evaluation (10.31)
$\mathrm{V}^{\to}$	light directions (10.27)
$V(1)$	absolute velocities (2.2), (5.4), (10.2)
$\boldsymbol{v}_{u'u} = \boldsymbol{u}' - \boldsymbol{u}$	nonrelativistic relative velocity (5.29)
$\boldsymbol{v}_{u'u} = \frac{u'}{-u \cdot u'} - \boldsymbol{u}$	standard relativistic relative velocity (10.34)
$\boldsymbol{\xi}_u = (\boldsymbol{\tau}, \boldsymbol{\pi}_u) : \mathbf{M} \to \mathbb{T} \times \mathbf{S}$	nonrelativistic vectorial splitting (5.32)
$\boldsymbol{\xi}_u = (\boldsymbol{\tau}_u, \boldsymbol{\pi}_u) : \mathbf{M} \to \mathbb{T} \times \mathbf{S}_u$	standard relativistic vectorial splitting (10.40)
$\xi_u : \mathrm{M} \to \mathrm{T}_a \times \mathrm{S}_u$	nonrelativistic spacetime splitting (5.54)
$\xi_u : \mathrm{M} \to \mathrm{T}_u \times \mathrm{S}_u$	standard relativistic spacetime splitting (10.59)
$\xi_{u,o} : \mathrm{M} \to \mathbb{T} \times \mathbf{S}$	nonrelativistic vectorized spacetime splitting (5.56)
$\xi_{u,o} : \mathrm{M} \to \mathbb{T} \times \mathbf{S}_u$	standard relativistic vectorized spacetime splitting (10.61)
$\boldsymbol{\xi}_{u'u} : \mathrm{T}^{\to} \times \mathbf{S} \to \mathbb{T} \times \mathbf{S}$	nonrelativistic vectorial transformation rule (5.36)
$\boldsymbol{\xi}_{u'u} : \mathrm{T}^{\to} \times \mathbf{S}_u \to \mathbb{T} \times \mathbf{S}_u$	relativistic vectorial transformation rule (10.44)
$\partial_{\mathbb{T}}$	differentiation with respect to time in splitting, Subsections 7.2 and 12.2
∇	nonrelativistic space-like differentiation, Subsection 7.2
∇_u	relativistic $\boldsymbol{u}$-space-like differentiation, Subsection 7.2
$\boldsymbol{\nabla}$	nonrelativistic space-like differentiation in vectorized splitting, Subsection 7.2
$\boldsymbol{\nabla}_u$	relativistic $\boldsymbol{u}$-space-like differentiation in vectorized splitting, Section 12.2

https://doi.org/10.1515/9783112219553-025

www.ingramcontent.com/pod-product-compliance
Lightning Source LLC
Chambersburg PA
CBHW061338210326
41598CB00035B/5812
9783119146067